The Wigner Function in Science and Technology

The Wigner Function in Science and Technology

David K Ferry
School of Electrical, Computer, and Energy Engineering, Arizona State University, USA

Mihail Nedjalkov
Institute for Microelectronics, Technical University of Vienna, Austria

IOP Publishing, Bristol, UK

Multimedia content is available from About Book online https://doi.org/10.1088/978-0-7503-1671-2.

ISBN 978-0-7503-1671-2 (ebook)
ISBN 978-0-7503-1669-9 (print)
ISBN 978-0-7503-1670-5 (mobi)

DOI 10.1088/978-0-7503-1671-2

Version: 20181101

IOP Expanding Physics
ISSN 2053-2563 (online)
ISSN 2054-7315 (print)

British Library Cataloguing-in-Publication Data: A catalogue record for this book is available from the British Library.

Published by IOP Publishing, wholly owned by The Institute of Physics, London

IOP Publishing, Temple Circus, Temple Way, Bristol, BS1 6HG, UK

US Office: IOP Publishing, Inc., 190 North Independence Mall West, Suite 601, Philadelphia, PA 19106, USA

Contents

Preface ix

Acknowledgments xii

Author biographies xiii

1 Introduction **1-1**

1.1 Classical mechanics 1-2

1.2 Rise of quantum mechanics 1-7

1.3 Eugene Wigner 1-13

1.4 Modern devices and simulation 1-16

1.5 Our approach 1-17

References 1-18

2 Approaches to quantum transport **2-1**

2.1 Modes and the Landauer formula 2-5

2.2 The scattering matrix approach 2-9

2.3 The density matrix 2-13

2.4 Green's functions 2-21

2.5 What are the relative advantages? 2-27

References 2-32

3 Wigner functions **3-1**

3.1 Preliminary considerations 3-2

3.2 The equations of motion 3-5

3.3 Generalizing the Wigner function 3-8

3.4 Other phase space approaches 3-11

3.5 Wigner–Weyl transforms 3-15

3.6 The hydrodynamic equations 3-18

References 3-21

4 Effective potentials **4-1**

4.1 Size of the electron 4-2

4.2 The Bohm potential 4-6

4.3 Bohm and the two-slit experiment 4-11

4.4 The Wigner potential 4-14
4.5 Feynman and effective potentials 4-17
References 4-23

5 Numerical solutions 5-1

5.1 The initial state 5-2
5.2 Numerical techniques 5-8
5.2.1 Stability and convergence 5-10
5.2.2 The boundary/contact 5-12
5.2.3 Artificial reflections 5-16
5.2.4 Spectral methods 5-17
5.3 The resonant tunneling diode: Wigner function simulations 5-21
5.4 Other devices 5-22
References 5-23

6 Particle methods 6-1

6.1 The classical Monte Carlo technique 6-2
6.1.1 The path integral 6-4
6.1.2 Free-flight generation 6-7
6.1.3 Final state after scattering 6-8
6.1.4 Time synchronization 6-10
6.1.5 Rejection techniques for nonlinear processes 6-11
6.2 Paths in quantum mechanics 6-14
6.2.1 Bohm trajectories 6-16
6.2.2 Feynman paths 6-17
6.2.3 Wigner paths 6-18
6.3 Using particles with the Wigner function 6-21
6.3.1 Weighted Monte Carlo 6-22
6.3.2 Introducing an *affinity* parameter 6-23
6.3.3 Signed particles 6-27
References 6-30

7 Collisions and the Wigner function 7-1

7.1 The interaction representation 7-2
7.2 The electron–phonon interaction 7-4
7.2.1 Acoustic phonons 7-4
7.2.2 Piezoelectric scattering 7-6

7.2.3 Non-polar optical and intervalley phonons 7-7
7.2.4 Polar optical phonons 7-8
7.2.5 A precautionary comment 7-9
7.3 The Wigner scattering integrals 7-9
7.4 Collisions in the Monte Carlo approach 7-11
References 7-18

8 Entanglement 8-1

8.1 An illustration of entanglement 8-3
8.2 Entanglement in harmonic oscillators 8-4
8.3 Measures of entanglement 8-10
8.4 Some illustrative examples 8-15
8.4.1 Photons 8-15
8.4.2 Condensed matter systems 8-17
References 8-20

9 Quantum chemistry 9-1

9.1 Quantum statistics 9-2
9.2 Reactions and rates 9-4
9.3 Tunneling 9-8
9.4 Spectroscopy 9-13
References 9-14

10 Signal processing 10-1

10.1 Signal propagation 10-2
10.2 Wavelets 10-6
References 10-7

11 Quantum optics 11-1

11.1 Propagation 11-2
11.2 The Jaynes–Cummings model 11-4
11.3 Squeezed states 11-8
11.4 Coherence I 11-11
11.5 Coherence II 11-14
11.6 Bell states 11-15
References 11-20

12 Quantum physics **12-1**

12.1 The harmonic oscillator 12-2

12.1.1 The driven oscillator 12-3

12.1.2 Qubits 12-6

12.2 Quantum physics 12-7

12.2.1 Parity again 12-8

12.2.2 Quantum Hall effect 12-9

12.2.3 Qubits 12-10

12.3 Superconductivity 12-13

12.3.1 Coupling to a resonator 12-14

12.3.2 SQUIDS 12-16

12.3.3 Qubits 12-17

12.4 Plasmas 12-21

12.4.1 Kinetics 12-22

12.4.2 High-density plasmas 12-24

12.4.3 Hydrogen 12-25

12.5 Relativistic systems 12-25

12.5.1 Waves and particles 12-29

12.5.2 The strong force 12-32

12.5.3 Other forces 12-33

12.6 Quantum cascade laser 12-34

References 12-36

Preface

The Wigner function is almost as old as quantum mechanics itself. It arose as a method in which to bring phase space representations into quantum mechanics. As most people know, quantum mechanics deals with operators that exist either in real space or in momentum space, but not in both. Heisenberg first developed the matrix form of quantum mechanics by the set of real space operators. Subsequently, Schrödinger gave us wave mechanics through his famous equation, which existed in only real space. Yet, both Madelung and Kennard rapidly developed the ideas of the phase space motion of particles from the Schrödinger equation, an innovation that continued to be important many years later with the developments of Bohm. In these latter developments, it was clear that there was an additional potential that arose quantum mechanically, and that the motion was then governed by both the normal (classical) potential appearing in the Hamiltonian and this new 'quantum' potential. Indeed, Wigner's development was not aimed precisely at a phase space representation, but by the desire to examine how thermodynamic equilibrium would incorporate any additional potentials that arose from the quantum description. In this pursuit, he developed his own form of a 'quantum' potential. We will discuss this and the various potentials met in both classical and quantum mechanics in chapter 1.

What Wigner triggered by his work was not only the recognition that phase space descriptions could exist in quantum mechanics, but that there were many types of additional potentials that arose in these quantum descriptions. Which type was important depended to a great extent upon how one formulated the phase space prediction. Notable advances in understanding the phase space approach were made by both Weyl and Moyal originally, and the interest has existed in the math community even up to today. At the same time, the ideas of an effective, or 'quantum' potential, also received a good deal of work, and continues to do so up to the present time. One form actually appears in the book by Feynman and Hibbs, which is important for its description of the path integral approach.

Zachos, Fairlie, and Curtright suggest that there are actually three important lines of development for quantum mechanics. The first is the original developments due to Heisenberg and Schrödinger. The second approach is the path integral approach, conceived by Dirac, and heavily developed, of course, by Feynman. They suggest that the phase space approach represents an equivalent third approach to quantum mechanics. This third approach relies upon not only Wigner, but the developmental work of many other interested parties, as mentioned above. Particularly in the path integral approach and in the Wigner approach, the role of the effective potential becomes quite important. So, in addition to trying to highlight the importance of the Wigner function today in science and technology, we will also try to highlight the various approaches to the effective potential and the part it plays in the use of Wigner functions.

Nevertheless, if advances and interest were only based upon the very interesting mathematics of the phase space formulation, Wigner functions would even today remain something of a backwater in both science in general and quantum mechanics in particular. What has drawn considerable interest to the Wigner function, as well as its cousins, in the last third of the twentieth century is a very enhanced interest in non-equilibrium statistical mechanics in the various sciences. In both chemistry and electrical engineering, a strong interest in transport physics drew attention to the Wigner function as a method of incorporating quantum effects into approaches based upon the Boltzmann transport equation. This made it natural to extend standard phase space approaches into the quantum regime. Then, work, for example by Carruthers and Zachariasen, pointed out the extreme usefulness of the Wigner function in examining collision dynamics in a variety of physical systems. The rise of quantum optics and quantum information theory, in a variety of physical guises, led to an intense interest in the use of Wigner functions, not only to explore various quantum regimes but also as a method to illustrate, at least mathematically, the presence of entanglement in such systems. This has highlighted the realization that the Wigner function gives a methodology to examine correlations in the quantum systems.

But, why has the Wigner phase space formulation thrived and actually become an important approach to a wide frame of science? It was proved early in the second half of the twentieth century that a Wigner function defining a single ground state wave packet had to have a Gaussian form, which meant absolutely no negative excursions of the Wigner packet. As a result, the study of quantum systems with Wigner functions could clearly identify non-classical behavior by the development of negative excursions of the function. This meant that it would be easy to study the quantum-classical transition and clearly identify behavior that was quantum mechanical in nature. Later, as one began to study the presence of entanglement in disparate quantum objects, it became clear that such a quantum effect was also suggested by the negative excursions of the Wigner function. Hence, any study in particular of the extreme interest in qubits (quantum bits) for quantum computing needed to be expressed in terms of the Wigner function: it was the only description which clearly identified quantum entanglement, and did so in a very easily understood pictorial manner. Even when the situation needed the more complicated non-equilibrium Green's functions, the results were quite often transformed into the Wigner representation in order to gain an understanding of the meaning of these results. And, perhaps just as importantly, experimental techniques now exist which can probe the Wigner function and demonstrate the existence of the non-Gaussian and negative parts of the function.

In recent years, the authors have been involved in an effort to bring more recognition to the importance of Wigner functions and to advertise their use in widely disparate fields of science and technology[1]. The central organizational point for this effort has been at the Technical University of Vienna, and owes a lot to the efforts of Joseph Weinbub and others in Vienna (links to more than 90 scientists around the world are provided on the aforementioned link). This has led to, so far, two international workshops on the wide applications of Wigner functions, one in

[1] http://www.iue.tuwien.ac.at/wigner-wiki/doku.php?id=start

Hawaii in 2015 and the second in the Lake District of England in 2017 (a third will be in Evanston, IL, in 2019). It was from these two workshops that the idea, or perhaps the need, for this book was born. The workshops have been enthusiastically supported by a wide group of scientists with interest in the Wigner function, and this has pointed out the need for such a book. Certainly, there are other books in the area, but none with the breadth and completeness that we hope we achieve in this work.

The book itself is designed to give a background on the origins and development of Wigner functions, as well as its mathematical underpinnings. Along the way, we will point out the connections (and differences) from the more popular non-equilibrium Green's function approaches. But the importance of the text lies in the discussions of the applications of the Wigner function to various fields of science. To be sure, we would be unable to do this justice if not for the large group of scientists involved with the Wigner Initiative mentioned above, the start of a set of Wigner workshops[2]. Without their work, and the role of the workshops in bringing this work to our attention, we would not be able to write this book. There are too many to list and thank individually, but the citations to their work are clearly our thanks to their efforts and collaborations. We are sure that there are others out there whose work has not been brought to our attention. To them, we can only offer our apologies, and to assure them that we look forward to learning of their work in future workshops.

David Ferry, Tempe, AZ
Mixi Nedjalkov, Vienna, Austria

[2] http://www.iue.tuwien.ac.at/iw22017

Acknowledgments

The authors would like to thank Josef Weinbub, who has dutifully read each and every word of the chapters and provided important corrections and feedback.

Author biographies

David K Ferry

David Ferry is Regents' Professor Emeritus in the School of Electrical, Computer, and Energy Engineering at Arizona State University. He was also graduate faculty in the Department of Physics and the Materials Science and Engineering program at ASU, as well as Visiting Professor at Chiba University in Japan. He came to ASU in 1983 following shorter stints at Texas Tech University, the Office of Naval Research, and Colorado State University. In the distant past, he received his doctorate from the University of Texas, Austin, and spent a postdoctoral period at the University of Vienna, Austria. He enjoys teaching (which he refers to as 'warping young minds') and research. The latter is focused on semiconductors, particularly as they apply to nanotechnology and integrated circuits, as well as quantum effects in devices. In 1999, he received the Cledo Brunetti Award from the Institute of Electrical and Electronics Engineers, and is a Fellow of this group as well as the American Physical Society and the Institute of Physics (UK). He has been a Tennessee Squire since 1971 and an Admiral in the Texas Navy since 1973. He is the author, co-author, or editor of some 40 books and about 900 refereed scientific contributions.

Mihail Nedjalkov

Mihail (Mixi) Nedjalkov received his doctoral (PhD) degree in Physics in 1990 and his doctor of science (DSc) degree in Mathematics in 2011 at the Bulgarian Academy of Sciences (BAS). He is Associate Professor with the Institute of Information and Communication Technologies, BAS. He held visiting research positions at the University of Modena (1994), Arizona State University (2004) and mainly at the Institute for Microelectronics, TU–Wien (1998–date), supported by national and European projects. He has served as a lecturer at the 2004 International School of Physics 'Enrico Fermi', Varenna, Italy. His research interests include physics and modeling of classical and quantum carrier transport in semiconductor materials, devices and nanostructures, collective phenomena, theory and application of Monte Carlo methods.

Chapter 1

Introduction

Quantum mechanics arrived on the scene on December 14, 1900, when Max Planck gave his talk on blackbody radiation to the German Physical Society [1]. In truth, he had presented a preliminary version two months earlier, but the theory was not fully ready at that time. However, at the earlier meeting he had found that Rubens and Kurlbaum [2] had new data, confirming very recent results of other groups, and that he could match his theory to this new data. So, now experiment and theory agreed, and this provided the full proof that his revolutionary ideas had merit. What was new and different with Planck's theory was the idea that light existed as small bundles of energy given by hf, where f was the frequency of the light and h was a new fundamental constant, which is now known as Planck's constant and is 6.6025×10^{-34} J s. The idea that light was composed of these small bundles of energy, which he called corpuscles, was revolutionary and completely at odds with the accepted theory that light was a wave and not a particle. The view that light should be a wave had existed for almost a century since Young's work on the diffraction of light and its interference when passed through a pair of slits [3]. Clearly, the suggestion that light was in fact made up of particles, each of which had a well-defined energy given by its frequency, was at odds with the wave picture and stunned the audience at his presentation.

But it did not take long for confirmation of Planck's thesis, for Einstein soon showed that experiments on photoemission—the use of light to cause electrons to be emitted from a solid—could be explained by the corpuscular nature of light [4]. Einstein's theory needed only to treat light as corpuscular particles, and then, after accounting for an energy barrier which kept the electrons within the solid, conservation of energy and momentum explained the experimental observations. Yet, if light was a particle, how was one to explain the diffraction and interference observed by Young? Could both views be right? The answer was apparently stimulated by Thomson, who was studying the ionization of gases under ultraviolet and x-ray light excitation [5]. Thomson noticed that only a small fraction of the gas

doi:10.1088/978-0-7503-1671-2ch1

molecules were ionizing. This lead him to conjecture that the energy of the light was localized in Planck's packets, which also were *physically small in dimension*, so that they missed the majority of the molecules. This led him to suggest that light striking a metal surface would actually cause a 'series of bright specks on a dark ground'. He supposed that if the plate was moved further from the source, 'we shall diminish the number of these [specks] ... but not the energy in the individual units.' This went beyond Planck's quantization by also suggesting that the photons were physically quite small. Thompson estimated that, with illumination of 1.9 eV photons with an intensity of 10^{-11} W cm^{-2}, he would get about 3.3×10^7 photons per cm^2 per second, for which there would be about one photon for each 1000 cm^3. He finally concluded that 'the structure of light would be exceedingly coarse'. This suggested to a colleague, Taylor at Trinity College, that one should measure very weak light. Taylor used a flame passing through a single slit as his source, and then diffracted the light around a needle [6]. He attenuated the light with various smoked glass windows, which he calibrated first to determine the attenuation of each window. With the weakest light, the photographic plate had to be exposed for 2000 h. He estimated that this corresponded to imaging a standard candle from 1 mile away with a 10 s exposure of the plate. Taylor states that he saw no diminution in the diffraction pattern for any of his exposures, thus confirming Thomson's ideas. Even though light arrived at each photographic plate as a single photon at a time, the net pattern formed over time matched the expected wave interference patterns. Consequently, the light in this experiment was acting as both particle (upon striking the plate) and wave (when passing the needle). And there we had it: light was paranoid. It behaved both as a particle and as a wave, and each characteristic could be observed with the right experiment.

It is hard today to understand fully the shock this incurred on the understanding of physics at the time. Naturally, after more than a century, we are inured to this. But it was a jolt at the time. It didn't help any when Einstein proposed his theories of relativity, producing still more jolts to the established world of physics. However, these new ideas merely opened the door to the revolution that was to follow. Our purpose here is to follow only one thread of this revolution, and that is the appearance, development, and usage of the Wigner distribution function. To understand this thread, we need first to take a look at the state of classical physics, particularly statistical mechanics, at the time (and which still exists today). We will do this in the next section, and then approach the arrival of the Wigner function.

1.1 Classical mechanics

It has been almost a third of a millennium since Isaac Newton formalized the mechanics of motion in his *Principia Mathematica* [7]. Almost a century and half after Newton, the mathematics of motion for a many-particle system were written down by Hamilton [8], who developed these equations from those of Lagrange from half a century earlier [9]. When we deal with the many-particle system, we traditionally use $6N$ variables in a multi-particle phase space, where N is the number of particles. Of the six variables for each particle, three give the position in normal space and three give us the momentum of the particle. Position and momentum are referred to as conjugate variables, as the momentum of a single particle is related to the time rate of change of its position. We shall take the time to give a brief review of Hamilton's equations of motion, as they will reappear when we talk about quantum dynamics below. The set of six variables for each particle define that particle's position in phase space: hence, the idea of phase space as the six-dimensional space of position and momentum. This idea carries over to the many-particle description.

To illustrate these points, let us consider a one-dimensional harmonic oscillator, such as a linear pendulum. The ball of the pendulum, which is attached via a long bar to its pivot point, oscillates back and forth (we will ignore the gradual rotation of the oscillation in the Earth's gravity). We take the position directly below the pivot point as the $x = 0$ point (or origin), so that x is the distance away from the origin (and is positive or negative depending upon one's choice of directions). The ball has a velocity v which measures the rate of change of the variable x. The velocity times the mass of the ball gives us the momentum p of the ball. This momentum and the position are the two coordinates of our single ball, and are known as conjugate coordinates. When the ball reaches its maximum excursion in x, the momentum goes to zero and the ball begins to move back toward the origin. As the ball passes the origin, the momentum has a maximum in its amplitude. At the maximum excursion, the entire energy of motion lies in the potential energy of the ball. On the other hand, as the ball passes the origin, the entire energy of motion is the kinetic energy of the ball. As the ball passes the origin, it has to give up energy to raising over its lowest point. At the maximum excursion, the ball lies at its highest point, and the kinetic energy has been converted to potential energy. If we perform a time average of the two energies, we find that the time average of the kinetic energy is precisely equal to the time average of the potential energy. Thus, the stable motion of the oscillating ball is achieved when the time averages of the two energies are equal. The kinetic energy is usually denoted by the symbol T, while the potential energy is denoted by the symbol V. In a sense, then, the potential energy was obtained by forcing the ball up to his maximum height above that of the origin. Thus, the potential energy is a measure of the work done to raise the ball to this height.

Such oscillators are found in many situations in the real world. For example, consider a smooth round stone situated at the peak of a hill, in which a surrounding valley separates the hill from others of like kind around it. The stone is in equilibrium in this position, and has potential energy deriving from its height above the surrounding valley and the existence of a gravitation attraction. Yet, the stone is not in a stable equilibrium, as any fluctuation of its position will cause it to roll down the hill. In the absence of any friction or scattering, the stone will acquire an oscillatory motion as it moves from one hill to another and back again. In many cases, various motion, such as that of the atoms in a solid, are described in terms of their normal modes (Fourier modes), and each of these modes is described by a harmonic oscillator. So, when we talk about a phonon, it is an excitation of the harmonic oscillator at the given value of phonon momentum.

Lagrange developed a methodology for obtaining the equations of motion from a constraint of least action and a variational principle. Both were quite advanced for the time. One typically gets the equations of motion from what we now call the Lagrangian, which is the difference in potential and kinetic energy, $L = T - V$. Hence, we see that in this formulation the time average of L for the oscillator is zero, so the motion is derived from the balance between the two energies. Typically, the kinetic energy is written as $p^2/2m$, where m is the mass of the object, and the dynamic equation of motion derives from the Lagrangian as

$$-\frac{d}{dt}\left(m\frac{\partial T}{\partial p}\right) = \frac{\partial V}{\partial x}. \tag{1.1}$$

Assuming that the mass is constant, this leads us to Newton's famous equation:

$$\frac{d}{dt}\left(m\frac{p}{m}\right) = m\frac{dv}{dt} = -\frac{\partial V}{\partial x} = F, \tag{1.2}$$

where v is the velocity of the particle and F is the classical force. Hence, we see that the force acting upon the particle to produce acceleration is the negative (spatial) derivative of the potential energy. It is also important to note that the two equations above can be combined into a single second-order differential equation for the position of the particle in time. While Newton's approach would lead us to believe that the force is the central source of the motion, we now can see that it is only derivative, and it is the potential that is centrally important for the motion. Now, it is relatively obvious that these equations can be immediately extended to a system of N particles, and to three-dimensional space. Then, there will be three directions of position and three dimensions of momentum given in vector form as $\mathbf{x}_i$ and $\mathbf{p}_i$, respectively, for the ith particle of the group. This then leads to the N points in six-dimensional phase space, while the N vector positions in real space give us a single point in the higher dimensional configuration space. That is, we can represent the position of N particles by N points in normal three-dimensional space or as a single point in N-dimensional configuration space. Each different space has important uses in both classical and quantum physics.

Hamilton noted that in the motion of the oscillator the total energy was a constant of the motion. While the time variant energy moved from kinetic energy to potential energy and back again, the sum of the two energies was constant. And, in linear steady motion of a free particle, the momentum was also a constant of the motion. In his approach, he preferred to work with the total energy, which was expressed as

$$H = T + V. \tag{1.3}$$

It is known that Lagrange also discussed the total energy, and when his book was reprinted and translated in the early nineteenth century (before Hamilton's publication), the form (1.3) was present in the work. So, historically, it is not known if the H refers to Hamilton, but one presumes that it does, and it is called the Hamiltonian today. Hamilton himself introduced the action into the problem of dynamics, with this quantity being the time integral of the total energy. We typically denote the action today with S (Hamilton's actual notation differed from that normally used today [8], so we will stay with the modern notation). We have seen the action in the last section, as the units on Planck's constant h are those of the action, joule-second. From the Hamiltonian (1.3), it is possible to arrive at (what are called) the Hamilton's equations of motion for the ith particle as [10]

$$\frac{d\mathbf{x}_i}{dt} = \frac{\partial H}{\partial \mathbf{p}_i},$$
$$\frac{d\mathbf{p}_i}{dt} = -\frac{\partial H}{\partial \mathbf{x}_i}. \tag{1.4}$$

Now, we clearly see an equation of motion of all of the six variables in phase space for each particle. Once again, we note that, since the first of these equations relates the postion to the momentum, these equations can be combined into a single second-order differential equation for the position as a function of the time.

Most modern approaches to classical mechanics have adopted the use of Hamilton's equations of motion for many reasons. First, of course, is that the full motion of both momentum and position are accounted for in the equations. This means that phase space is the natural scenario in which to monitor the motion of the particles. Indeed, if we scale the position and momentum of the pendulum oscillator discussed above properly, and then plot the motion on a graph of position and momentum as the coordinate axes, the motion is a simple circle (as long as the oscillation remains described by the harmonic oscillator equations). Second, it is easier to understand these equations without fully understanding the calculus of variations intrinsic to Lagrange's equations. Yet both sets of equations have the underlying basis that the motion constitutes an extremum of the action, rather than of the energy. Instead, for conservative systems, the energy is a constant of the motion. But it is absolutely clear that phase space is the best canvas upon which to paint the motion of the particles in our system, and it is heavily used in the study of a wide variety of physical systems.

We will see later that Hamilton's equations are quite useful to us in describing our statistical Wigner distribution in terms of an ensemble of particles. The distribution of these particles in phase space is a representation of the statistical distribution, and the motion of each particle in phase space is described by Hamilton's equations of motion. The movement of the particles is determined by integrating the equations over a very small time step. Hence, we can in fact follow the trajectory of each of these particles as it moves through phase space. Importantly, the equations are as useful for quantum transport as for classical transport. An important aspect of this motion of the particles through phase space is Liouville's theorem, which requires that the volume of phase space occupied by the particles in a conservative system must remain constant. If we have a non-conservative system, the energy we impart from an external potential leads to an increase of the phase space volume, while the loss of energy to collisions yields a loss of phase space volume. Hence, even in a non-conservative system, the steady state (if it exists) evolution obtains a phase space volume that is a balance between the driving forces derived from the applied potential and the dissipative forces derived from the collisions.

In equilibrium, both the external potential and the dissipative interactions are zero, the latter required to do so by detailed balance (between, for example, the emission and absorption of phonons). In general, the phase space function of N particles can be described in terms of a one-particle distribution function in phase space; we will return to this point below. Then, we can write for a time-independent phase space function, such as the statistical distribution of the particles,

$$\frac{df}{dt} = \frac{\partial f}{\partial \mathbf{x}} \cdot \frac{d\mathbf{x}}{dt} + \frac{\partial f}{\partial \mathbf{p}} \cdot \frac{d\mathbf{p}}{dt} = \frac{\partial f}{\partial \mathbf{x}} \cdot \frac{\partial H}{\partial \mathbf{p}} - \frac{\partial f}{\partial \mathbf{p}} \cdot \frac{\partial H}{\partial \mathbf{x}} = \{f, H\}_p, \tag{1.5}$$

where the last bracket is called the Poisson bracket, and indicated by the subscript. In the equilibrium situation, this bracket and the time derivative are equal to zero. In the non-equilibrium steady-state situation, the Hamiltonian would deliver the driving potential and the time derivative would be balanced by decay terms due to the collisional interactions. In the case of a steady-state phase space function, which is also a time varying function, (1.5) would be altered as

$$\frac{df}{dt} = \frac{\partial f}{\partial \mathbf{x}} \cdot \frac{\partial H}{\partial \mathbf{p}} - \frac{\partial f}{\partial \mathbf{p}} \cdot \frac{\partial H}{\partial \mathbf{x}} + \frac{\partial f}{\partial t}, \tag{1.6}$$

and it is this total derivative that vanishes. Then, the right-hand side may be rearranged to yield the Liouville equation:

$$\frac{\partial f}{\partial t} = -\frac{\partial f}{\partial \mathbf{x}} \cdot \frac{\partial H}{\partial \mathbf{p}} + \frac{\partial f}{\partial \mathbf{p}} \cdot \frac{\partial H}{\partial \mathbf{x}} = -\{f, H\}_p. \tag{1.7}$$

The amplitude of this distribution function, in a six-dimensional phase space box, is given by the number of particles in the box. Hence, the normalization of the distribution is that the integral of it over the phase space must produce the total number of particles (or, more often, the number of particles per unit volume). In actual fact, we can achieve this one-particle distribution only for non-interacting particles. More properly, in the case of interacting particles the equation of motion for the distribution must have a correction arising from the pair-wise interactions between the particles. This result is given by what is known as the Born–Bogoliubov–Green–Kirkwood–Yvon hierarchy, and in which the two-particle distribution depends upon a three-particle interaction term, and so on [11]. Various approximations have arisen for describing this pair-wise interaction, such as the Coulomb interaction between two particles. One such approximation is to just ignore the interactions, as Boltzmann did with his molecular chaos assumption. Indeed, Boltzmann's one-particle distribution function then satisfies the form of (1.5) as

$$\frac{\partial f}{\partial t} + \frac{\partial f}{\partial \mathbf{x}} \cdot \mathbf{p} + \frac{\partial f}{\partial \mathbf{p}} \cdot \frac{d\mathbf{p}}{dt} = 0 \tag{1.8}$$

in equilibrium. The Boltzmann distribution function has become the canonical classical phase space distribution in a large part of statistical physics. Consequently, it is widely used, and experience has given us a large background of historical developments of its use in a wide range of scientific fields. If for no other reason, one would desire a phase space description in quantum mechanics just to leverage this rich understanding as a guide to the interpretation of quantum mechanics.

In the above, we have gone from the *N* particles moving independently in either configuration space or phase space to a one-particle distribution function in phase space. In classical mechanics, this distribution function is usually the Boltzmann distribution. As we will see below, quantum mechanics is usually not done in phase space. This is the rationale that has driven many, one of whom is Wigner, to seek a method for quantum mechanics in phase space. If this is achieved, the reduction to a one-particle distribution may be reversed to consider the *N* particles once again. One rationale for doing so lies in the fact that quantum mechanics, like statistical mechanics, is a probabilistic domain. Hence, it is then profitable is adapt the Monte Carlo procedure, which has been so successful in classical transport [12, 13], to the evaluation of the temporal motion of the Wigner distribution in the presence of external potentials and scattering/collision potentials.

1.2 Rise of quantum mechanics

As remarked above, it was established fairly early in the development of quantum mechanics that light was paranoid. It behaved as both a wave and a particle, and one or the other of these characteristics would be selected by the experiment. But this was light, and not really particles. The quantum mechanics of particles was first considered by Bohr, who adapted Planck's light quanta to the classical motion of the electrons in the hydrogen atom [14]. By doing so, he found that he could explain the observed spectra of absorption and emission of light by this quantization. In his calculations, he also found that the angular momentum of an electron in its atomic orbit was quantized, but he could give no reason for this to occur. The answer would come from elsewhere.

Louis de Broglie, the son of a nobleman, did his thesis in Paris on waves and particles. He was able to show that the stability of the orbits in Bohr's atom depended upon wave properties. That is, he showed that the quantization of the angular momentum of one of the particle orbits required it to contain an integer number of wavelengths of the equivalent wave [15]. This was a dramatic step forward. Not only did light behave as a particle or a wave, now particles also behaved as a wave, a view that later experiments would prove to be correct. Hence, both light and normal particles were paranoid: they both could have wave-like properties and particle-like properties. De Broglie would go on to suggest that the wave was the central feature and would guide the particle onto a given path. So, in his view, it was not wave *or* particle, but wave *and* particle. With this work, waves and particles were to be considered as two interpretations, or observations, of a single quantity. But de Broglie had highlighted another problem: if electrons were waves, where was the equation which governed their motion? Bohr could not give the answer, as he had only applied Planck's quantum of light to the *classical* mechanics of the atom. Light waves had Maxwell's equations, which provided a litany of governing equations. So far, electron waves had zero equations. Progress would soon come rapidly.

The first would come from Werner Heisenberg. Heisenberg was a young associate of Max Born in Göttingen, but also had a fellowship with the Bohr Institute in Copenhagen. In 1925, he was suffering severely from allergies and retired to the island of Heligoland in the North Sea. The first task would be relieving his allergies, and this took a few days; but, this also gave him time to think, and it was here that he formulated his new view of quantum mechanics. Today we think of this as the new quantum mechanics, but he thought of it as just a new advance in the single flow of quantum mechanics that had begun with Planck, Einstein, and Bohr. It is not clear whether he viewed the new ideas as a break from the old views. Having come to Heligoland in early May, by June he was both recovered from the allergies and ready to discuss his new approach. He first discussed the results with Wolfgang Pauli in Hamburg. He knew Pauli already from his student days in Munich and considered him almost as an older brother. Thus, he wanted to run his ideas past Pauli to make sure that they were not total rubbish. Pauli was rather encouraging, so Heisenberg gave the manuscript to Born in early July to read. Born tended to love difficult and involved mathematics. Heisenberg's paper spends considerable time comparing the classical approach and its formulas with the new quantum-mechanical formulas, but with little advanced math. In addition, there were some conditions on the new derivations, and the paper in general was somewhat devoid of the detailed mathematics Born loved. Perhaps it was due to the fact that Heisenberg was not familiar with the mathematics he himself demanded for the paper. It is generally believed that he had no acquaintance with matrix mathematics, and this could explain his reluctance. Although he worked for Born, he did not approach Born for help with the mathematics. Nevertheless, the paper was submitted and was received by the journal on July 29, and appeared in print in the December issue of *Zeitschrift für Physik* [16].

From the time of Planck, the properties of atoms and solids had been discussed in terms of harmonic oscillators, such as those discussed in the previous section. Radiation was emitted and, in turn, was absorbed. In Heisenberg's treatment, with his new formulation, he found that the energy levels of the oscillators were not given by $n\hbar\omega$, as Bohr had asserted, but by $(n + 1/2)\hbar\omega$. Since $n = 0$ is allowed in this new expression, it seems that the available classical states are symmetrically compressed into a quantum state; for example, states from both above and below a quantum state are quantized into the particular energy level. Hence, each quantum level has the same statistical weight. Heisenberg was able to show that the dispersion of the optical emission and absorption in atoms was related to the squared magnitude of the transition probability (or amplitude), but that the phase was, in principle, still measurable. In the end, he recovered the same energy transitions for the hydrogen atom that Bohr had found, even with the correction to the oscillator energies. Yet in this approach he introduced another symbolic problem. In making the transition that was induced by either the emission or absorption of radiation, the electron in the atom would have to 'jump' from one energy level to the other. There was still no chance in the theory for a gradual transition in this process. The emission or absorption of a light corpuscle required this jump to occur. Heisenberg introduced another critical property. In this new quantum mechanics, variables such as the position x and the momentum p were no longer simple variables. Rather, they now became mathematical operators. Consequently, their order in an expression was crucial. This lead to his famous introduction of the commutator relationship, much like the classical Poisson brackets. Now, the two operators were required to satisfy

$$[x, p] = xp - px = i\hbar, \tag{1.9}$$

where they were presumed, for example, to operate on the state function of one of the energy levels in the atom. Here, $\hbar = h/2\pi$ is the *reduced* Planck's constant.

Pascual Jordan had become Born's assistant in mid-1925, and this was to lead to fruitful work between them. After seeing Heisenberg's manuscript, Born turned to Jordan to begin the mathematical efforts required to create equations underscoring the new ideas on quantum mechanics. It was clear that, although he allowed Heisenberg to submit his paper without what Born considered important changes, Born himself would work to make sure that these changes appeared in the literature as soon as practicable. Jordan provided him with the aid he needed to get this accomplished. Within a few days of beginning to work on the task, Jordan brought to Born the first major result of the fact that, in general, Heisenberg's work related to matrices, which do not commute. That is, if we take two (non-diagonal) matrices A and B, the products AB and BA usually give different results. Jordan had approached the problem in a manner to first assure that Heisenberg's notation was consistent with matrix operations, and also consistent with Hamilton's canonical equations of motion, which would lead to classical behavior in the right limits. As a result, he achieved the important commutator relation that Heisenberg had proposed. Now, for the system of quantum states in the atom, both x and p were to be matrices and the natural non-commutation of these matrices would explain the new relationship (1.9).

By early September, Jordan had completed the first two chapters of the resulting paper. In order to keep Heisenberg informed of the progress, Jordan sent these to him in Munich. The later chapters served to fill out the full mathematical treatment and to address a few examples. Moving quickly, Jordan and Born were able to complete their paper in good time. Their paper [17] was received by the journal on September 27, just two months after the Heisenberg paper; it was further published in December 1925. While Born worked with Jordan on the mathematical paper, he still felt that he needed to get Heisenberg back into the loop. Thus, he put the two, Heisenberg and Jordan, together to work on developing the theory further, primarily as Jordan was an expert in the mathematics of matrices and so would aid in Heisenberg's learning about this methodology. This new effort would take them until the end of the year. The three of them were working hard on a longer exposition to follow up on the first two papers. This new paper was referred to by Heisenberg as the *dreimännerarbeit* (which roughly translates as the 'work of three men'), a name which has stuck to it even with the passage of time. By November they had more or less completed the paper, and it was accepted by the journal on 16 November 1925, though it did not appear until August [18] of the following year.

From the very name of the paper, it was clear that this was a sequel to the Born–Jordan paper and extended the work to an arbitrary finite number of degrees of freedom. By this approach, they were able to work with the many-particle picture. In addition, they worked through the entire formulation of perturbation theory so as to apply the results to more complicated problems. In the matrix formulation, the perturbation series was described via a scattering matrix, often referred to as the S-matrix. The use of a preferred set of basis functions was augmented by a transformation theory which utilized unitary matrices to implement basis set transformations. This means that one can always find a system in which the Hamiltonian approach yields a diagonal matrix for the energies, and therefore the energy eigenvalues can be used to find an orthonormal basis set of eigen-states for these energies. This paper became the cornerstone of the most extended understanding of the new quantum mechanics.

By 1925, Erwin Schrödinger was almost 40 years old, certainly beyond the age one was expected to make major scientific breakthroughs. He was already well recognized for his color theories. He had recently moved to the University of Zurich, turning down a position in his native Vienna due to the economic situation in post-World War I Austria. In Zurich, he could work closely with the physics group at the Technical University, especially with Hermann Weyl. When de Broglie's paper appeared with his wave and particle concept, it was agreed among the group that they would ask Schrödinger to give a joint colloquium on the topic. Schrödinger had become familiar with the de Broglie work through its mention in a paper by Einstein in the popular press, where he indicated that perhaps de Broglie's approach was quite promising; Schrödinger generally attributed this as the beginning of his interest in wave mechanics. During the winter vacation period, Schrödinger spent his time at a resort cabin with his *l'amour du jour*, in Arosa, Switzerland. During this stay, he developed the meat of four important papers establishing a new quantum mechanics based on continuous wave theory, as put forward by de Broglie. Herman Weyl would categorize this massive achievement as a [19] 'late erotic outburst in his life'.

In the first paper, Schrödinger developed the basis of his wave equation and showed how the atomic spectra of the single-electron atom was a natural eigenvalue problem in mathematics [20]. His famous equation in one spatial dimension is

$$-i\hbar\frac{\partial\psi(x, t)}{\partial t} = \frac{\hbar^2}{2m}\frac{\partial^2\psi(x, t)}{\partial x^2} + V(x)\psi(x, t), \tag{1.10}$$

where $\psi(x, t)$ is the wave function, which is a complex quantity. Just a few weeks later, the second paper appeared, in which he continued the treatment showing how the energies of the oscillator used by Planck would occur [21]. In the third paper, he established that the matrix mechanics of the Göttingen crowd was fully equivalent to his wave equation approach [22]. While he had introduced the wave equation in his first paper, he returned to it in the fourth paper and showed how perturbation theory could be applied to problems that were more difficult to solve [23]. In this perturbation approach, there was a continuous transition between states over time. Schrödinger's work crashed upon the physics community as a tidal wave smashing the shore. Here was an approach that produced all the same results Heisenberg had achieved for his new quantum mechanics, but was eminently easier to understand than the complex mathematics of the matrix theory. The continuity of the waves provided a more understandable approach to the new theories, and essentially all physicists were accepting of waves as they had experience with them in acoustics and electromagnetics. This was something they could get their heads around. There were, however, some subtleties underlying the extension from classical physics. For Schrödinger, his theory held that it was the waves that were fundamental and particles could only exist as localized regions of the waves (essentially what we would call a wave packet).

But Schrödinger's wave mechanics was not without its problems. The most important of these is that half of the information about dynamics that classical mechanics possessed appears to have been lost. We recall that in classical mechanics, each particle had six variables, three for position and three for momentum. In Schrödinger's wave equation, only the position appeared. Even Heisenberg had retained momentum, although it had become a differential operator. In the wave approach, the momentum was described as the differential operator

$$p = -i\hbar\frac{\partial}{\partial x}, \tag{1.11}$$

and it was hard to see how this would be an independent quantity in the sense of classical mechanics. (In classical mechanics, position and momentum are considered to be conjugate variables, but independent from one another. Here, that is no longer the case as momentum is no longer independent in the wave function.) In addition, the Schrödinger equation contains the potential directly. There are a vast number of different potential functions $V(x)$ which arise not only in one dimension, but also in three dimensions. The solution of the equation thus depends upon the exact form of the potential function as well as the boundary values that are imposed. Finally, the equation contained directly the factor $i = \sqrt{-1}$, which made it different from the normal form of an eigenvalue equation in classical mechanics.

Generally, the eigenvalue equation in classical mechanics has the second derivative with respect to time, rather than the first derivative as in (1.10). The general form of the classical equation is often called the Sturm–Liouville equation after two French mathematicians, Jacques Sturm and Joseph Liouville, who did their work in the mid-nineteenth century. Various forms that arise from the specific coordinate system and the detailed potential function have been found, and give rise to a wide library of special functions, usually named after the discoverer of that particular function. Wave equations usually couple spatial derivatives to the time derivatives. Schrödinger's equation is different in that the form is that of a diffusion equation. Classically, a diffusion equation does not produce waves, but the presence of the factor of i in the time-derivative term changes the nature and allows waves to exist. The Schrödinger equation (1.10) clearly, then, is different than what one finds in classical mechanics. Born was highly complimentary of Schrödinger's new wave equation, and formulated his probability conjecture that the magnitude squared of the wave function gave the probability of finding the 'particle' at a point in space [24]. Schrödinger had already shown that the magnitude squared of the wave function could be related to charge density and had developed the probability current [23]. Born went further with his probability interpretation. He recognized that, because the wave function was a complex quantity, the superposition of two wave functions would lead to an interference effect, by which the quantum probability would be significantly different than the classical probability. He pointed out that the wave function did not give the exact position of any particle, only that it contributed to the probability density of where the particle may be found. This statistical approach is significant, as we now know that the Schrödinger equation can be transformed into the classical Fokker–Planck equation of statistical physics and probability theory [25].

Erwin Madelung took careful note of Schrödinger's work, and immediately noted that the probability density had all the appearances of a fluid flow [26]. To investigate the complex nature of the wave function in (1.10), he suggested that one make the substitution

$$y = a\, e^{iS/\hbar}. \tag{1.12}$$

This yields two new equations, an equality relating the real parts of the equation and a second equality relating the imaginary parts of the equation. After some effort in rearranging the various terms of the imaginary parts, he arrived at

$$\begin{aligned} \frac{\partial a^2}{\partial t} + \nabla \cdot \left[a^2 \nabla \left(\frac{S}{m} \right) \right] = 0 \\ \mathbf{v} = \frac{1}{m} \nabla S. \end{aligned} \tag{1.13}$$

The first of these equations is a continuity equation for the probability density, while the second defines the flow velocity in terms of the action S. These equations are today known as the Madelung equations, or the quantum Euler equations in connection with the fluid-like behavior. David Bohm, apparently unaware of Madelung's work, rediscovered this approach and brought these equations to the attention of a wider audience some decades later [27].

The real parts of the Schrödinger equation can be cast in terms of an equation for the energy as

$$\frac{\partial S}{\partial t} = -\left[\frac{1}{2m} (\nabla S)^2 + V + Q \right], \tag{1.14}$$

which is referred to as the quantum Hamilton–Jacobi equation. The first term in the square brackets is the kinetic energy T and V is the potential energy. The factor Q is a new term found by Madelung (and Bohm), which is

$$Q = -\frac{\hbar^2}{2m} \left(\frac{1}{a} \nabla^2 a \right), \tag{1.15}$$

and may be called a quantum potential, although it is commonly called the Bohm potential these days.

Bohm would go on to note that, since the velocity or momentum appeared only in terms of the action, which itself appeared only in the imaginary phase of the wave function, it should be considered as a *hidden* variable. But both he and Madelung realized that the wave function could be a many-particle wave function. The interaction between these particles would appear not only in their possible contribution to the potential energy, but most certainly would appear in the quantum potential. Through this latter quantity, one particle would feel the presence of all the other particles through their contribution to the total wave function and therefore to the quantum potential.

Earle Kennard, of Cornell University, was on sabbatical in Göttingen in 1925–26, so he quickly learned of the new developments in quantum mechanics. At that time, he was interested in hydrodynamics, but quickly became intrigued with the new developments. In 1928, he published his work on the quantum mechanics of a system of particles, showing that the dynamics came directly from the Schrödinger equation without any need for the use of matrices [28]. His starting point was actually the expansion (1.12) introduced by Madelung. He immediately rederived the above equations and this led him to the result

$$m_j \frac{d\mathbf{v}_j}{dt} = -\nabla_j (V + Q). \tag{1.16}$$

Here, as in classical mechanics, the gradient operation is taken with respect to the position of the jth particle. It is clear that the quantum potential acts as any other potential to produce a force acting upon the particular particle. As he pointed out [28], '*each element of the probability moves in the Cartesian space of each particle as that particle would move according to Newton's laws under the classical force plus a "quantum force".*' (italics his). He also pointed out that one could replace the n-dimensional wave function by n-separate wave packets, one for each particle. Thus, to him, the relationships between quantum and classical mechanics stood out very clearly, and the response to an external field would be exactly that of classical mechanics augmented by the quantum force in (1.16). The position of any single particle could be described by taking its classical position and smoothing it over the wave packet for that particle. He then went on to show that the conservation of quantities such as energy and momentum were natural extensions from the classical mechanics.

1.3 Eugene Wigner

Jenô Pál Wigner was born in Budapest (the name was Anglicized to Eugene when he moved to Princeton in the early 1930s). He did his doctoral work in Berlin at the Technical University, and was a regular at the weekly seminars of the German Physical Society, where he was introduced to all the new physics occurring at this time. He became a great colleague of Leo Szilárd, another transplanted Hungarian, and they worked together on many projects. Concerned with the problem of a phase space representation, Wigner and Szilárd found a way to transform the wave function to introduce the missing momentum variables. Wigner published the result in 1932 [29]. He began with the concept of the probability of a position and momentum (for many particles, although we will limit the discussion to a single particle for now) in the classical world, based upon the work of Gibbs and Boltzmann. Although he could not extend the momentum idea to the quantum world, he noted that the mean value of any physical quantity was given by a formula from von Neumann [30] as

$$Qe^{-\beta H}, \tag{1.17}$$

where Q is the matrix operator of the dynamic variable under consideration and H is the Hamiltonian of the system. Here, $\beta = 1/k_B T$ is the inverse temperature of the system. Since the diagonal sum is invariant under any basis transformations, any operator representation can be used. Wigner noted that this led to a somewhat difficult approach to gain the statistical behavior, and then suggested that one can consider the following expression for the probability in phase space:

$$P(x, p) = \frac{1}{\pi\hbar}\int_{-\infty}^{\infty} dy\, \psi^*(x + y)\psi(x - y)e^{2ipy/\hbar}, \tag{1.18}$$

which today is termed the Wigner function. Here, we have recovered a phase space distribution function which is based upon the quantum wave function. However, as Wigner noted, this function is real but is not always positive definite. If integrated over the momentum, it yields the correct probability density of the wave function. More importantly, as expectation values are additive, he pointed out that any function of the coordinates and/or the momentum operator (1.11) would result in the proper form with (1.18) if it were placed between the two wave functions in the normal manner of the Schrödinger wave approach.

While the result for the Wigner function is very suggestive, Wigner himself was quite clear that (1.18) cannot be considered to be the simultaneous probability for a position and a coordinate. This was clear from the fact that the function takes negative values for just this situation. We know that these regions of negative value are connected with the uncertainty relationship of Heisenberg. But, this minor difficulty cannot hinder the advantage of using such a phase space function in many quantum calculations.

Wigner clearly pointed out that his actual development was valid for a wave function for the many-particle case. That is, the wave function in (1.18) can be a function of the position of the N particles. Hence, we will develop a Wigner function which is also a function of the momentum of the N particles. As long as we have a single (many-body) wave function, no confusion arises. Kennard [28] has shown us that the many-electron wave function for N electrons can be decomposed into N single-particle wave functions. But one needs to understand whether the composite wave function is the product of the N single-particle wave functions or the sum of these individual wave functions. Since he asserts that the probability is the product of the individual probabilities, we may assume that the composite wave function is the product of the individual wave functions. Then, in this case, the Wigner function is a product of the Wigner functions of the individual single-particle wave functions. This leads to a straightforward interpretation of the many-electron Wigner function, and any interactions between the individual particles is considered in the total potential energy (and any quantum potential contribution).

Were the many-particle wave function a sum of the individual wave functions, then the Wigner construction would lead to interference terms between the individual packets. Even if the many-particle wave function were constructed on a basis set, with each particle on a different basis function, the presence of the exponential term in (1.17) would lead to failure of the orthogonality principle. This is not necessarily a bad conclusion, especially when we consider the Pauli principle for distinguishable particles. One construction for the many-electron wave function, in which the Pauli principle is included, is the Slater determinant [31]. Expansion of the determinant leads to a sum of products of individual wave functions. The Wigner transformation of the sums will lead to interference terms which are representative of the Pauli interaction. Hence, there is a meaningful interpretation of these interferences. They are not simply mathematical artifacts.

While the Wigner function was an interesting introduction, Wigner was primarily interested in the corrections to the description of thermodynamic equilibrium that would arise from the presence of quantum mechanics. To this end, he also found a correction to the energy, specifically to the potential energy. For example, if we note that, as in classical mechanics, the local probability density varies exponentially with the potential, then this probability is modified as [29]

$$e^{-\beta V}\left[1 - \frac{\hbar^2\beta^2}{12m}\frac{\partial^2 V}{\partial x^2} + \frac{\hbar^2\beta^3}{24m}\left(\frac{\partial V}{\partial x}\right)^2\right] \tag{1.19}$$

for a single particle. This can obviously be extended to the many-particle case so that this probability is affected by the positions of all of the particles in the system. If we suggest that the probability density be written as $P \approx \alpha^2 \approx \exp(-\beta V)$ in the bracketed terms, this will allow a combination of the latter two terms. In addition, this relates Wigner's construction with the Bohm potential. Using this expression in the bracket, we find that the correction term in (1.19) can be approximated as

$$e^{-\beta V}\left[1 - \frac{\hbar^2\beta}{6m\alpha}\frac{\partial^2 \alpha}{\partial x^2}\right] \sim e^{-\beta(V+Q/3)}. \tag{1.20}$$

All that this suggests is that the Wigner approach also leads to a quantum correction to the potential that has similar form to that found by Madelung and Bohm. The difference is in the magnitude of this correction, but the form appears to be very similar. We will explore this further when we address the general nature of effective potentials, which accounts for various quantum corrections.

The important result of Wigner's introduction of the phase space distribution is that it is now possible to treat classical mechanics and thermodynamics on an equal footing with quantum mechanics and thermodynamics. This introduces an additional window into the quantum world and makes it easier to discover the novel new effects of quantum mechanics. Such a phase space approach has become far more useful in recent years where one wants to see the quantum corrections and to make them visible in the analysis of experiments.

1.4 Modern devices and simulation

A *nanodevice* is a functional structure with nanoscale dimensions which performs some useful operation, for example a nanoscale transistor. We almost have to laugh at this comment, as the transistors in modern microprocessors have been made with nanoscale dimensions for more than a decade. Today, we are near the 10 nm (or even the 7 nm) *node*, so-called for the description of a particular design dimension as the 'design node' in Moore's law. Whether we are dealing with a research structure or a modern nanoscale metal-oxide-semiconductor field-effect transistor (MOSFET), the *device* can be considered to have an active region coupled to two contacts, left and right, which serve as a source and sink (drain) for electrons. Here the contacts may be heavily-doped metallic-like reservoirs, characterized by chemical potentials μ_S and μ_D, and are separated by an external bias, $qV_{DD} = \mu_S - \mu_D$. The current flowing through the device is then a property of the chemical potential difference and the transmission properties of the active region itself [32]. A separate gate electrode serves to change the transmission properties of the active region, and hence controls the current. This gate coupling may be through electric fields (field effect), another junction (bipolar operation), or even chemical control. This separation of a nanodevice into ideal injecting and extracting contacts, an active region which limits the transport of charge, and a gate(s) which regulates current flow, is a common way of visualizing the transport properties of nanoscale devices. However, it clearly has limitations: the contacts themselves form/become a part of the active system, and are driven out of equilibrium due to current flow, as well as coupling strongly to the active region through the long-range Coulomb interaction of charge carriers. Disorder, driven by impurities or other random potentials, can disrupt the transport properties and localize a fraction of the carriers.

The nature of transport in a nanodevice depends on the characteristic length scales of the active region of the device. If scattering events are frequent as carriers traverse the active region of the device, carrier transport is *diffusive* in nature, and is reasonably approximated by the semi-classical Boltzmann transport equation (BTE). Energy dissipation occurs throughout the device, and the contacts are simply injectors and extractors of carriers near equilibrium. In contrast, if little or no scattering occurs from source to drain, transport is said to be *ballistic*, and the wave nature of charge carriers becomes important in terms of quantum-mechanical reflection and interference *from the structure itself*, and the overall description of transport is in terms of quantum-mechanical fluxes and transmission, which means quantum transport. Energy is no longer dissipated in the active region of the device, rather it is dissipated in the contacts themselves. Nevertheless, the key element in the modern MOSFET is the quantum potential well at the oxide interface. For quite some time, the use of the effective potential to account for the initial features of quantization has been used in a great many simulations. If we need to discuss the quantum effects that may occur in the transport, then we need to solve something like the Schrödinger equation. We will discuss different approaches to quantum transport in chapter 2, but the Wigner function is highly desirable because of its similarity to the semi-classical BTE.

Beyond MOSFETs, there have been numerous studies over the past two decades of alternatives to classical CMOS at the nanoscale. As dimensions become shorter than the phase-coherence length of electrons, the quantum-mechanical wave nature of electrons becomes increasingly apparent, leading to phenomena such as interference, tunneling, and quantization of energy and momentum, as discussed earlier. Indeed, for a one-dimensional wire, the system may be considered a waveguide with 'modes', each with a conductance less than or equal to a fundamental constant $2e^2/h$. While various early schemes were proposed for quantum interference devices based on analogies to passive microwave structures (see for example [33–35]), most suffer from difficulty in controlling the desired waveguide behavior in the presence of unintentional disorder. As mentioned above, this disorder can arise from the discrete impurity effects as well as the necessity for process control at true nanometer-scale dimensions.

As mentioned, the role of discrete impurities is an undesirable element in the performance of nanoscale FETs due to device-to-device fluctuations. However, the discrete nature of charge in individual electrons and the control of the charge motion of single electrons has in fact been the basis of a great deal of research in single-electron devices and circuits (see for example [36]). The understanding of single-electron behavior is most easily provided in terms of the capacitance C of a small tunnel junction, and the corresponding change in electrostatic energy, $E = e^2/2C$, when an electron tunnels from one side to the other. When the physical dimensions are sufficiently small, the corresponding capacitance (which is a geometrical quantity in general) is correspondingly small, so that the change in energy during the tunneling process is greater than the thermal energy, resulting in the possibility of a 'Coulomb blockade', or suppression of tunnel conductance due to the necessity to overcome this electrostatic energy. This Coulomb blockade effect allows the experimental control of electrons as they tunnel one by one across a junction in response to a control gate bias. As in the case of quantum interference devices, the present-day difficulties arise from fluctuations due to random charges and other inhomogenieties, as well as the difficulty in realizing lithographically defined structures with sufficiently small dimensions to have charging energies approaching kT and above. The larger scale case of such tunneling behavior is the resonant-tunneling diode, which has been treated for a couple of decades as the prototypical quantum device [37]. Indeed, it is important here, because it is the device preferred for many simulations based upon the Wigner function.

So, there are a large number of what one can call nanoscale devices and/or structures, and there are several different approaches to quantum transport. The Wigner function provides only one such approach. However, it basically is the only one that is represented in phase space, and therefore it has natural connections to semi-classical transport and the BTE, as mentioned above. One definitive advantage of this connection is the ability to study the classical to quantum transition in the device physics.

1.5 Our approach

In chapter 2, we begin by discussing the need for quantum transport. Here, we discuss the differences between quantum and classical behavior. Then we turn to various approaches to quantum transport and look at the strengths and weaknesses of each. This will lead to a discussion of the relative advantages and disadvantages of each of these approaches. Then, in chapter 3, we return to the question of the effective potential. There have been many approaches to this concept through the years, beginning with the Madelung–Bohm potential and Wigner's effective potential. In many cases, the effective potential may be all that is needed to bring the first consequences of quantum effects into the semi-classical world. However, they can be coupled with Wigner functions quite effectively; in fact, Wigner himself suggested that they were necessary for the full treatment of quantum thermodynamics.

In chapter 4, we introduce a more extensive introduction to the Wigner function and derive its equations of motion, as well as the equivalent equation for determining the stationary state at any initial time. But other phase space forms have appeared over the years, as variants of the Wigner approach. We discuss their relation to the normal Wigner form and how some differences can be observed. The role of scattering, simplified hydrodynamic approaches and particle simulation approaches are discussed in chapters 5–7. Scattering is important as it provides the dissipation that balances any driving forces in transport. Scattering with the Wigner function can be more complicated than what is found in classical transport. In many cases, the equation of motion for the Wigner function provides more detail than necessary, and effective (i.e. faster) solution techniques arise from taking moments of this equation. This leads to the hydrodynamic approach due to the similarity with other approaches to the Boltzmann equation. Then, in chapter 7, we return to particle simulations as representations of the Wigner function. These are often done via the Monte Carlo approach [12, 13], which has added complexities in the Wigner case. Chapter 8 deals with an important part of the Wigner function, and that is its ability to clarify interference effects and entanglement between correlated particle motion.

Finally, in the remaining chapters, we examine the use of the Wigner function to illustrate important properties like entanglement in the explanation of experiments carried out in a variety of disparate fields. It is this broad usage of the Wigner function in different fields that is the basis upon which it was decided to write this volume. More importantly, however, it is this broad usage of the Wigner function which has brought the latter into the mainstream of quantum transport in the last few years.

References

[1] Planck M 1900 Protokollbuch https://www.dpg-physik.de/veroeffentlichung/archiv/index.html

[2] Rubens H and Kurlbaum F 1900 *Sitzungber. Imp. Pruss. Acad. Sci.* **pt. 2** 929

[3] Young T 1807 *Course of Lectures on Natural Philosophy and Mechanical Arts*vol 1 (London: J. Johnson) lecture XXXIX

[4] Einstein A 1905 *Ann. Phys.* **322** 132

[5] Thompson J J 1907 *Proc. Camb. Phil. Soc.* **14** 417

[6] Taylor G I 1909 *Proc. Camb. Phil. Soc.* **15** 114

[7] Newton I 1686 *Philosophiae Naturalis Principia Mathematica* (London: S. Pepys)

[8] Hamilton W R 1834 *Phil. Trans. R. Soc.* **124** 247
[9] Lagrange J-L 1788 *Mécanique Analytique* (Paris: Chez la Veuve Desaint)
[10] Goldstein H 1950 *Classical Mechanics* (Reading, MA: Addison-Wesley)
[11] See, e.g. ed Ferry D K 1991 *Semiconductors* (Reading, MA: Addison-Wesley) sec 5.11.1, *and references contained therein.*
[12] Jacoboni C and Lugli P 1989 *The Monte Carlo Method for Semiconductor Device Simulation* (Wien: Springer)
[13] Jacoboni C 2010 *Theory of Electron Transport in Semiconductors* (Heidelberg: Springer)
[14] Bohr N 1913 *Phil. Mag.* **26** 1
[15] de Brogle L 1923 *Nature* **112** 540
de Brogle L 1923 versions in French were published in Comptes Rendus **177** 507–548
[16] Heisenberg W 1925 *Z. Phys.* **33** 879
Tr. in ed van der Waerden B L 1967 *Sources of Quantum Mechanics* (Amsterdam: North-Holland) pp 261–76
[17] Born M and Jordan P 1925 *Z. Phys.* **34** 858
Tr. in ed van der Waerden B L 1967 *Sources of Quantum Mechanics* (Amsterdam: North-Holland) pp 277–306
[18] Born M, Heisenberg W and Jordan P 1926 *Z. Phys.* **35** 557
Tr. in ed van der Waerden B L 1967 *Sources of Quantum Mechanics* (Amsterdam: North-Holland) pp 321–85
[19] Gribbon J 2013 *Erwin Schrödinger and the Quantum Revolution* (New York: Wiley)
[20] Schrödinger E 1926 *Ann. Phys.* **79** 361
[21] Schrödinger E 1926 *Ann. Phys.* **79** 489
[22] Schrödinger E 1926 *Ann. Phys.* **79** 734
[23] Schrödinger E 1926 *Ann. Phys.* **80** 437
[24] Born M 1926 *Z. Phys.* **37** 863
[25] Debosscher A 1991 *Phys. Rev.* A **44** 908
[26] Madelung E 1927 *Z. Phys.* **40** 322
[27] Bohm D 1952 *Phys. Rev.* **85** 166
Bohm D 1952 *Phys. Rev.* **85** 180
[28] Kennard E H 1928 *Phys. Rev.* **31** 876
[29] Wigner E P 1932 *Phys. Rev.* **40** 749
[30] von Neumann J 1927 *Gött. Nachr.* 273 (Proc. Meeting of the Göttingen Acad. of Sciences for 11 November 1927, full text available at http://gdz.sub.uni-goettingen.de/dms/load/img/?PPN=PPN252457811_1927&DMDID=DMDLOG_0020)
[31] Slater J C 1929 *Phys. Rev.* **34** 1293
[32] Landauer R 1957 *IBM J. Res. Develop.* **1** 223
Landauer R 1970 *Phil. Mag.* **21** 863
[33] Sols F, Macucci M, Ravaioli U and Hess K 1989 *J. Appl. Phys.* **66** 3892
[34] Datta S 1989 *Superlatt. Microstruct.* **6** 83
[35] Weisshaar A, Lary J, Goodnick S M and Tripathi V K 1991 *IEEE Electron. Device Lett.* **12** 2
[36] Likharev K K 1999 *Proc. IEEE* **87** 606
[37] Ferry D K 2017 *An Introduction to Quantum Transport in Semiconductors* (Singapore: Pan Stanford Publishing)

IOP Publishing

The Wigner Function in Science and Technology

David K Ferry and Mihail Nedjalkov

Chapter 2

Approaches to quantum transport

From the past chapter, and the desire to learn more about Wigner functions, it is clear that a more detailed modeling of the quantum contributions in modern physics and technology is desired. These contributions appear in many forms: (1) changes in the statistical thermodynamics within the structures themselves as well as in their connection and interaction with the external world, (2) new critical length scales, (3) an enlarged role for ballistic transport and quantum interference, and (4) new sources of fluctuations which will affect performance. Indeed, many of these effects have already been studied at low temperatures where the quantum effects appear more readily in such devices [1]. But there are many approaches to quantum transport, from the simple to the very complex. In the present chapter, we want to briefly introduce the main approaches and to set the realm of Wigner functions properly in the landscape defined by these various approaches. In that sense, we hope to get a feeling for both the advantages and the disadvantages of the Wigner function methods. First, however, it is probably fruitful to describe just what we mean by quantum transport and what it entails.

One of the hallmarks of classical physics is the strong connection between energy and momentum. For a free particle, it is clear that the relationship $E = p^2/2m$ is a fundamental relationship. Even when we move to the quantum mechanics of the energy bands in semiconductors, we still hold to the tenet that the energy is well defined at each and every value of momentum, in this case the crystal momentum. But then we have to remind ourselves that the normal calculation of energy bands is done for the so-called empty lattice, in which we seek the allowed energy levels for a single electron [2]. This is the reason that the approach called density functional theory (DFT) does not always give us nice results. Calculations for the energy bands of semiconductors using DFT is notorious for not getting the energy gap (between the conduction and valence bands) correct. The usual approach is to use a linear density approximation for the exchange and correlation energies, but without really correcting the single electron approximation, since it basically is a random-phase

doi:10.1088/978-0-7503-1671-2ch2

approximation with a local density. What has to be done is to account for the interactions among the electrons, usually through what is known as a GW approximation, where G is a Green's function and W an interaction self-energy. This lowers the entire valence band through the cooperative interactions and provides a correction to the energy gap. How this is performed is not the subject we want to address here, but the phrase 'self-energy' is a topic we want to begin to discuss, especially as it is connected with going beyond the simple relationship $E = p^2/2m$. In three-dimensional momentum space, this latter relationship defines a spherical shell of the given energy, and all states with this energy lie on this shell. In any computation, this connection between energy and momentum is introduced by the delta function

$$\delta(E - p^2/2m). \tag{2.1}$$

When we move to the quantum world, however, this constraining relationship is no longer either valid or required. As we will see later, energy and momentum are separate variables, although they do have a connection, just one not so strict as (2.1).

The relationship (2.1) can be thought of as a resonance condition, and we can actually see how it is modified with a classical analog. Let us consider an R, L, C circuit from undergraduate circuit theory. It does not matter whether this is a parallel or a series circuit in principle, although the exact relationships will differ. So, let us consider the series case. This circuit has a resonant frequency in the absence of the resistance, which is given by

$$\omega_0 = \frac{1}{\sqrt{LC}}. \tag{2.2}$$

However, when the resistance is present, then the oscillation of the circuit is damped exponentially in time with the time constant $2L/R$, and the frequency of the oscillation is shifted to

$$\omega = \omega_0\sqrt{1 - \frac{R^2}{4L^2\omega_0^2}} \equiv \omega_0\sqrt{1 - \frac{\omega_X^2}{\omega_0^2}}, \tag{2.3}$$

where we have defined $\omega_X = R/2L$. The resistance leads to a quality factor $Q = \omega_0 L/R$, which tells us how good the resonant circuit is at its task. But, if the resistor has too large a value, then the second term in the bracket can be larger than 1, and the circuit will no longer exhibit the resonance. In order to connect to our energy and momentum relationship, we note that the resistance moves the frequency away from the shell value of ω_0.

Two things happened when the resistance was added to the resonant circuit. First, damping appeared, and, second, the frequency was shifted by the finite Q of the circuit. In the quantum world, the resistor represents interactions which take the system from a one-electron picture to what is called the many-body picture. These interactions can be among the many electrons that are present, or among the electrons and the lattice vibrations, or among the electrons and the impurities in the system, and so on. But, the interactions lead to the same two effects: energy broadening and shift of the peak. Generally, though, we define them with a single term: the self-energy. In the world of Schrödinger wave mechanics, the time variation for an energy eigenstate is the simple exponential

$$e^{i\omega t} = e^{iEt/\hbar}, \tag{2.4}$$

where we have used the Planck relationship between frequency and energy. In the absence of interactions, the energy is given by the momentum according to (2.1), but when the interactions are present, the energy is shifted to

$$E = \frac{p^2}{2m} - \Sigma(p). \tag{2.5}$$

The last term in the above equation is the self-energy. The real part of the self-energy corresponds to a shift downward of the energy (frequency shift). The imaginary part of the self-energy provides a damping of the state in time, which is the analog of the resistive damping of the resonant circuit. So, in situations when the self-energy is non-zero, momentum states which do not lie on the energy shell can be important in the transport. The impact of these states is cleverly described as off-shell effects, thereby obfuscating their role to anyone who is not one of the cognoscenti. But, how do we account for these off-shell effects? The answer lies in (2.1). The so-called delta function becomes a broadened function which peaks at the value of energy given by the momentum term, although it is now like a probability function which yields the probability of states off the shell being important. The form of this function does not concern us here, but it is often called the spectral density function.

Now, it has to be pointed out that not all formulations of quantum transport make clear the role of this new spectral density function. Indeed, in many formulations it is sometimes obscure as to whether or not the fuzziness actually occurs, or is even being considered. Nevertheless, a key distinction between quantum and classical transport is just this existence of the states off the energy shell and their contribution to the transport. Indeed, it will be important to estimate the importance of this term in the Wigner formulation of quantum transport. This is because the Wigner function is explicit in phase space, and the energy is not an explicit variable in the equations. Then, we expect the spectral density will be quite subtle in its presence.

In general, any physical level of description of many-body effects is done with the help of macroscopic parameters that are related to length and time scales in the problem. One of the most important lengths in quantum transport arises from processes that can break the phase coherence of the wave function. Such processes are not necessarily energy relaxing, although nearly all energy relaxation processes also break phase coherence. For example, phase-breaking processes which do not relax energy can arise from elastic interactions that are sufficiently strong they introduce localization. In general, we distinguish between different situations by the discussion of appropriate *lengths*. Hence, we can define a phase-breaking length as

$$l_\varphi = \sqrt{D\tau_\varphi}, \tag{2.6}$$

where D is the diffusion constant related to the mobility as

$$D = \frac{\mu k_B T}{e} \tag{2.7}$$

for non-degenerate carrier statistics. In the case in which the statistics are degenerate, there is a correction to (2.7) which involves the ratio of a pair of Fermi–Dirac integrals whose specific order depends upon the dimensionality of the system. In addition to this important length, we also have various lengths that are associated with the scattering processes. One of these is the mean free path:

$$l = v\tau_m, \tag{2.8}$$

where v is usually either the thermal velocity or the Fermi velocity and τ_m is the mean free time. Note that this is not the actual scattering time τ, but is usually thought of as the momentum relaxation time. The second is the inelastic length:

$$l_{in} = v\tau_{in}. \tag{2.9}$$

Many other lengths have been introduced for various situations, and these will be discussed as the need arises. But it is important to note that equations (2.6) through (2.9) are not independent of one another.

As we consider physical systems, we must also be aware that there exists a hierarchy in (length) scale, as we move from the macro-scale through the micro-scale to the nano-scale. At the lowest scale, the basic concepts of transport in nano-scale quantum systems, in the presence of localized scatterers or tunneling barriers, can be traced to Landauer [3]. From many approaches, it is also clear that the onset of inelastic scattering will suppress many of the quantum effects that are of interest, primarily through broadening of the density of states as discussed above, but also broadening the important nature of the quantum states themselves. It has often been assumed that the Landauer formula is only applicable in cases of ballistic transport, but this is just not true. The Landauer approach can be applied in nearly any situation and with nearly any scattering process as long as it is done in a smart manner that incorporates the specific (and detailed) physics of the system. That is, the Landauer approach is quite fundamental as its extensions to the use of the distribution functions in the contact regions is all that is required. We will discuss this in the next section. From the wave functions that are given by the Schrödinger equation, one can generate a quite general quantum statistical quantity, whether this is the density matrix or the Green's function. These are related to each other, and also lead to the Wigner function, which is the phase space version of them. The Green's function can also be extended to a non-equilibrium form, which is necessary when the system is far from equilibrium. That is to say, all of the approaches we describe work quite well in the non-equilibrium, or far from equilibrium, world. However, the Landauer approach is more usually found in the near-equilibrium situation, but this is not a real limit to its use. These various approaches will be discussed in the subsequent sections of this chapter.

2.1 Modes and the Landauer formula

Let us now expand the principles that are inherent in the Landauer formula mentioned above. For this, we consider a quasi-two-dimensional semiconductor system that is of finite extent. We assume that this semiconductor system has a narrow width and a much longer length, so that it appears more as a ribbon. The confining potential in the transverse (narrow) direction leads to quantization and a set of transverse modes. Each transverse energy level will produce one mode propagating in each direction if this energy lies below the Fermi energy (see figure 2.1). Hence, the ribbon can be considered as a waveguide. Many times, the transverse confinement is created by a experimental situation, such as by the use of gates which impose an electrostatic potential upon the overall two-dimensional system. Then, Schrödinger's equation can be written as

$$-\frac{\hbar^2}{2m}\left(\frac{\partial^2}{\partial x^2} + \frac{\partial^2}{\partial y^2}\right)\psi(x, y) + V(x, y)\psi(x, y) = E\psi(x, y) \tag{2.10}$$

with

$$V(x, y) = V_c(x, y) + V_a(x, y) \tag{2.11}$$

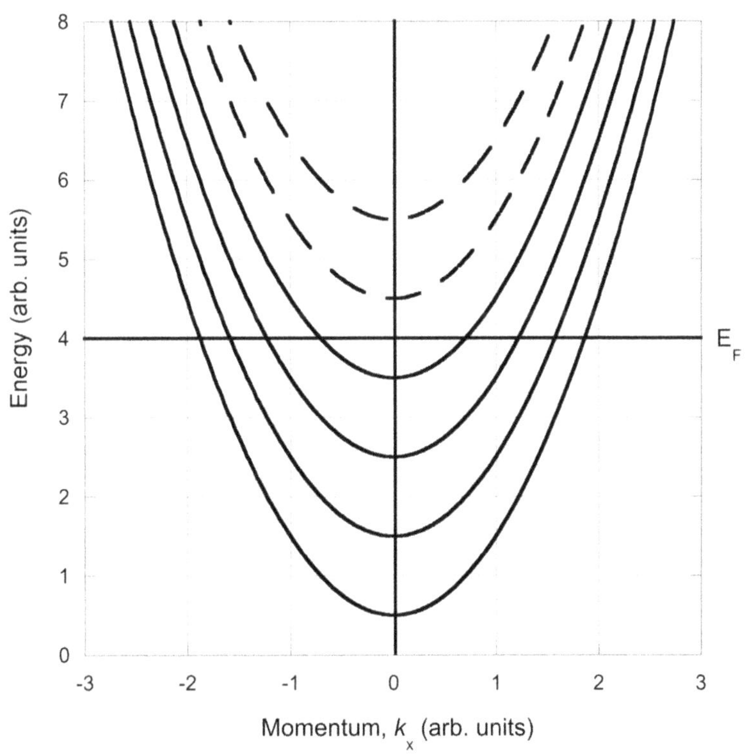

Figure 2.1. Momentum as a function of total energy for various modes in the transverse direction that arise from (2.13). The abscissa is the longitudinal momentum.

and V_a can be any potential describing impurities, bias, or phonon scattering through an imaginary self-energy, as well as the applied potential V_a, and V_c defining the confinement potential. The confinement potential is taken as the transverse potential that leads to the mode creation.

The general solution of the wave function within the waveguide region, which is defined by the confinement potential, may be written as a product form in terms of various modes as

$$\psi(x, y) = \sum_n \varphi_n(x)\chi_n(y), \tag{2.12}$$

where, in general, we take the y-direction as the transverse confinement direction. Hence, we may write the transverse modes as

$$\chi_n(y) = \sqrt{\frac{2}{W}} \sin\left(\frac{n\pi y}{W}\right). \tag{2.13}$$

If the transverse potential is a soft wall potential, then the wave function will assume another form, but this one is sufficient for our present purposes. The longitudinal modes are described, in general, by a combination of forward and backward waves, as

$$\varphi_n(x) = a_n e^{\gamma_n x} + b_n e^{-\gamma_n x}, \tag{2.14}$$

where γ_n is the propagation constant. If the mode is a propagating mode, which usually corresponds to an eigen-energy below the Fermi energy, then this propagation constant is imaginary and describes the wave nature of the mode, through $\gamma_n = k_n = (p_n/\hbar)$. If, on the other hand, the mode is evanescent, which usually corresponds to an eigen-energy above the Fermi energy, then the propagation constant is real. It is these evanescent modes that create the problem with simple matching of modes at interfaces in the waveguide region. One solution of (2.14) is always growing in either the forward or backward direction, and this leads to an instability in which the amplitude of the wave grows without limit. It would be simple to just ignore the evanescent modes, but they must be included in the matching problem as they will correspond to reactive elements at the boundaries, just as in microwave waveguides. They can also represent tunneling transfer across short regions.

We can see the nature of the modes by referring to figure 2.1. Here, we plot the kinetic energy of each of the lowest few transverse modes. This kinetic energy is given by the wave number k_n ($\gamma_n = \mathrm{i}k_n$) for the modes where the minimum of the curves lie below the Fermi energy. Here, we show four modes below the Fermi level, and two of the many modes which have quantized energies above the Fermi level. Thus, there will be four channels flowing through the waveguide. The picture in figure 2.1 is for an unbiased channel. When bias is applied the left-going modes will shift relative to the right-going modes due to the bias. Generally, we may assume that various modes do not cross along the length of the channel, which is generally valid for low bias.

As remarked above, Rolf Landauer presented an approach to transport and the calculation of conductance that was dramatically different from the microscopic kinetic theory based on the previously utilized Boltzmann equation (and which is still heavily utilized in macroscopic conductors) [3, 4]. He suggested that one could compute the conductance of low-dimensional systems simply by computing the transmission of a mode from an input reservoir to a similar mode in an output reservoir. The transmission of this probability from one mode to the other was then very similar to the computation of the tunneling probability, except that there was no requirement that the process be one of tunneling. The only real constraint was that of lateral confinement so that the two reservoirs could be discussed in terms of their transverse modes (2.13). The key property of the two reservoirs is that they are in equilibrium with any applied potentials. That is, the electrons in the reservoirs are to be described by their intrinsic Fermi–Dirac distributions, with any externally applied potentials appearing only as a shift of the relative energies (which would shift one Fermi level relative to the other). While he originally considered that the transport was ballistic, this is not required. Rather, the requirement is that we can assign a definitive mode to the electron when it is in either of the two reservoirs, which means that if scattering is present, it must be described specifically as a transfer of the electron from one internal mode to another within the active region and not in the reservoir. If we consider a potential applied to the right (output) reservoir relative to the left (input) reservoir, then the right reservoir emits carriers into the active region with energies up to the local Fermi level plus the applied bias, $E_F + eV_a$ (note that the energy eV_a will be negative for a positive voltage). The left reservoir emits electrons into the active region with energies only up to E_F. In the following discussion, we will assume that the applied voltage is quite small, although this also is not a stringent requirement.

For simplicity, we make the initial assumption that no scattering takes place within the active region of the structure, so that we can make a definitive association between the energy of the carriers and their direction of propagation. That is, electrons injected into the active region from the left reservoir have a positive momentum and travel from left to right. On the other hand, electrons injected into the active region from the right reservoir have a negative momentum and thus travel from the right to the left. Because of the applied bias, there are more electrons traveling to the right than are traveling to the left, and this gives us a net current through the constriction. The imbalance between left-going and right-going electrons means that the constriction is in a condition of non-equilibrium, which is required to support a net current.

In order to compute the current through the active region, we need first to determine the charge that is occupied in each transverse mode in the energy range between E_F and $E_F + eV$. As we remarked above, if the applied bias is small, then we can estimate the excess charge as

$$\delta Q = e \cdot eV_a \cdot \frac{1}{2}\rho_{1d}(E_F) = e^2 V_a \sqrt{\frac{m^*}{2\pi^2\hbar^2 E_F}}, \tag{2.15}$$

where ρ_{1d} is the one-dimensional density of states for the semiconductor with effective mass m^*. Hence, we are treating each mode as quasi-one-dimensional in the longitudinal direction. It should be noted that we use only one-half of the density of states since we are only interested in those electrons that are traveling to the right (those with positive momentum). Also, because of our definition of the density of states, (2.15) is the charge per unit length. In order to compute the current, we need only multiply (2.15) by the group velocity of the carriers at the Fermi energy, which is

$$v_F = \frac{1}{\hbar}\frac{dE}{dk_x}\bigg|_{E=E_F} = \frac{\hbar k_F}{m^*} = \sqrt{\frac{2E_F}{m^*}}, \tag{2.16}$$

and this leads to the current:

$$I = \delta Q \cdot v_F = \frac{2e^2}{h}V_a. \tag{2.17}$$

A very interesting effect has occurred in the above derivation of the current, and that is the energy has dropped out of the equation. Hence, the current is completely independent of the energy, and this is true for each quasi-one-dimensional channel that flows through the ribbon. The result (2.17) has no dependence on the channel index; the current is identical in each channel. This result has been called the equipartition of the current, i.e. the current is divided equally among the available channels. Since the current in each channel is equal, the total current carried in the constriction is simply the number of such channels N that are occupied and the current per channel. Hence,

$$I_{total} = NI = N\frac{2e^2}{h}V_a. \tag{2.18}$$

From this last expression, we obtain the conductance through the constriction as

$$G = \frac{I_{total}}{V_a} = \frac{2e^2}{h}N. \tag{2.19}$$

This last expression is called the Landauer formula. Thus, the conductance of the quantum wire (our constriction) is quantized in units of $2e^2/h$ (~77.28 μS) with a resulting magnitude that depends only upon the number of occupied channels. However, this value is obtained for spin degeneracy; the factor of 2 arises for the two spin-degenerate channels. In the case of some materials, such as Si, there can also be a valley degeneracy, usually 2. This will double the conductance given in (2.19).

We note from (2.19) that, as one raises the Fermi energy, the conductance increases through a set of steps. These steps in the conductance occur as individual channels are occupied or emptied (depending upon a rising or decreasing potential). As each new channel is occupied, the conductance jumps upward by $2e^2/h$, and this increase is the same for each and every channel. To clearly see the steps in the conductance requires high-mobility materials so that the transport is almost ballistic.

The presence of significant scattering will wash out the steps by introducing transitions in which electrons jump from one channel to another and this will upset the balance in the channels. The conductance steps were observed experimentally in what are known as quantum point contacts (QPC), which are narrow channels defined in a quasi-two-dimensional semiconductor by a set of gates [5, 6].

2.2 The scattering matrix approach

A normal approach to solving the Schrödinger is to match the wave function and its derivative across interfaces, but this is generally unstable when a large number of such interfaces is present. However, stability can often be restored by modifying the approach to a recursion technique based upon the scattering matrix. Long a staple in the treatment of microwave systems [7], the scattering matrix entered quantum mechanics through the Lippmann–Schwinger equation [8]. Microwave systems, as well as confined quantum systems, give rise to *modes*, as discussed in section 2.1. What we mean by modes is the following. In the simplest case, we are seeking to solve the Schrödinger equation for a two-dimensional ribbon of semiconductor. This ribbon has a finite width and is connected to two reservoirs at the ends. One of these reservoirs is considered to be the source (usually taken to be the left-hand reservoir) and the other is considered to be the drain. Because the ribbon has a well-defined width, as we described in the previous section, the particles are constrained to remain in the ribbon because of potential barriers on the two sides of the ribbon. This potential constraint causes a quantization in the transverse direction. Hence, the solutions of the Schrödinger equation consist of a set of modes, precisely as described in (2.12). The difference here, however, is that we will discretize the Schrödinger equation on a two-dimensional grid. Like the Landauer equation of the previous section, each transverse mode will propagate from the source toward the drain. As it passes through the active ribbon region, part of it will be reflected and part will be transmitted, just as for the Landauer equation, except that this is all found numerically. There are two approaches. First, the entire Hamiltonian matrix may be set up for the overall two-dimensional grid, and the wave functions found by matrix diagonalization. This usually limits one to a small grid due to the matrix inversion problems for large matrices. The second approach is to use a recursion algorithm to start with the solutions in one reservoir, and generate the general solution by recursion, slice by slice as one propagates through the ribbon. This is the approach used here. The strength of this approach lies in the fact that modal solutions of the Lippmann–Schwinger equation maintain their orthogonality through the scattering process [8]. Even in the non-recursion mode, the use of the scattering states approach, described here, provides a viable method of building up an orthogonal ensemble, even with weighting by, for example, a Fermi–Dirac distribution [9]. This approach allows one to determine the transmission probability from one excited mode to a second mode in the output end of the structure, and therefore to use the well-known Landauer formula [3, 4].

The procedure generally begins with discretizing the ribbon so that the equations of the previous section are evaluated on a grid in the two-dimensional space. The wave function of each transverse slice thus is given in the *site* representation; that is, it is evaluated at each grid point in the lateral slice. As there are different modes, there will be different wave functions whose values on the sites vary significantly. However, the procedure must be initiated, and this is done in the so-called *contact* layer, usually the left reservoir. In this layer, it is assumed that equilibrium conditions exist, and for the present we consider that the system is at a zero-temperature state (non-zero temperature is easily added, as we point out below). That is, the first step is to determine the allowed solutions in the transverse direction, e.g. the solutions $\chi(y)$ in (2.12). As there are forward and reverse modes and eigenvalues, we have to first find these modes and eigenvalues in the contact layer. This is done through the equation [10]

$$\det\,[\mathbf{T}_0 - \lambda \mathbf{I}] = 0, \tag{2.20a}$$

where $\mathbf{T_0}$ is a matrix whose rank is twice that of the Hamiltonian, and is given as

$$\mathbf{T}_0 = \begin{bmatrix} \mathbf{U}_+ & \mathbf{U}_- \\ \lambda_+ \mathbf{U}_+ & \lambda_- \mathbf{U}_- \end{bmatrix}. \tag{2.20b}$$

Here, $\mathbf{U}_\pm$ are matrices whose rank is equal to that of the Hamiltonian in its discretized mode, and that represent the modes of the structure. In essence, these eigenvalues are found by solving the equation

$$\begin{bmatrix} \mathbf{0} & \mathbf{I} \\ -\mathbf{H}_{0,1}^{-1}\mathbf{H}_{0,-1} & \mathbf{H}_{0,1}^{-1}(E\mathbf{I} - \mathbf{H}_0) \end{bmatrix} \begin{bmatrix} \mathbf{u}_m^\pm \\ \lambda_m^\pm \mathbf{u}_m^\pm \end{bmatrix} = \lambda_m^\pm \begin{bmatrix} \mathbf{u}_m^\pm \\ \lambda_m^\pm \mathbf{u}_m^\pm \end{bmatrix} \tag{2.21}$$

at the zero slice. Here, $\mathbf{H}_0$ is the bare transverse Hamiltonian (the second subscript refers to the transverse slice number). The set of values for λ are the eigenvalues and describe the propagation constants in (2.14). These have both positive and negative signs, which correspond to the subscripts in (2.21). The positive subscript is for modes propagating from the left contact to the right (and to the other contact), while the negative one is for modes going in the opposite direction. The matrix $\mathbf{U}$ is generated during the solution of (2.20), and the corresponding diagonalization problem. Each column of $\mathbf{U}$ provides the wave function for a particular mode with the elements representing the value on each grid point of the slice. In essence, these matrices also represent the needed mode to site transformation matrices. As remarked, the $\boldsymbol{\lambda}_\pm$ are the diagonal eigenvalue matrices whose elements are the values of the propagation constants that appear in (2.14). Associated with this zero slice is a value of the Fermi energy, so that modes whose eigenvalues are below the Fermi energy have propagating wave vectors, while those whose eigenvalues are above the Fermi energy are evanescent modes.

We now introduce two matrices $\mathbf{C}_1^s$ and $\mathbf{C}_2^s$ which represent the amplitudes of the forward (positive) and backward (negative) modes, respectively, at slice s (here, this slice defines the values on the lattice points in the transverse direction; s refers to the longitudinal point). It is these matrices with which we will work in the recursion procedure. Now, the scattering matrix recursion relation may be expressed as [11]

$$\begin{bmatrix} \mathbf{C}_1^{s+1} & \mathbf{C}_2^{s+1} \\ \mathbf{0} & \mathbf{I} \end{bmatrix} = \begin{bmatrix} \mathbf{0} & \mathbf{I} \\ -\mathbf{I} & t^{-1}(\mathbf{H}_{0,s} - E\mathbf{I}) \end{bmatrix} \begin{bmatrix} \mathbf{C}_1^s & \mathbf{C}_2^s \\ \mathbf{0} & \mathbf{I} \end{bmatrix} \begin{bmatrix} \mathbf{I} & \mathbf{0} \\ \mathbf{P}_{1,s} & \mathbf{P}_{2,s} \end{bmatrix}, \tag{2.22}$$

where $\mathbf{H}_{0,s}$ is the slice Hamiltonian (with the values for the local potentials that exist on that slice). If we left off the last matrix, the recursion would be no more than mode matching, which is unstable. The last matrix is the transformation to scattering matrix form, but we still need to determine some values. The matrices $\mathbf{P}_1$ and $\mathbf{P}_2$ are unknowns, but are required to satisfy this recursion equation. The quantity $t = \hbar^2/2ma^2$ is the 'hopping energy' between grid points and a is the spacing of the grid points. The $\mathbf{P}$ matrices may be found by expanding the matrix products, as

$$\begin{aligned} \mathbf{C}_1^{s+1} &= \mathbf{P}_{1,s} = -\mathbf{P}_{2,s}\mathbf{T}_{21,s}\mathbf{C}_1^s \\ \mathbf{C}_2^{s+1} &= \mathbf{P}_{2,s} = (\mathbf{T}_{21,s}\mathbf{C}_2^s + \mathbf{T}_{22,s})^{-1} \end{aligned}. \tag{2.23}$$

Here, the $\mathbf{T}$ matrices refer to the various entries in the first block matrix on the right-hand side of (2.22). Hence, $\mathbf{T}_{22}$ refers to the Hamiltonian and energy matrix, which is the first large matrix on the right-hand side of (2.22). The second line of (2.23) is a form of the Dyson equation, while the first line is the transformation of mode amplitudes into scattering matrix form. The iteration is started by the initial conditions $\mathbf{C}_1{}^0 = \mathbf{I}$ and $\mathbf{C}_2{}^0 = \mathbf{0}$ (for non-zero temperature, the value of unity in $\mathbf{C}_1{}^0$ is replaced by the value of the Fermi function at the energy of that particular mode). The equation (2.23) then becomes the recursion relation. At each step, the local values for the $\mathbf{P}$ matrices are determined at a slice, and then the recursion is applied to reach the next slice. Once the last slice is reached, the final transmission matrix is found using the mode to site transformation as

$$\mathbf{t} = -(\mathbf{U}_+\lambda_+)^{-1}[\mathbf{C}_{1,N+1} - \mathbf{U}_+(\mathbf{U}_+\lambda_+)^{-1}]^{-1}. \tag{2.24}$$

It is important here to explain that the Landauer equation (2.19) is written in terms of modes, whereas the internal computations here are in terms of values on the discrete grid points. The mode to site transformations provide the conversion from one picture to the other. The numerical stability of this method in large part stems from the fact that the iteration implied by (2.22) and (2.23) involves products of matrices with *inverted* matrices. Taking such products tends to cancel out most of the troublesome exponential factors that arise in straight mode matching approaches.

Once the transmission is determined, we can also find the wave functions in the interior of the structure by back propagating from the output contact. At the last slice, it can be shown that the wave function is just

$$\psi_N = \mathbf{P}_{2,N}. \tag{2.25}$$

Of course, this wave function is a matrix whose columns are the mode wave functions, and in turn whose elements are the values on each grid point. Propagating backwards, we do the iteration

$$\psi_s = \mathbf{P}_{1,\,s} + \mathbf{P}_{2,\,s}\,\psi_{s+1}, \tag{2.26}$$

and the density at each grid point can be found from a sum over the various mode contributions at that site, as

$$n(x,\,y) = n(i,\,j) = \sum_{k=1}^{M} |\psi_k(i,\,j)|^2, \tag{2.27}$$

where k is the mode index (there are as many modes as there are transverse grid points). Because each mode may have a different velocity associated with it, Landauer's formula must be modified to account for this as

$$G = \frac{2e^2}{h}\sum_{n,m}\frac{v_n}{v_m}\,|t_{nm}|^2, \tag{2.28}$$

where t_{nm} is the transmission from mode n in the input contact to mode m in the output contact. The summation in (2.28) is carried out only over the propagating modes, and the velocities are, of course, those in the longitudinal direction of the device.

The magnetic field may be included in the solutions to the Schrödinger equation once one decides upon an appropriate gauge for the vector potential. Generally, the magnetic field is oriented normal to the surface defined by the ribbon coordinates. That is, if the ribbon is defined as in (2.12), then the magnetic field is taken to lie in the z-direction. But, we have to face the fact that it is the vector potential, and not the magnetic field, that appears in the Hamiltonian. There are various gauges that are commonly used, but we almost always work in the overall Coulomb gauge (or electrostatic gauge), where the divergence of the vector potential is set to zero. This then allows us to uniquely connect the magnetic field to the vector potential through

$$\mathbf{B} = \nabla \times \mathbf{A}. \tag{2.29}$$

When we study the magnetic field effect in various mesoscopic devices, there are two usual further choices for the gauge that are made. These arise from the manner in which we can force the magnetic field and the vector potential to satisfy (2.29). One of these is the *Landau* gauge:

$$\mathbf{A} = Bx\mathbf{a}_y, \tag{2.30}$$

where the last quantity is a unit vector in the direction transverse to the current flow direction in the approach of the last section. The other choice often used is the *symmetric* gauge:

$$\mathbf{A} = \frac{1}{2}(-By\mathbf{a}_x + Bx\mathbf{a}_y). \tag{2.31}$$

In quantum mechanics, it is quite general to make the wave function gauge invariant, especially if we want to talk about the wave function as a field. Normally, we invoke gauge invariance through the condition that we create a new function Λ, which we use to change the vector potential through

$$\mathbf{A} \rightarrow \mathbf{A} + \nabla\Lambda. \tag{2.32}$$

This now also requires that the scalar potential be modified, and the wave function is modified according to

$$\psi(\mathbf{r}) \rightarrow e^{ie\Lambda/h}\psi(\mathbf{r}). \tag{2.33}$$

If we make a connection between this gauge change and the vector potential itself, we then arrive at the Peierls' substitution which changes the momentum via the vector potential as

$$\mathbf{p} \rightarrow \mathbf{p} - e\mathbf{A}. \tag{2.34}$$

Now, as we move around each small square of the discretized grid, we couple phase to the wave function from the presence of a magnetic field normal to the two-dimensional plane. In the Landau guage, the amount of phase change is given by

$$\delta\varphi = \frac{e}{h}\oint \mathbf{A} \cdot d\mathbf{l} = \frac{e}{h}\iint B_z dx\, dy = 2\pi\frac{\delta\Phi}{\Phi_0}, \tag{2.35}$$

where $\delta\Phi$ is the flux coupled through that square and $\Phi_0 = h/e$ is the flux quantum. This is accomplished by adding a phase factor to the hopping energy t, in which the phase is given by

$$t \rightarrow te^{i\varphi}, \quad \varphi = \frac{eBa^2 j}{\hbar}, \tag{2.36}$$

where a is the grid spacing (uniform in this case) and j is the transverse row index. One point worth noting is that the simulations are often more stable if the number of rows is odd, and the j index above is taken to be zero at the center line of the simulation space. Hence, this would give positive and negative values for the vector potential, although this has no effect on the resulting magnetic field or the overall results of the simulation. The above approach is easily extendable to three dimensions [12], and scattering can be included through the introduction of the self-energy in the site representation [13].

2.3 The density matrix

Generally, one can solve the Schrödinger equation by assuming an expansion of the wave function in a suitable basis set so that each basis function is an energy eigenfunction according to

$$H\varphi_n = E_n\varphi_n, \tag{2.37}$$

in which E_n is the energy level corresponding to the particular basis function. Then, the total wave function can be written as

$$\psi(\mathbf{r}, t) = \sum_n c_n \varphi_n(\mathbf{r}) e^{-iE_n t/\hbar}. \tag{2.38}$$

If we now multiply this equation with the complex conjugate of another basis function, say $\varphi_m^*(\mathbf{r})$, and integrate over the volume, we can determine the coefficients of the expansion as

$$\int d\mathbf{r} \varphi_m^*(\mathbf{r}) \psi(\mathbf{r}, t) = \sum_n c_n \delta_{nm} e^{-iE_n t/\hbar} = c_m e^{-iE_m t/\hbar}, \tag{2.39}$$

and the coefficient can be inserted into (2.38), using different time and position, to give

$$\psi(\mathbf{r}, t) = \sum_n \int d\mathbf{r}' \varphi_n^*(\mathbf{r}') \varphi_n(\mathbf{r}) \psi(\mathbf{r}', t_0) e^{-iE_n(t-t_0)/\hbar}. \tag{2.40}$$

Now, this equation can be rewritten as

$$\psi(\mathbf{r}, t) = \int d\mathbf{r}' K(\mathbf{r}, t; \mathbf{r}', t_0) \psi(\mathbf{r}', t_0), \tag{2.41}$$

which defines the generalized propagator or kernel K. This quantity is a function of two positions and two times. It is often called a Green's function, but we will define this more efficiently below.

The kernel describes the general propagation of any initial wave function at time t_0 to any arbitrary time t (which is normally greater than t_0, but not always). There are a number of methods of evaluating it, either by differential equations or by integral equations known as path integrals [14]. In general, the usefulness of the present form (2.41) lies in the existence of the entire set of basis functions, which are characteristic of the problem at hand. Normally, the diagonal elements of the density matrix, which is the magnitude squared basis function, is used to compute the average of a physical quantity. Here, the kernel generally represents an admixture of states, which differs from the average by the inclusion of the off-diagonal terms which infer correlations between the wave function at different positions and elements of the basis set. We define the kernel as

$$K(\mathbf{r}, t; \mathbf{r}', t_0) = \sum_{n,m} c_{nm} \varphi_m^*(\mathbf{r}, t) \varphi_n(\mathbf{r}', t_0). \tag{2.42}$$

The indication of (2.42) is that the basis functions are now time dependent, but this could arise merely through the exponential functions of the energy eigenvalues for these functions. We will see this form again when we discuss Green's functions below in the next section. For now, however, we concentrate on the density matrix form of these equations.

In a great many situations, the separation into two times is not necessary, as the processes which lead to these separate times are incredibly fast. For the general situation, one may then reduce the kernel to a single time variable, and we then have the density matrix

$$\rho(\mathbf{r}, \mathbf{r}', t) = \sum_{n,m} c_{nm}\varphi_m^*(\mathbf{r}, t)\varphi_n(\mathbf{r}', t) \equiv \psi^*(\mathbf{r}, t)\psi(\mathbf{r}', t). \tag{2.43}$$

There are many different definitions of the density matrix, but these differ only in the fine detail. One can also introduce the imaginary time $t \to -i\hbar\beta$, where $\beta = 1/k_BT$ is the inverse temperature. Then, the density matrix can be written in the thermal equilibrium form:

$$\rho(\mathbf{r}, \mathbf{r}', T) = \sum_{n} e^{-\beta E_n}\varphi_n^*(\mathbf{r})\varphi_n(\mathbf{r}'). \tag{2.44}$$

Now, we should remark that going to a single time description will hinder our consideration of the functional broadening of the spectral density, but we address this below.

The temporal differential equation for the density matrix can be developed by the standard approach for the time variation of any operator, which involves its commutator with the Hamiltonian

$$i\hbar\frac{\partial\rho}{\partial t} = [H, \rho] \tag{2.45}$$

and this leads to the form

$$i\hbar\frac{\partial\rho}{\partial t} = \left[-\frac{\hbar^2}{2m^*}\left(\frac{\partial^2}{\partial\mathbf{r}^2} - \frac{\partial^2}{\partial\mathbf{r}'^2}\right) + V(\mathbf{r}) - V(\mathbf{r}')\right]\rho(\mathbf{r}, \mathbf{r}', t), \tag{2.46}$$

which is known as the *Liouville equation.* While no dissipation is indicated here, it is quite easy to incorporate a dissipative term [15–17] through incorporating the self-energy into the total Hamiltonian. A different form of the equation arises when we introduce the imaginary time used for (2.44). If we use the same definition in (2.46), we arrive at

$$-\frac{\partial\rho}{\partial\beta} = \left[-\frac{\hbar^2}{2m^*}\left(\frac{\partial^2}{\partial\mathbf{r}^2} - \frac{\partial^2}{\partial\mathbf{r}'^2}\right) + V(\mathbf{r}) - V(\mathbf{r}')\right]\rho(\mathbf{r}, \mathbf{r}'), \tag{2.47}$$

which is known as the *Bloch equation.* In a sense, the Bloch equation is a quasi-steady-state, or quasi-equilibrium, form in which the time variation is either nonexistent or sufficiently slow as to not be important in the form of the statistical density matrix. It is important to note that the Bloch equation possesses an adjoint equation, which arises from the anti-commutator form as

$$-\frac{\partial\rho}{\partial\beta} = \left[-\frac{\hbar^2}{2m^*}\left(\frac{\partial^2}{\partial\mathbf{r}^2} + \frac{\partial^2}{\partial\mathbf{r}'^2}\right) + V(\mathbf{r}) + V(\mathbf{r}')\right]\rho(\mathbf{r}, \mathbf{r}'). \tag{2.48}$$

The importance of the adjoint equation is that it allows one to find the equilibrium form of the density matrix, and this can be used as the initial condition for the time evolution of (2.46). This initial condition holds also for (2.45).

Let us take the form of the density matrix that appears above as (2.43). Then, if we integrate over the spatial coordinate, we have

$$\rho(\mathbf{r}, \mathbf{r}, t) = \int_{-\infty}^{\infty} |\psi(\mathbf{r}, t)| d^3\mathbf{r} = 1. \tag{2.49}$$

This has certain constraints if we expand the wave function in a basis set as done in (2.38), we obtain

$$\begin{aligned}\int_{-\infty}^{\infty} \rho(\mathbf{r}, \mathbf{r}, t)\, d^3r &= \sum_{n,m} c_n c_m^* \int_{-\infty}^{\infty} \varphi_m^\dagger(\mathbf{r}, t)\varphi_n(\mathbf{r}, t)\, d^3\mathbf{r} \\ &= \sum_{n,m} c_n c_m^* \delta_{nm} = \sum_n |c_n|^2 = 1\end{aligned} \quad . \tag{2.50}$$

The first line of this latter equation requires that the density matrix be Hermitian. This implies that we can use the density matrix quite generally to evaluate the expectation values of various operators. For example, we can write this expectation value as

$$\begin{aligned}\langle A \rangle &= \int_{-\infty}^{\infty} \psi^\dagger(\mathbf{r}, t) A \psi(\mathbf{r}, t)\, d^3\mathbf{r} = \sum_{n,m} c_n^* c_m \int_{-\infty}^{\infty} \varphi_n^\dagger(\mathbf{r}, t) A \varphi_m(\mathbf{r}, t)\, d^3\mathbf{r} \\ &= \sum_{n,m} c_n^* c_m A_{nm} = \sum_{n,m} A_{nm}\rho_{nm},\end{aligned} \tag{2.51}$$

where ρ_{mn} are the matrix elements for the density matrix coupling the various basis functions.

The above equations (2.46)–(2.48) give the solution to the density matrix for an arbitrary potential that is imposed upon the system. But the equations do not contain any dissipation. On the other hand, the Boltzmann equation is a kinetic equation which describes the dynamic evolution of a distribution function under the influence of external fields and collisional forces. The resulting transport is a balance between the driving forces of the fields and the relaxation forces of the collisions. Here, we want to develop an analogous kinetic equation (to the Boltzmann equation) that is derived from the density matrix and describes the transport in equivalent fields and relaxation forces. We will also introduce a more extensive approach which gives a generic prescription for the general treatment. We do this now, so that the development of a transport equation for the Wigner function may be interpreted in terms of both the classical Boltzmann equation and the quantum equation for the density matrix.

We describe a semiconductor system composed of electrons and phonons, and the interaction between these two sub-systems, although we neglect the interactions among the electrons (which of course could be added). The system is described by the Hamiltonian

$$H = H_0 + H_L + H_F + H_{eL}, \tag{2.52}$$

where the terms on the right-hand side describe the electrons, the lattice, the external fields, and the electron–lattice interaction, respectively. The external field will be taken in the scalar potential gauge, and the electron interaction could contain all of the appropriate many-body interactions if we so chose. The third term in the Hamiltonian is thus expressed, for a homogeneous field, as

$$H_F = -\mathbf{E} \cdot \mathbf{r}. \tag{2.53}$$

The reader will note that this bold faced **E** is the field and not the energy.

The total density matrix ρ is defined over the entire system. If there were no interactions between the electrons and the lattice, then the density matrix could be written as a tensor product of the density matrix for the electrons and the density matrix for the lattice. In this situation, a trace Tr_L over the lattice coordinates would yield just the density matrix for the electrons. Similarly, a trace Tr_e over the electron coordinates would yield just the density matrix for the lattice. As we will see, when the interactions are present, then the simple divide into two distinct density matrices will not be possible. Nevertheless, we can still define the relevant density matrix for the electrons by the partical trace over the lattice variables as

$$\rho_e = \mathrm{Tr}_L\{\rho\}. \tag{2.54}$$

In the interacting case, this will not be the simple density matrix for the electrons alone described above. Instead, it will incorporate the effect that the lattice has on the electrons due to the interactions between the two systems. There are, of course, some limitations upon this approach and the resulting (2.54). These will not concern us here, but they have been addressed in a more general approach dealing in particular with the effect the field may have on the lattice as well [18]. In the above equations, we have invoked the effective mass approximation, so that transport within a single band is being studied.

The approach with the Hamiltonian (2.52) is followed by introducing this into the Loiuville equation (2.45), and then tracing over the lattice variables to produce the partial density matrix for the electrons, as

$$i\hbar\frac{\partial\rho_e}{\partial t} = [H_0 + H_F, \rho_e] + Tr_L\{[H_{eL}, \rho]\}. \tag{2.55}$$

Clearly, the first term on the right-hand side is the electronic motion within the effective mass approximation under the influence of the applied field. The second term represents the electron–phonon interaction and the effect that this has upon the resulting electronic density matrix. In the subsequent developments, this trace over the lattice variables corresponds to the summation over the phonon momentum vector **q** that describes the change in momentum during a scattering event (by phonons, of course).

To proceed beyond (2.55), we will introduce the use of projection superoperators which will project the one-electron density matrix out of the many-body multi-electron density matrix [19]. This approach is fully equivalent to the BBGKY hierarchy of equations, but is somewhat simpler to work through [20–23]. To proceed, we Laplace transform (2.55) to give

$$\left(s + \frac{i}{\hbar}\hat{H}_e\right)\tilde{\rho}_e = -\frac{i}{\hbar}\mathrm{Tr}_L\{\hat{H}_{eL}\tilde{\rho}\} + \rho_e(0), \tag{2.56}$$

where $\hat{H}_e = \hat{H}_0 + \hat{H}_F$, in superoperator notation, and the tilde over the density matrix refers to the Laplace transformed form. Note that there is no tilde over the last term as it is an initial condition. We now introduce the projection superoperator [24–26] as

$$\tilde{\rho}_1 = \hat{P}\tilde{\rho}_e, \qquad \hat{P}^2 = \hat{P}, \qquad \hat{Q} = 1 - \hat{P}, \qquad \hat{P}\hat{Q} = \hat{Q}\hat{P} = 0. \tag{2.57}$$

The last two expressions tell us that the projection superoperator is idempotent and that there exists a projection operator onto the complement space. Now, what do we mean with this projection operator and the one-electron density matrix? In general, there is a very large number of electrons in the semiconductor, say N. Then, the total phase space, or configuration space, has $6N$ coordinates (plus time). What we want is a typical single electron density matrix which has six coordinates plus time. This is achieved via the BBGKY hierarchy with one complication in which the one-electron density matrix results with one term that involves a two-electron function, just as in chapter 1. In the present approach, this projection from $6N$ coordinates to six coordinates is achieved by the projection operator, but we will still have the more complicated term to treat. To proceed, one must make some approximations to this two-electron function. With the projection operator defined above, we do not miss this term, but it is buried within the scattering function. This particular projection superoperator commutes with the trace operation in (2.56), so that we can introduce the scattering superoperator through the definition

$$\hat{\Sigma}\tilde{\rho}_1 \equiv \mathrm{Tr}_L\{\hat{P}\hat{H}_{eL}\tilde{\rho}\}. \tag{2.58}$$

We will explore this latter quantity more deeply below. With these definitions, (2.56) can now be rewritten in the form

$$\tilde{\rho}_1 = \hat{P}\frac{1}{i\hbar s - \hat{H}_e - \hat{\Sigma}}i\hbar\rho_e(0) = \hat{P}\frac{1}{i\hbar s - \hat{H}_e - \hat{\Sigma}}i\hbar[\hat{P}\rho_e(0) + \hat{Q}\rho_e(0)]. \tag{2.59}$$

It is clear at this point that one needs only products of various projections of the resolvent operator (the term following the first $\hat{P}$ on the right-hand side of the equation; we will later call this term a Green's function).

At this point, we need to develop an expansion identity for the resolvent operator, which we write as [27]

$$R(s) = \frac{1}{i\hbar s - \hat{H}_e - \hat{\Sigma}} = \frac{1}{i\hbar s - \hat{H}'}. \tag{2.60}$$

To begin, we note that the denominator can be rewritten as

$$\begin{aligned} i\hbar s - \hat{H}' &= i\hbar s - (\hat{P} + \hat{Q})\hat{H}'(\hat{P} + \hat{Q}) \\ &= i\hbar s - \hat{P}\hat{H}'\hat{P} - \hat{P}\hat{H}'\hat{Q} - \hat{Q}\hat{H}'\hat{P} - \hat{Q}\hat{H}'\hat{Q}. \end{aligned} \tag{2.61}$$

This result can now be used to expand (2.60) in two different ways:

$$\begin{aligned} \hat{R}(s) &= \frac{1}{i\hbar s - \hat{P}\hat{H}'\hat{P}}[1 - (\hat{P}\hat{H}'\hat{Q} + \hat{Q}\hat{H}'\hat{P} + \hat{Q}\hat{H}'\hat{Q})\hat{R}(s)] \\ &= \frac{1}{i\hbar s - \hat{Q}\hat{H}'\hat{Q}}[1 - (\hat{P}\hat{H}'\hat{Q} + \hat{Q}\hat{H}'\hat{P} + \hat{P}\hat{H}'\hat{P})\hat{R}(s)]. \end{aligned} \tag{2.62}$$

We note that this is an iterative equation for the resolvent $\hat{R}(s)$, and we need to try to separate with the projection operators. We operate on the first of these equations with $\hat{P}$ and on the second of these equations with $\hat{Q}$. This gives us the two equations

$$\begin{aligned} \hat{P}\hat{R}(s) &= \frac{1}{i\hbar s - \hat{P}\hat{H}'\hat{P}}[1 - (\hat{P}\hat{H}'\hat{Q})\hat{Q}\hat{R}(s)] \\ \hat{Q}\hat{R}(s) &= \frac{1}{i\hbar s - \hat{Q}\hat{H}'\hat{Q}}[1 - (\hat{Q}\hat{H}'\hat{P})\hat{P}\hat{R}(s)]. \end{aligned} \tag{2.63}$$

We insert the second equation into the first to eliminate the irrelevant part, and then use this in the second to isolate the irrelevant part. Then, we can then recombine in (2.62) to give us a new formulation for the resolvent, which is

$$\begin{aligned} \hat{R}(s) = &\left(\hat{P} + \hat{Q}\frac{1}{i\hbar s - \hat{Q}\hat{H}\hat{Q}}\hat{Q}\hat{H}\hat{P}\right)\frac{1}{i\hbar s - \hat{P}\hat{H}\hat{P} - \hat{C}}\left(\hat{P} + \hat{P}\hat{H}\hat{Q}\frac{1}{i\hbar s - \hat{Q}\hat{H}\hat{Q}}\hat{Q}\right) \\ &+ \hat{Q}\frac{1}{i\hbar s - \hat{Q}\hat{H}\hat{Q}}\hat{Q} \end{aligned} \tag{2.64}$$

where the collision operator is

$$\hat{C} = \hat{P}\hat{H}\hat{Q}\frac{1}{i\hbar s - \hat{Q}\hat{H}\hat{Q}}\hat{Q}\hat{H}\hat{P}. \tag{2.65}$$

This is an almost standard form, as the leading and trailing parts connect the projected density matrix to the so-called 'off diagonal' part, and vice versa. In the simplest case, this just generates the magnitude squared matrix element for the collision event. This is just the type of term which leads to the Fermi golden rule for scattering, and the leading term in the electron–phonon interaction must be of this type. Thus, our crude approximation (2.58) is found to be exactly this more correct version. Let us probe this a little further. We can rewrite the expression (2.58), using the result (2.65). To do this, we use only the Hamiltonian terms that involve scattering. Then, (2.58) becomes

$$\begin{aligned}\hat{\Sigma}\tilde{\rho}(s) &= \mathrm{Tr}_L\{\hat{C}(\hat{H}_{eL} + \hat{H}_e)\tilde{\rho}\}\\ &= \mathrm{Tr}_L\left\{\hat{P}(\hat{H}_{eL} + \hat{H}_e)\hat{Q}\frac{1}{i\hbar s - \hat{Q}(\hat{H}_{eL} + \hat{H}_e)\hat{Q}}\hat{Q}\tilde{\rho}\right\}.\end{aligned} \tag{2.66}$$

This complicated structure tells us many things. In the first line, we see that collisions via the electron–lattice, or electron–phonon, process as well as the electron–electron interaction, are present. We include the electron–electron interactions in the Hamiltonian even though we are projecting out the single electron density matrix. The electron–phonon interactions can lead to real energy shifts, which are the self-energy corrections to the single particle energies. This is in addition to the normal dissipative scattering processes. Thus, we see that this approach brings the off-shell dynamics and the broadening introduced by the self-energy into the description in a correct manner.

We can now use these various operators, together with the form of the collision operator, to rewrite (2.56) as

$$\begin{aligned}i\hbar s\tilde{\rho}_1(s) &= i\hbar\rho_1(0) + \hat{P}\hat{H}_e\hat{P}\tilde{\rho}_1 + \hat{\Sigma}\tilde{\rho}_1\\ &\quad + \hat{P}(\hat{H}_{eL} + \hat{H}_e)\hat{Q}\frac{1}{i\hbar s - \hat{Q}(\hat{H}_{eL} + \hat{H}_e)\hat{Q}}\hat{Q}\tilde{\rho}_e(0).\end{aligned} \tag{2.67}$$

The last term has been thought to produce a number of effects, including the random force used for the Langevin equation at the lowest order of the electron–phonon interaction [28] and a screening of the driving field in higher order [27]. The temporal equation can now be obtained by retransforming this last equation, which results in

$$\frac{\partial\rho_1}{\partial t} = -\frac{i}{\hbar}\hat{P}\hat{H}_e\hat{P} - \frac{1}{\hbar^2}\int_0^t \hat{\Sigma}(t - t')\rho_1(t')\,dt', \tag{2.68}$$

and the last term of (2.67) has been ignored. If the system is homogeneous, the only contribution from the first term on the right-hand side is from the accelerative electric field, which produces

$$\frac{1}{\hbar}[H_F, \rho_1] = e\mathbf{E}\cdot\frac{\partial\rho_1}{\partial\mathbf{p}}, \tag{2.69}$$

so that the final quantum kinetic equation for the homogeneous system is just

$$\frac{\partial\rho_1}{\partial t} + \frac{e\mathbf{E}}{\hbar}\cdot\frac{\partial\rho_1}{\partial\mathbf{k}} = -\frac{1}{\hbar^2}\int_0^t \hat{\Sigma}(t - t')\rho_1(t')\,dt'. \tag{2.70}$$

This form of the transport equation was first derived by Barker [29], but its form is quite analogous to the Prigogine–Resibois equation [30]; it is also in the same form as derived by Levinson [31, 32]. Except for the convolution form of the collision term, this equation is essentially the same as the Boltzmann transport equation, which means that some quantum effects, such as quantization of the states, is buried in the form of the one-electron density matrix. Moreover, the convolution form of the collision term means that we are accounting for evolution of the density matrix *during the collision process.*

From the discussion in chapter 1, and the form (2.43) for the density matrix, it is apparent that the Wigner function is obtained by the simple transformation of the density matrix, in which we first introduce the center-of-mass and difference coordinates through

$$\mathbf{R} = \frac{\mathbf{r} + \mathbf{r}'}{2}, \quad \mathbf{s} = \mathbf{r} - \mathbf{r}'. \tag{2.71}$$

We can now introduce the phase-space Wigner function as the Fourier transform on the difference coordinate as [33]

$$f_W = \frac{1}{h^3}\int d\mathbf{s}\, \rho(\mathbf{R}, \mathbf{s}, t)e^{i\mathbf{p}\cdot\mathbf{s}/\hbar} = \frac{1}{h^3}\int d\mathbf{s}\, \rho\left(\mathbf{R} + \frac{\mathbf{s}}{2}, \mathbf{R} - \frac{\mathbf{s}}{2}, t\right)e^{i\mathbf{p}\cdot\mathbf{s}/\hbar}. \tag{2.72}$$

Thus, any equation of motion for the Wigner function will have certain similarities to that for the density matrix, although it will, of course, be in phase space. We defer the derivation of the equation of motion for the Wigner function until chapter 4.

2.4 Green's functions

In the preceeding section, we briefly introduced the kernel or propagator for the solutions to the Schrödinger equation. Replacing the t_0 we had in (2.41) with an arbitrary t', we can write the wave function as

$$\psi(\mathbf{x}, t) = \int d\mathbf{x}'\, K(\mathbf{x}, \mathbf{x}'; t, t')\psi(\mathbf{x}', t'), \tag{2.73}$$

where

$$K(\mathbf{x}, \mathbf{x}'; t, t') = \sum_n \varphi_n^*(\mathbf{x}')\varphi_n(\mathbf{x})e^{-iE_n(t-t')/\hbar}. \tag{2.74}$$

It would be very tempting to write the Green's function directly as the kernel, or propagator. But this would be a mistake (in terms of common usage), especially for the many-body case. It is convenient to use field operators that maintain the symmetry properties of the many-body state. With field operators, we want to write the retarded and advanced Green's functions as commutator products, as [34]

$$G_r(\mathbf{x}, \mathbf{x}'; t, t') = -i\vartheta(t - t')\langle\{\hat{\psi}(\mathbf{x}, t), \hat{\psi}^\dagger(\mathbf{x}', t)\}\rangle \tag{2.75a}$$

$$G_a(\mathbf{x}, \mathbf{x}'; t, t') = i\vartheta(t' - t)\langle\{\hat{\psi}^\dagger(\mathbf{x}', t'), \hat{\psi}(\mathbf{x}, t)\}\rangle. \tag{2.75b}$$

Here, we have introduced the use of field operators, in which we turn the wave function itself into an operator, primarily because users of Green's functions tend to like to use such a notation. The curly brackets indicate the anti-commutator of the two field operators (discussed just below) while the angle brackets mean an expectation value over the ground state of the system [34]. Normally, we have no trouble referring to the wave function as a field, since mathematically the description of a quantity as a *field* means only that the quantity has a defined value at every point in space. But, if we start with a basis set, as in the previous chapters, then we can use our operators to create excitations into the electron basis set, and write the field operator for the wave function as

$$\begin{aligned}\hat{\psi}(x, t) &= \sum_i \alpha_i c \varphi_i(x, t) \\ \hat{\psi}^\dagger(x, t) &= \sum_i \alpha_i c^\dagger \varphi_i^*(x, t).\end{aligned} \tag{2.76}$$

That is, for each basis function of interest, there is an operator which creates or destroys an excitation in that state, with an amplitude of probability of $|\alpha_i|^2$. First, the use of the basis set distributes the probability from the total wave function into parts assigned to each basis vector. Secondly, we have used fermion creation and annihilation operators, as we chose to deal with electrons. When a fermion state, which may be defined by its position and spin, is full, a second fermion cannot be created in this state. This is the Pauli exclusion principle. Hence, we require that (and we now use the letter c to indicate fermion operators)

$$c^\dagger \mid 1\rangle = 0, \quad c \mid 0\rangle = 0, \tag{2.77}$$

where the last expression indicates that the empty state cannot have its occupancy set to zero as it is already zero. The first term is new for fermions. These two operations lead to the existence of an anti-commutator relationship for the fermion operators, which may be expressed as

$$\{c, c^\dagger\} = cc^+ + c^+c = 1. \tag{2.78}$$

If the state is empty, only the first term works to create a particle in the empty state and then destroys this particle, returning it to the empty state. If the state is full, then the second term first destroys this particle and then recreates a particle in the state. In either case the result leaves the state unchanged, thus the unity of the anti-commutator. We have used the curly brackets to indicate the anti-commutator, but other notations are commonly used, such as square brackets with a subscript '+' attached to them.

Let us now proceed. For the retarded Green's function, we need to first create an excitation from the ground state (or vacuum state), and then destroy this excitation at a later time. The Green's function tells us how this excitation propagates. To see this, let us expand the right-hand side of (2.75*a*) as

$$\begin{aligned}\langle\{\hat{\psi}(\mathbf{x}, t), \hat{\psi}^\dagger(\mathbf{x}', t')\}\rangle = &\langle 0 \mid \hat{\psi}(\mathbf{x}, t)\hat{\psi}^\dagger(\mathbf{x}', t')\mid 0\rangle \\ &+ \langle 0 \mid \hat{\psi}^\dagger(\mathbf{x}', t')\hat{\psi}(\mathbf{x}, t)\mid 0\rangle.\end{aligned} \tag{2.79}$$

In the second term, the $\hat{\psi}$ operating on the $|0\rangle$ state (the empty or ground state) produces zero. But the adjoint creation operator working on this state in the first term produces an electron at $\mathbf{x}'$. This electron is then annihilated at a later time t at position $\mathbf{x}$. The two positions and times cannot be the same, since that would lead to self-annihilation of the excitation. This is given by the field operator version of the anti-commutator:

$$\{\hat{\psi}(\mathbf{x}, t), \hat{\psi}^{\dagger}(\mathbf{x}', t')\} = \delta(\mathbf{x} - \mathbf{x}')\delta(t - t'). \tag{2.80}$$

To see how we formulate a transport equation for these two functions, let us begin with the general non-interacting ground state Green's function that must satisfy two forms of the Schrödinger equation, in terms of the different sets of variables. These equations are

$$\left[i\hbar\frac{\partial}{\partial t} - H_0(\mathbf{x}) - V(\mathbf{x})\right]G_0(\mathbf{x}, \mathbf{x}'; t, t') = \hbar\delta(\mathbf{x} - \mathbf{x}')\delta(t - t') \tag{2.81a}$$

$$\left[-i\hbar\frac{\partial}{\partial t'} - H_0(\mathbf{x}') - V(\mathbf{x}')\right]G_0(\mathbf{x}, \mathbf{x}'; t, t') = \hbar\delta(\mathbf{x} - \mathbf{x}')\delta(t - t'). \tag{2.81b}$$

If we Fourier transform in both the position difference (the Fourier variable is $\mathbf{k}$) and the time difference (the Fourier variable is ω), and assume that the Green's function is a function of only the difference in the two positions and the two times, this Fourier version becomes, when the potential is neglected,

$$G_0(\mathbf{k}, \omega) = \frac{\hbar}{\hbar\omega - E(\mathbf{k})}, \tag{2.82}$$

where we note that the bare Hamiltonian H_0 produces the energy eigenvalue, and we have assumed that the potential term is zero. Strictly speaking, this Fourier transform is a little sloppy, for reasons we will address. Neglecting the potential means that we are looking at a free propagating Green's function for a homogeneous system with no boundaries. Now, if we want to invert this Fourier transform, we first have to address the proper form of the transform, and this requires us to introduce a convergence factor. we can recover the retarded and advanced Green's functions by choosing the sign of the convergence factor properly. Hence, we modify (2.82), and write the retarded and advanced bare (or ground state) Green's functions as

$$G^0_{r,a}(\mathbf{k}, \omega) = \frac{\hbar}{\hbar\omega - E(\mathbf{k}) \pm i\eta}. \tag{2.83}$$

For example, let us do the inverse transform in time, so that we have

$$G^0_{r,a}(\mathbf{k}, t) = \int\frac{d\omega}{2\pi}G^0_{r,a}(\mathbf{k}, \omega)e^{-i\omega t}. \tag{2.84}$$

In general, the integration proceeds along the real frequency axis. But, as for any integration involving a complex variable, we have to decide how to close the contour to obtain a full contour integration. We can close the contour in either the upper or lower half of the complex plane. On the contour, we can rewrite the frequency as

$$\omega = Ae^{i\theta} = A\cos\theta + iA\sin\theta, \tag{2.85}$$

and the exponential term in (3.22) becomes

$$e^{-i\omega t} = e^{-iA\cos\theta t}e^{A\sin\theta t}. \tag{2.86}$$

For stability, we require that the argument of the last exponential be less than zero, so that the integral converges as we let A go to infinity. So, for $t > 0$, we require that the sine term be negative, which means that we have to close in the lower half plane. On the other hand, for $t < 0$, we require the sine term to be positive, which means that we have to close in the upper half plane. The importance of this closure lies in the convergence factor that was added to (2.83). We require that the pole of this latter equation lie within the closed contour for there to be a solution to the integral. Hence, when we close in the lower half plane, for positive time, we take the upper sign in (2.83), and this leads to the retarded Green's function as

$$G_r^0(\mathbf{k}, \omega) = \frac{\hbar}{\hbar\omega - E(\mathbf{k}) + i\eta}. \tag{2.87}$$

Similarly, when we close in the upper half plane, we require the pole to lie in this upper half plane, and this requires the lower, negative sign in (2.83). This leads us to the advanced Green's function as

$$G_a^0(\mathbf{k}, \omega) = \frac{\hbar}{\hbar\omega - E(\mathbf{k}) - i\eta}. \tag{2.88}$$

These two definitions now give us an important result that relates the two functions via the property

$$G_r^0(\mathbf{k}, \omega) = [G_a^0(\mathbf{k}, \omega)]^*. \tag{2.89}$$

This property is readily apparent from their respective equations, but carries forward beyond just the bare Green's functions. Indeed, when we replace later the convergence factor by the full self-energy, this property will continue to be obeyed, since the imaginary parts of the self-energy play the convergence roles in the full Green's functions.

The integral (2.84) leads to a number of important properties that we associate with the Green's functions. First, when η is vanishingly small, as it is for these two functions where it serves only as a convergence factor, we can write the functions as

$$\frac{1}{x \pm i\eta} = P\frac{1}{x} \mp i\pi\delta(x), \tag{2.90}$$

where P denotes the principal part of the function (and the resulting integral). We recognize the delta function as giving us the on-shell relationship between the energy $\hbar\omega$ and the momentum dependent kinetic energy $E(\mathbf{k})$. This delta function assures us that the energy is given by the classical value. The delta function arises from the imaginary part of the Green's function. But, let us be absolutely correct in the use of (2.90). This equation pre-dates the use of Green's functions, and is a property of the mathematics of generalized functions, or distributions, developed in the middle of the last century [35]. In the present case, we get a delta function (which is a form of a generalized distribution) because we have no interactions in the system as yet. When these arise, then the delta function will be broadened into a spectral function and account for the off-shell contributions to the integrals. Consequently, this will give us the interacting Green's functions. Hence, for our non-interacting Green's functions, we can write

$$2\pi\delta(\hbar\omega - E(\mathbf{k})) = -\mathrm{Im}\{G_r^0(\mathbf{k}, \omega) - G_a^0(\mathbf{k}, \omega)\}, \tag{2.91}$$

and the corresponding spectral density for the interacting system will be (for historical reasons, we incorporate the 2π into the definition of the spectral density)

$$A(\mathbf{k}, \omega) = -\mathrm{Im}\{G_r(\mathbf{k}, \omega) - G_a(\mathbf{k}, \omega)\} = i\{G_r(\mathbf{k}, \omega) - G_a(\mathbf{k}, \omega)\}. \tag{2.92}$$

In condensed matter physics, an important quantity is the density of states, which is often written as

$$\rho(E) = \sum_{\mathbf{k}} \delta(E - E(\mathbf{k})), \tag{2.93}$$

where the states are assumed to be of plane wave form. We can extend this to the spectral density if we account for the actual basis states in the definition, and

$$A(E) = 2\pi \sum_{\mathbf{k}} \psi_{\mathbf{k}}^{\dagger} \psi_{\mathbf{k}} \delta(E - E(\mathbf{k})). \tag{2.94}$$

In a sense, the wave functions retain the position information, and some useful normalizations of the spectral function can be found by rewriting (2.94) as [36]

$$A(\mathbf{x}, \mathbf{x}', E) = 2\pi \sum_{\mathbf{k}} \hat{\psi}(\mathbf{x}) \hat{\psi}_k^{\dagger}(\mathbf{x}') \delta(E - E(\mathbf{k})). \tag{2.95}$$

If we now integrate this over the energy, we find

$$\int \frac{dE}{2\pi} A(\mathbf{x}, \mathbf{x}', E) = \sum_{\mathbf{k}} \hat{\psi}_k(\mathbf{x}) \hat{\psi}_k^{\dagger}(\mathbf{x}') = \delta(\mathbf{x} - \mathbf{x}'), \tag{2.96}$$

which is another way of writing the completeness of the basis set. If we now integrate the diagonal form over the position, we get

$$\int d\mathbf{x}\, A(\mathbf{x}, \mathbf{x}, E) = 2\pi \sum_{\mathbf{k}} \delta(E - E(\mathbf{k})) = 2\pi\rho(E), \tag{2.97}$$

which is the density of states per unit energy defined in (2.93). While these normalizations are useful, the key element is the spectral density (2.92), which exactly determines the broadening of the off-shell effects that arise as a result of the interactions within the many-body system.

In the above discussion, we have encountered the retarded and advanced Green's functions. These two functions bring us into the world of off-shelf interactions, as they describe the evolution of the traditional, classical delta function $\delta(E - \hbar^2k^2/2m^*)$, for a single parabolic band, into the broadened spectral density $A(E, k)$. However, when we go to a 'device' operating under significant bias potentials, we must go beyond these two Green's functions. This is because the device is driven significantly far from equilibrium. In the non-equilibrium world, densities are no longer well defined with the Fermi–Dirac distribution, just as for the classical case, where the distribution is no longer a nice Maxwellian distribution, as in the equilibrium case. This requires new, real-time Green's functions which can give us the proper distribution function. Hence, we now have a necessity to be able to solve for the non-equilibrium distribution of the carriers under the applied forces. As remarked, in the semi-classical world, this is achieved through solutions of the Boltzmann transport equation. Now, we need to find new Green's functions that will be described by quantum evolution equations, and which will give us the quantum distribution functions. We will not be concerned with these evolution functions, but only with the concepts of the addition of the correlation functions. Those interested in a deeper study of these functions are referred to [37].

The additional Green's functions, which are proper correlation functions, seems to have first been used by Martin and Schwinger [38]. Many of us were then introduced to them through the book by Kadanoff and Baym [39], although useful advances were then introduced by Keldysh [40]. These new Green's functions do not involve the commutator products, but do involve the wave function and its adjoint. They may be defined as [36]

$$\begin{aligned} G^<(\mathbf{x}, \mathbf{x}'; t, t') &= i\langle\hat{\psi}^\dagger(\mathbf{x}', t')\hat{\psi}(\mathbf{x}, t)\rangle \\ G^>(\mathbf{x}, \mathbf{x}'; t, t') &= i\langle\hat{\psi}(\mathbf{x}, t)\hat{\psi}^\dagger(\mathbf{x}', t')\rangle. \end{aligned} \tag{2.98}$$

These two functions are usually referred to as the 'less than' Green's function and the 'greater than' Green's function, respectively. They are related to the non-equilibrium distribution function $f(\omega)$ through

$$\begin{aligned} G^<(\mathbf{k}, \omega) &= f(\omega)A(\mathbf{k}, \omega) \\ G^>(\mathbf{k}, \omega) &= [1 - f(x)]A(\mathbf{k}, \omega), \end{aligned} \tag{2.99}$$

where the spectral density is given by (2.92), and is related to the retarded and advanced Green's functions. One further point of interest is to introduce the center-of-mass coordinates (2.71) in both position and time as

$$\begin{aligned} \mathbf{R} &= \frac{1}{2}(\mathbf{x} + \mathbf{x}'), \quad T = \frac{1}{2}(t + t'), \\ \mathbf{s} &= (\mathbf{x} - \mathbf{x}'), \quad \tau = (t - t'). \end{aligned} \tag{2.100}$$

Then, a Fourier transform may be taken of the functions in (2.98) in the $\mathbf{s}$ and τ variables as before. It is recognized that (2.99) is thus written only in terms of the difference variables, which have been Fourier transformed as in (2.82). This allows us to arrive at a connection between the less than function and the Wigner distribution, which may be written as

$$f_W(\mathbf{R}, \mathbf{k}, T) = \lim_{\tau \to 0} \int d^3\mathbf{s}\, G^<(\mathbf{R}, \mathbf{s}, T, \tau) e^{-i\mathbf{k}\cdot\mathbf{s}}$$
$$= \int \frac{d\omega}{2\pi i} G^<(\mathbf{R}, \mathbf{k}, T, \omega). \tag{2.101}$$

Here, the frequency that arises from the difference in time coordinate continues to play the role of the energy in the system. The fact that this energy is integrated out to obtain the Wigner function has certain implications. The foremost is that we lose correlations in time, as the frequency relates to the difference in the two times of the original Green's function. The second implication is that the momentum must now be related to the energy through an on-shell delta function, just as in the classical case. This does not mean that off-shell effects cannot arise, but these must be treated carefully as part of the scattering process. The result of (2.101) is that the critical kinetic variable in the Wigner function is the momentum and not the energy, although the two are certainly related. In a sense, the integration has removed the spectral density, as the Wigner function is the distribution function for the system. In that sense, solving the Wigner equation of motion is equivalent to solving a similar equation for the less than Green's function.

2.5 What are the relative advantages?

If we want to understand the relative advantages of the various approaches to quantum transport, we need to couch these in terms of the standard approach used in semi-classical transport. So, we will begin with this, but not for an unspecified structure. Instead, we are concerned about simulating real semiconductor devices, which are far more complex than a simple nano-ribbon of a single material. In particular, the structure possesses a very inhomogeneous carrier density and a corresponding inhomogeneous electric potential. Current flows by virtue of both the motion of the carriers and, in the time-varying situation, displacement current. During the switching process, in which the device becomes either conducting or non-conducting, both the local carrier density at each point in the device and the local electric field or potential at each point in the device are time varying. Hence, it is a complicated and difficult problem that has to be solved self-consistently at each time step of the solution. For example, we may start the simulation from the unbiased device [41], and determine the carrier density throughout the structure. As may be seen in figure 2.2, this leads to determination of the local potential and electric field through the solution of Poisson's equation. Once the local potential is found, the transport of the carriers is found from a solution of the Boltzmann transport equation. In modern devices, the carrier heating from the motion in the electric field is sufficient to scatter carriers to almost all parts of the Brillouin zone, so that a full-band Monte Carlo technique is used to solve the Boltzmann transport equation [42]. From this solution, the new carrier density is determined throughout the device, and

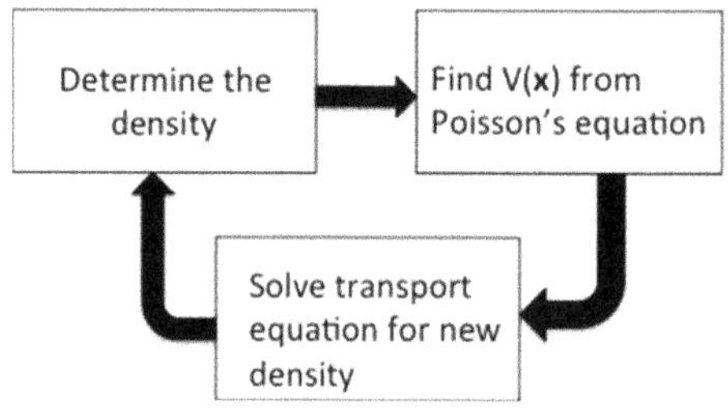

Figure 2.2. The self-consistent loop for solving the various parameters in a semiconductor device simulation. All of these processes are dependent upon the boundary conditions and doping variations.

the cycle is repeated. Once this self-consistent loop reaches a consistent solution, we advance the time and start again.

There are two steps in this self-consistent approach that are important and involve serious levels of approximation. The first is the assumption of Poisson's equation. The second is the many-body corrections. We start with the second problem. As we discussed earlier, the one-electron distribution function is obtained from the many-electron distribution by the BBGKY hierarchy. Yet, with this approach, the resulting distribution is still coupled to a two-electron distribution. This has many names, and we will refer to it as a two-electron correlation function or a two-electron Green's function. While this many-body term is usually ignored when discussing semi-classical simulations, this can be a costly error.

For example, in finding the band structure to be used for the full-band Monte Carlo, one can use first-principles pseudo-potential approaches or empirical pseudo-potential methods for the atomic potentials. The first-principles approach, known as density functional theory (DFT), needs a functional for the many-body interactions [43], and this will be a function of the density (hence the name). This many-body term usually can be separated into three terms. The first term assumes that one electron interacts with the average background potential of the other electrons, and is termed the Hartree term. The Hartree term is the density that appears in Poisson's equation. The second term is the exchange interaction, while the third term takes all the remaining interactions and is called the correlation energy. Often, it is the exchange and correlation terms that are expressed as, for example, the linear density approximation [2]. In the empirical pseudo-potential method, the computed bands are fit to experimental results. Hence, it is assumed that all many-body interactions are accounted for when the fit is made to the experimental data. Nevertheless, both approaches are approximations, and these approximations may become especially important in nano-scale devices.

As we noted, though, the Poisson equation incorporates only the Hartree approximation. Worse, even in this case, it involves another approximation. That is, the Poisson equation is the low-frequency approximation that arises in the electrostatic case of the Coulomb gauge where the divergence of the vector potential is set to zero. Thus, in high-frequency cases, the Poisson equation may be inadequate, and one needs to consider whether the exchange and correlation energies lead to significant differences in the solution. Here is where the effective potential, first mentioned in section 1.5, can be important. For sure, it brings quantum corrections into the picture. But the effective potential itself can include corrections for the exchange and correlation energies [44]. The most common approach is to add the effective potential effects to the potential found from Poisson's equation, and proceed from that point.

What happens when we move to quantum transport? With the density matrix and the Wigner function, the only major change in the self-consistent loop is to replace the Boltzmann transport equation with the evolution equation for either the density matrix or the Wigner function. The former was obtained in section 2.3, and the evolution equation for the Wigner function will be found in the next chapter. So, in these two situations, the mean velocity of the carriers at each point in space is found from an appropriate average of the momentum over the distribution function. The current is then found from this spatial varying velocity and the spatial varying density. Yet there remains a constraint in that the current flowing through the device must be conserved according to Kirchoff's current law. Nevertheless, the computation of either the density matrix or the Wigner function from their evolution equations ensures the full and proper treatment of scattering on the carrier density and velocity.

When we deal with the scattering matrix approach, we are dealing with the wave function itself. As we saw in section 2.2, we can evaluate this wave function throughout the structure and determine the local density at each point. However, we do not determine an average velocity directly, but include this in determining the transmission probability for each mode in the contact regions. This determines the conductance from (2.28). In reality, the transmission coefficient is included in a modified form of the Landauer formula, according to

$$I = \frac{2e}{h}\int dE \cdot T(E)[f_s(E) - f_d(E - eV)], \tag{2.102}$$

which assumes that the reservoirs for source s and drain d are described by Fermi–Dirac distributions, which may be modified to parameterized forms [45]. Finding any such parameters is part of the self-consistency demanded by the loop in figure 2.2. The result (2.101) is actually a reduction of the Duke tunneling formula [46] in which the transmission includes a summation over the modes (the factor of N in section 2.1) which replaces integration over transverse momentum. In this approach, scattering may easily be included in the computation of the transmission through the self-energy, and this will account for all back-scattering of carriers as part of that process [13]. This is a strong advantage of this approach over some others.

Now, let us turn to the advantages and disadvantages of Green's functions. Many in the field will say that the Green's function approach is the most complete and basic approach, if for no other reason than that both coordinates and times are kept for both of the two wave functions. But, this can also be a disadvantage as it leads to extensive complexities in the approach. What is worse, however, is that the approach is being incorrectly applied to the area of semiconductor device simulation by a vast majority of the people using it! But, we will come to that. Let us consider the first box in figure 2.2, in which we need to find the density at every position within the structure being simulated. With non-equilibrium Green's functions, we find the density from the so-called less than function given in the first line of (2.98). To get to the density as a function of position, we have to collapse this less than function to a single time (corresponding to the time step) and the single center-of-mass coordinate **R** in (2.100). This means a double integral over the energy (the time difference) and the momentum (the position difference). The first of these is shown in (2.101) where we obtain the Wigner function. Then, this result is integrated again to obtain the density as a function of position at the particular time step. In particular, the energy integral in (2.101) is very sensitive, as the gridding can lead to the integral missing some eigen-energies. Many approaches have been suggested to ease this integral, and the best seems to be to introduce a complex energy, which broadens all the states, do the integral, and then let the broadening go to zero. Of course some of this is naturally included via the self-energy. The second block, determining the potential from the density, is essentially the same in every approach we have talked about, so it is no more difficult with Green's functions than with other approaches.

Solving the transport equation is considerably more difficult for the Green's function approach, especially in the presence of scattering. First, there are two Green's functions that we need: the less than function and the retarded function. However, the equations for these two functions generally involve all four Green's functions, and are further complicated by the fact that there are four self energies, one for each of the four Green's functions. Hence, solving the equations of motion are far more complicated with the Green's functions than with other approaches. But this doesn't end the increased difficulty, because we now need to determine the conductance, and this is far more difficult with Green's functions. This step is where most people make a significant error.

In general, the conductance may be found with Green's functions in a manner quite similar to that in (2.102). In the Green's function form, it has been given for *ballistic* transport as [47]

$$I = \frac{2e}{h}\int dE\, [f_s(E) - f_d(E - eV)]\mathrm{Tr}\{G^a\Gamma^d G^r\Gamma^s\}, \tag{2.103}$$

where the Γ are the imaginary parts of a self-energy-like connection between the active channel and the two reservoirs of source and drain. The trace term is the summation over the modes in the source and drain and give the transmission that appears in (2.102). This result is the source of most errors in using the Green's function. While it is absolutely correct in the ballistic case, it cannot be applied in the case of scattering in the active region. The fact that this is not widely recognized may be seen by its use in a main stream paper by reputable engineers, even in the case of scattering [48], where the importance of the Bethe–Salpeter equation (discussed below) is ignored. The problem with this formula is the two Green's functions. In fact, these must be replaced by the proper two-particle Green's function, which is termed the polarization. From the Kubo formula, we understand that the current is a result of the correlation between the electrons and their flows. This correlation is represented by a diagram depicting the two-particle Green's function. The Green's functions in (2.103) give a polarization as shown in figure 2.3, which is only the lowest order approximation to the two-particle Green's function.

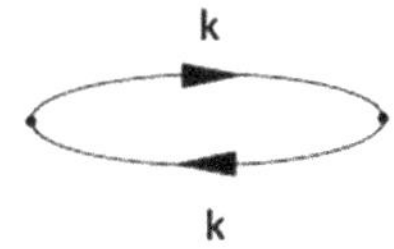

Figure 2.3. The polarization bubble for the two Green's functions of (2.103). The upper line represents G^r while the lower line represents G^a.

We can illustrate the source of this error with the simplest of scatters: the impurities. If we first assume that we use only the results of (2.103), we find that the conductivity may be found to be of the form

$$\sigma = \frac{e^2}{m^*}\int dE(k)E(k)\tau(E)\rho(E)\left(-\frac{\partial f}{\partial E}\right), \tag{2.104}$$

where we have included the density of states, which arises from an integration of the spectral density over the momentum, and τ is the total scattering time inferred from the self-energy itself. If we compare this with the equivalent form that arises in semi-classical transport with the Boltzmann equation, we immediately see that something is missing. In the semi-classical case, the scattering integral gives an additional factor that accounts for the conversion of the total scattering time to the momentum relaxation time through

$$\sigma = \frac{e^2}{m^*}\int dE(k)E(k)\tau_m(E)\rho(E)\left(-\frac{\partial f}{\partial E}\right), \tag{2.105}$$

where the momentum relaxation involves this additional factor in the integration over momentum as

$$\tau_m\rho(E) = \int d^3\mathbf{k}\ \tau(k)(1 - \cos\vartheta)A^2(\mathbf{k}, E). \tag{2.106}$$

That is, the form used in (2.103) has lost the $1 - \cos\theta$ that is found in the semi-classical approach with the Boltzmann transport equation. This term is important because of the anisotropy in impurity scattering and accounts for the fact that it takes several actual collisions to fully destroy the momentum. So, we have to ask whether an approach that does not lead, in the semi-classical limit, to the correct answer is, in fact, a valid approach. The proper form for the conductivity may be written as

$$\sigma = \frac{2e^2\hbar^2}{3m^{*2}}\int\frac{d^3\mathbf{k}}{(2\pi)^3}\int\frac{d\omega}{2\pi}\left(-\frac{\partial f}{\partial\omega}\right)\Pi(\mathbf{k}, \omega), \tag{2.107}$$

where we have retained the energy as ω and the two-particle polarization is given as

$$\begin{aligned}\Pi(\mathbf{k}, \omega) &= G^r(\mathbf{k}, \omega)G^a(\mathbf{k}, \omega)\\ &\times\left\{k^2 + \int\frac{d^3\mathbf{k}}{(2\pi)^3}\mathbf{k}\cdot\mathbf{k}'\Lambda(\mathbf{k}', \omega)G^r(\mathbf{k}', \omega)G^a(\mathbf{k}', \omega)+\ldots\right\}\end{aligned} \tag{2.108}$$

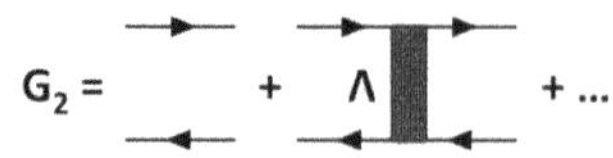

Figure 2.4. Diagrammatic expansion of the two-particle Green's function, known as the polarization of (2.108).

We show this diagrammatically in figure 2.4. It is clear that the cosine term arises from the first term under the integral of the second term in parentheses. Actually this expansion is an extended series approximation for an integral, known as the Bethe–Salpeter equation, for the polarization. The scattering kernel Λ accounts for scattering which begins on one of the Green's functions and ends upon the other one. It is this scattering across the Green's functions which is omitted in (2.103). In the case of impurity scattering, the set of impurity diagrams is referred to as the ladder diagrams [34]. From the complexity of (2.107) and (2.108), it is clear that computing the conductance is more complicated with the use of Green's functions. But if one is going to use the non-equilibrium Green's functions, then the necessity to utilize the Bethe–Salpeter equation is evident [49]. Indeed, this added complexity is one of the disadvantages to the use of non-equilibrium Green's functions. Yet if you don't see a discussion of it in any treatment of device simulation, then you can't trust the results.

References

[1] Ferry D K, Goodnick S M and Bird J P 2009 *Transport in Nanostructures* 2nd edn (Cambridge: Cambridge Univ. Press)

[2] Kohanoff J 2006 *Electronic Structure Calculations for Solids and Molecules* (Cambridge: Cambridge Univ. Press)

[3] Landauer R 1957 *IBM J. Res. Develop.* **1** 223

[4] Landauer R 1970 *Philos. Mag.* **21** 863

[5] Wharam D A, Thornton T J, Newbury R, Pepper M, Ahmed H, Frost J E, Hasko D G, Peacock D C, Ritchie D A and Jones G A C 1988 *J. Phys.* C **21** L325

[6] van Wees B J, van Houten H, Beenakker C W J, Williamson J G, Kouwenhouven L P, van der Marel D and Foxon C T 1988 *Phys. Rev. Lett.* **60** 848

[7] Collin R E 1960 *Field Theory of Guided Waves* (New York: McGraw-Hill)

[8] Merzbacher E 1970 *Quantum Mechanics* 2nd edn (New York: Wiley)

[9] Kriman A M, Kluksdahl N C and Ferry D K 1987 *Phys. Rev.* B **36** 5953

[10] Akis R, Bird J P, Vasileska D, Ferry D K, de Moura A P S and Lai Y-C 2003 On the influence of resonant states on ballistic transport in open quantum dots *Electron Transport in Quantum Dots* ed J P Bird (Boston: Kluwer) pp 209–76

[11] Usuki T, Saito M, Takatsu M, Kiehl R and Yokoyama N 1995 *Phys. Rev.* B **52** 8244

[12] Gilbert M J and Ferry D K 2004 *J. Appl. Phys.* **95** 7954

[13] Gilbert M J, Akis R and Ferry D K 2005 *J. Appl. Phys.* **98** 094303

[14] Feynman R P and Hibbs A R 1965 *Quantum Mechanics and Path Integrals* (New York: McGraw-Hill)

[15] Kohn W and Luttinger J M 1957 *Phys. Rev.* **108** 590

[16] Siegel J L and Argyres P N 1969 *Phys. Rev.* **178** 1016

[17] Argyres P N and Siegel J L 1973 *Phys. Rev. Lett.* **31** 1397

[18] Krieger J B and Iafrate G J 1986 *Phys. Rev.* B **33** 5494
Krieger J B and Iafrate G J 1987 *Phys. Rev.* B **35** 9644
[19] Ferry D K 1991 *Semiconductors* (New York: Macmillan) section 15.3
[20] Bogoliubov N N 1946 *J. Phys. Sowjet. U.* **10** 256
[21] Born M and Green H S 1946 *Proc. Roy. Soc. London* A **188** 10
[22] Kirkwood J G 1946 *J. Chem. Phys.* **14** 180
[23] Yvon J 1937 *Act. Sci. Ind.* 542–3
[24] Nakajima S 1958 *Prog. Theor. Phys.* **20** 948
[25] Zwanzig R 1960 *J. Chem. Phys.* **33** 1338
[26] Mori H 1965 *Prog. Theor. Phys.* **33** 423
[27] Barker J R 1980 *Physics of Nonlinear Transport in Semiconductors* ed D K Ferry, J R Barker and C Jacoboni (New York: Plenum) pp 127–51
[28] Pottier N 1983 *Physica* A **117** 243
[29] Barker J R 1978 *Sol.-State Electron.* **21** 197
[30] Balescu R 1975 *Equilibrium and Nonequilibrium Statistical Mechanics* (New York: Wiley)
[31] Levinson I B 1970 *Sov. Phys. JETP* **30** 362
[32] Nedjalkov M, Vasileska D, Ferry D K, Jacoboni C, Ringhofer C, Dimov I and Palankovski V 2006 *Phys. Rev.* B **74** 035311
[33] Wigner E P 1932 *Phys. Rev.* **40** 749
[34] Fetter A L and Walecka J D 1971 *Quantum Theory of Many-Particle Systems* (New York: McGraw-Hill)
[35] Schwartz L 1950–51 *Théorie des Distributions* 2 vols (Paris: Hermann & Cie)
[36] Rammer J 1998 *Quantum Transport Theory* (Reading, MA: Perseus)
[37] Ferry D K 2017 *Introduction to Quantum Transport in Semiconductors* (Singapore: Pan Stanford)
[38] Martin P C and Schwinger J 1959 *Phys. Rev.* **115** 1342
[39] Kadanoff L P and Baym G 1962 *Quantum Statistical Mechanics* (Reading, MA: Benjamin/ Cummings)
[40] Keldysh L V 1964 *J. Exper. Theor. Phys.* **47** 1515
Tr. in 1965 *Sov. Phys. JETP* 20 1018
[41] Ayoub-Moak J S, Ferry D K, Goodnick S M, Akis R and Saraniti M 2007 *IEEE Trans. Electron Dev.* **54** 2327
[42] Shichijo H and Hess K 1981 *Phys. Rev.* B **23** 4197
[43] Ferry D K 2013 *Semiconductors—Bonds and Bands* (Bristol: Institute of Physics Publishing)
[44] Ando T, Fowler A B and Stern F 1982 *Rev. Mod. Phys.* **54** 437
[45] Zubarev D N 1974 *Nonequilibrium Statistical Thermodynamics* (New York: Consultants Bureau) translated from the Russian by P J Shepard.
[46] Duke C B 1969 *Tunneling in Solids* (New York: Academic)
[47] Meir Y and Wingreen N S 1992 *Phys. Rev. Lett.* **68** 2512
[48] Anantram M P, Lundstrom M S and Nikonov D E 2008 *Proc. IEEE* **96** 1511
[49] Vasileska D, Edlgridge T, Bordone P and Ferry D K 1998 *VLSI Des.* **6** 21

IOP Publishing

The Wigner Function in Science and Technology

David K Ferry and Mihail Nedjalkov

Chapter 3

Wigner functions

In chapter 2, we introduced a number of quantum functions, including the density matrix and the Wigner function. Alone among the various approaches to quantum transport we have discussed so far, the Wigner function has a convenient phase space formulation. Wigner gave us his formulation already in 1932 [1], and demonstrated its usefulness particularly when we are interested in spanning the transition from the classical to the quantum world, as discussed in chapter 1. The Wigner function has been studied for more than eight decades. Historically, it was developed from the Schrödinger wave equation, although it could have been developed in its own right. The Wigner function approach is felt to offer a number of advantages for use in modeling the behavior of physical processes in many fields of science [2–4]. First, it is a phase space formulation, similar to classical formulations based upon the Boltzmann equation. By this, we mean that the Wigner function involves both real space and momentum space variables, distinctly different from the Schrödinger equation. In this regard, modern approaches with the Wigner function provide a distinct formulation that is recognized as equivalent, though a different alternative, to normal operator-based quantum mechanics [5]. Because of the phase space nature of the distribution, it is now conceptually possible to identify where quantum corrections enter a problem by comparing with the classical version, an approach that has been used to provide an effective quantum potential that can be used as a correction term in classical simulations, as discussed in the previous chapter.

When one is considering a nano-scale device, it is usually divided into three regions. These are a central active region, often called the channel, and two reservoirs that incorporate the physical contacts to the structure. At the ends of the active region, well into the reservoirs, the phase space distribution allows one to separate the incoming and outgoing components of the distribution, which now allows us to model both the contact itself and the entire open quantum system. Moreover, the Wigner function is entirely real, which simplifies both the calculation and the interpretation of results. For this reason, it was a natural choice for

doi:10.1088/978-0-7503-1671-2ch3

simulation of quantum transport in devices such as the resonant tunneling diode [6–8]. But the Wigner function is not the only possible phase space distribution, although it is certainly relatable to other such approaches. We will examine the similarities and differences between the Wigner function and other approaches in phase space later in this chapter.

3.1 Preliminary considerations

Let us begin the discussion by introducing the Wigner center-of-mass coordinates, discussed in the last chapter. These involve introducing the average and the difference coordinates as follows:

$$\begin{aligned} \mathbf{x} &= \frac{\mathbf{r} + \mathbf{r}'}{2}, \quad \mathbf{s} = \mathbf{r} - \mathbf{r}' \\ \mathbf{r} &= x + \frac{\mathbf{s}}{2}, \quad \mathbf{r}' = \mathbf{x} - \frac{\mathbf{s}}{2}. \end{aligned} \tag{3.1}$$

We can now introduce the phase space Wigner function as the Fourier transform on the difference coordinate as [1]

$$f_W(\mathbf{x}, \mathbf{p}, t) = \frac{1}{h^3} \int d\mathbf{s}\, \rho\left(\mathbf{x} + \frac{\mathbf{s}}{2}, \mathbf{x} - \frac{\mathbf{s}}{2}, t\right) e^{-i\mathbf{p}\cdot\mathbf{s}/\hbar}, \tag{3.2}$$

where the density matrix ρ is expressed in terms of the wave function as in (2.43). Coupling to this new function is an equation of motion, which has great similarities with the Boltzmann transport equation. If we incorporate the coordinate transformations of (3.1) into (2.46), we arrive at a new equation of motion for the density matrix as (we will derive this more formally below)

$$i\hbar \frac{\partial \rho}{\partial t} = \left[-\frac{\hbar^2}{m^*} \frac{\partial^2}{\partial \mathbf{x} \partial \mathbf{s}} + V\left(\mathbf{x} + \frac{\mathbf{s}}{2}\right) - V\left(\mathbf{x} - \frac{\mathbf{s}}{2}\right) \right] \rho(\mathbf{x}, \mathbf{s}, t). \tag{3.3}$$

The transform in (3.2), although introduced by Wigner, is often called the Weyl transform [1, 9–11]. When this transformation is applied to the above equation, we get a new equation of motion for the Wigner function as

$$\frac{\partial f_W}{\partial t} + \frac{\mathbf{p}}{m^*} \cdot \frac{\partial f_W}{\partial \mathbf{x}} - \frac{1}{i\hbar} \left[V\left(\mathbf{x} + \frac{i\hbar}{2} \frac{\partial}{\partial \mathbf{p}}\right) - V\left(\mathbf{x} - \frac{i\hbar}{2} \frac{\partial}{\partial \mathbf{p}}\right) \right] f_W = 0. \tag{3.4}$$

In the absence of any dissipative processes, this can be rewritten in a more useful form as

$$\frac{\partial f_W}{\partial t} + \frac{\mathbf{p}}{m} \cdot \frac{\partial f_W}{\partial \mathbf{x}} - \frac{1}{h^3} \int d^3\mathbf{p}'\, W(\mathbf{x}, \mathbf{p}') f_W(\mathbf{x}, \mathbf{p} + \mathbf{p}') = 0, \tag{3.5}$$

where

$$W(\mathbf{x}, \mathbf{p}') = \int d^3\mathbf{s} \sin\left(\frac{\mathbf{p}' \cdot \mathbf{s}}{\hbar}\right) \left[V\left(\mathbf{x} + \frac{\mathbf{s}}{2}\right) - V\left(\mathbf{x} - \frac{\mathbf{s}}{2}\right) \right]. \tag{3.6}$$

In this last form, we can clearly see the nonlocal behavior of the potential that is applied to the system. If the potential is of quadratic or lower behavior, then the results are no more than that found with the Boltzmann equation. That is, a simple quadratic potential yields only the first derivative of the potential in (3.6). This is the electric field, which is the principal driving force for transport, even in the classical case. On the other hand, if the potential is sharp, such as a step in potential, then this is felt to all orders of derivatives and can lead to difficulties in numerical simulations.

The problem with even the equation of motion (3.4) is that one must know the value of the Wigner function, or the wave function, at $t = 0$. Hence, one must also solve the adjoint equation for the initial condition [12], which we will do below. The importance of the initial conditions is especially critical for what is known as the bound state problem. For example, if the confining potential in (3.6) is just a quadratic potential, it represents a simple harmonic oscillator, and the resulting force from (3.6) is purely classical. The quantum bound states of the potential will not result from equation (3.4). The initial states from the adjoint equation are required to provide the quantization and, therefore, these bound states [12].

The nonlocal behavior of the potential is reflected in just how the wave function at one point senses the behavior of the wave function at a distant point. In other words, we may ask if there is some correlation in the wave function between two points if these two points are widely separated. It turns out that this effect can be significant in the absence of scattering, and the correlation length can be quite large. Ferrari *et al* [13] have investigated the role of device length of a resonant tunneling device. The main effect studied is the dependence of the tunneling current through the device as a function of the distance from the barriers to the contacts. It is found that the tunneling current is affected by the proximity of the contacts, and the closer the contacts are to the barrier, the higher the tunneling current becomes. In the presence of scattering, this long-range correlation is broken up by the scattering, and the potential can be cut-off at some length. But this length has to be determined by the efficacy of the scattering and may be different in each and every problem. Another approach that has been discussed is to separate the classical, quadratic potential from (3.6), and then treat the difference between the real potential and the classical potential as an effective quantum potential, which may be analyzed primarily as an additional scattering potential [14]. We will deal with this in more detail in later chapters; and, more information can be found in [4]. As one may infer from the above, the use of the Wigner function is particularly important in scattering problems [15], and clearly shows the transition to the classical world.

An important consideration is that the Wigner function satisfies the normalizations that are required on either real space wave functions or momentum space wave functions. We can show this with the Wigner function (3.2). If we integrate the Wigner function over all momentum, then we obtain

$$\begin{aligned}\int d^3\mathbf{p} f_W(\mathbf{x}, \mathbf{p}) &= \frac{1}{h^3}\int d^3\mathbf{p}\int d^3\mathbf{s}\psi^*(\mathbf{x} + \mathbf{s}/2)\psi(\mathbf{x} - \mathbf{s}/2)e^{i\mathbf{p}\cdot\mathbf{s}/\hbar} \\ &= \frac{1}{\hbar^3}\int d^3\mathbf{s}\psi^*(\mathbf{x} + \mathbf{s}/2)\psi(\mathbf{x} - \mathbf{s}/2)\delta(\mathbf{s}/\hbar) \\ &= \psi^*(\mathbf{x})\psi(\mathbf{x}).\end{aligned} \tag{3.7}$$

That is, we obtain the probability of finding the 'particle' at any position in space as given by the magnitude squared of the wave function itself. Similarly, if we introduce the Fourier transform of the wave function,

$$\varphi(\mathbf{p}) = \frac{1}{h^{3/2}} \int d^3\mathbf{x}\, \psi(\mathbf{x}) e^{i\mathbf{p}\cdot\mathbf{x}/\hbar}, \tag{3.8}$$

we may use the analogy to (3.1) and (3.2) to write the Wigner function in terms of these momentum space wave functions as

$$f_W(\mathbf{x}, \mathbf{p}, t) = \frac{1}{h^3} \int d^3\mathbf{q} \varphi^*\left(\mathbf{p} - \frac{\mathbf{q}}{2}\right) \varphi\left(\mathbf{p} + \frac{\mathbf{q}}{2}\right) e^{i\mathbf{p}\cdot\mathbf{x}/\hbar}. \tag{3.9}$$

In (3.9), of course, the position $\mathbf{x}$ is actually the time-shifted position $\mathbf{x} - \mathbf{v}_g t$, which yields the motion of the packet in phase space. The coherent wave packet (3.7) is often used to illustrate tunneling through barriers in simulations or in cases where the packet can be considered as a typical basis function. We note that (3.9) is a positive-definite function, which is appropriate for what is basically a ground state function. This is a general rule for Wigner functions: the ground state wave function is always positive-definite. Now, integration over all position also conserves the normalization in the momentum space, as

$$\int d^3\mathbf{x}\, f_W(\mathbf{x}, \mathbf{p}) = \varphi^*(\mathbf{p})\varphi(\mathbf{p}). \tag{3.10}$$

This is an important result, as not all suggested phase space formulations of quantum transport necessarily preserve this normalization. This has an important generalization. If we have a function $F(\mathbf{x}, \mathbf{p})$, which is either a function of the position operator alone or of the momentum operator alone, or of any *additive* combination of these two operators, the expectation value of this function is given by

$$\langle F \rangle = \int d^3\mathbf{x} \int d^3\mathbf{p} F(\mathbf{x}, \mathbf{p}) f_W(\mathbf{x}, \mathbf{p}). \tag{3.11}$$

This is completely analogous to the equivalent classical expression for such an average. We return to the question of taking these averages in section 3.5, where we show that this approach gives the same results as the quantum expectation value as well. One of the interesting aspects of the Wigner function is the ability to transfer many of the results of classical transport theory into quantum approaches merely by replacing the Boltzmann distribution with the Wigner function. However, there is a caveat that must be understood in this regard. This is the fact that the Wigner function may not be a positive-definite function; that is, there are cases in which the Wigner function possesses negative values, and these values can extend over a phase space region of the extent of Planck's constant. If one smooths over a region of this size, the negative excursions will go away. Clearly, these negative excursions represent the appearance of uncertainly in the quantum realm, but also have other important properties with which we shall deal in due time.

3.2 The equations of motion

So far, we have developed above the idea for the Wigner function from the density matrix, which itself is the product of a wave function and a complex conjugate wave function, in which the two are at different positions:

$$\rho(\mathbf{r}, \mathbf{r}', t) = \sum_{n,m} c_{nm}\varphi_m^*(\mathbf{r}', t)\varphi_n(\mathbf{r}, t) \equiv \psi^*(\mathbf{r}, t)\psi(\mathbf{r}', t). \tag{3.12}$$

The temporal differential equation for the density matrix can be developed by the standard approach for the time variation of any operator, which involves its commutator with the Hamiltonian

$$i\hbar\frac{\partial\rho}{\partial t} = [H, \rho]. \tag{3.13}$$

We consider that the Hamiltonian is written as a sum of a kinetic energy term and a potential energy term. For the Wigner function, we introduce the center-of-mass coordinates (3.1) and/or their conjugate forms for the momentum space wave functions. It turns out that the kinetic energy term is best analyzed using the momentum space wave functions (3.8), for which the Wigner function term becomes

$$\int\frac{d^3\mathbf{q}}{h^3}e^{i\mathbf{q}\cdot\mathbf{x}/\hbar}\left[\frac{\left(p+\frac{\mathbf{q}}{2}\right)^2}{2m} - \frac{\left(\mathbf{p}-\frac{\mathbf{q}}{2}\right)^2}{2m}\right]\varphi^*\left(\mathbf{p}-\frac{\mathbf{q}}{2}, t\right)\varphi\left(\mathbf{p}+\frac{\mathbf{q}}{2}, t\right). \tag{3.14}$$

The expansion of the two quadratic terms leads to the result

$$\int\frac{d^3\mathbf{q}}{h^3}e^{i\mathbf{q}\cdot\mathbf{x}/\hbar}\frac{\mathbf{p}\cdot\mathbf{q}}{m}\varphi^*\left(\mathbf{p}-\frac{\mathbf{q}}{2}, t\right)\varphi\left(\mathbf{p}+\frac{\mathbf{q}}{2}, t\right) = -i\hbar\frac{\mathbf{p}}{m}\cdot\nabla f_W(\mathbf{x}, \mathbf{p}, t). \tag{3.15}$$

For a free particle, we would have only this term and the time derivative on the left-hand side of (3.13). In this case, the equation of motion becomes

$$\frac{\partial f_W}{\partial t} + \mathbf{v}\cdot\nabla f_W = 0, \tag{3.16}$$

and the free particle hence moves as a function of $\mathbf{x} - \mathbf{v}t$, where $\mathbf{v} = \mathbf{p}/\mathrm{m}$. Hence, the Wigner function does not spread with time for a fixed momentum, even though the underlying wave function does so [12]. That is, in normal quantum mechanics, a Gaussian wave packet that moves with such a velocity spreads out in a diffusive manner, as the Schrödinger equation is technically a diffusion equation. In some sense, the spread in velocities leads to an increasing uncertainty for the normal Gaussian packet, as we show below. This does not seem to be the case for the Wigner function, however. Moreover, the corresponding uncertainties satisfy $\Delta x\Delta p_x \geqslant \hbar/2$ for each of the directions in space. Thus, the Wigner function for a single free particle is already different from the classical delta function as a result of the quantum uncertainty relation.

Because the uncertainty relation means that we really can't have a single fixed momentum, or position, the spread in momentum will lead to a spreading of the Wigner function. The sharper the spread in momentum, the broader the spread in position will be, as these are coupled. To see how this naturally follows, let us take a single Gaussian wave packet in momentum space (in one dimension) as

$$\varphi(k) = \left(\frac{2\sigma^2}{\pi}\right)^{1/4} e^{-\sigma^2(k-k_0)^2}. \tag{3.17}$$

For the moment, we have suppressed the time variation, and consider this Gaussian in one-dimensional momentum space. It is centered around an average value of momentum $p_0 = \hbar k_0$, with a mean spread of $1/\sigma$. This momentum space wave function transforms into a real space wave function:

$$\psi(x) = \frac{e^{-(x-v_g t)^2/4\sigma^2 - ik_0(x-v_g t)}}{(2\pi)^{1/4}\sqrt{\sigma}}, \tag{3.18}$$

where $v_g = \hbar k_0/m^*$ is the velocity corresponding to the mean momentum wave number. In addition, the wave packet has a general broadening with time in which the standard deviation σ increases by the factor

$$\sigma(t) = \sqrt{1 + \frac{\hbar^2 t^2}{4m^{*2}\sigma^4}}, \tag{3.19}$$

as is well known in quantum mechanics. For a reasonably sized wave packet of a few nanometers in semiconductors, such as GaAs, the packet doubles in the size extent every picosecond or so. Nevertheless, we will not discuss this broadening to any extent. To explore the spread further, we will actually use the Fourier transform version (3.9), where the wave functions are in momentum space, so that we use the function (3.17). Using this approach, we find that the Wigner function is now given by

$$f_W(x, k) = \frac{2}{h} e^{-2\sigma^2(k-k_0)^2 - x^2/2\sigma^2}, \tag{3.20}$$

where, once again, the time variation has been suppressed. If we had actually used (3.18) directly in (3.2), we would obtain precisely the same result. We get the same phase space representation whether we begin with the real space or the momentum space version of the wave function. In (3.20), of course, the position x is actually the time-shifted position. The spread in momentum space, $1/\sigma$, now appears as a spread in the real space of σ. Hence, the smaller one makes σ, the more confined the real space Gaussian becomes, but the broader the momentum space Gaussian becomes. So, these two are tied together intimately by the uncertainty in the wave packets.

Let us now turn our attention to the remaining terms in the normal Liouville equation (2.46), which are the two potential terms. If we insert the center-of-mass coordinate transformations (3.1), we can write the equation, using our results (3.16) as

$$\left(\frac{\partial}{\partial t}+\mathbf{v}\cdot\nabla\right)f_W(\mathbf{p},\mathbf{x},t)=\frac{1}{i\hbar}\int\frac{d^3\mathbf{s}}{h^3}e^{i\mathbf{p}\cdot\mathbf{s}/\hbar}\left[V\left(\mathbf{x}+\frac{\mathbf{s}}{2}\right)-V\left(\mathbf{x}-\frac{\mathbf{s}}{2}\right)\right]\times\psi^*\left(\mathbf{x}+\frac{\mathbf{s}}{2}\right)\psi\left(\mathbf{x}-\frac{\mathbf{s}}{2}\right). \tag{3.21}$$

To proceed, we invert the transform (3.2) in order to replace the two wave functions with the Wigner function. This leads to

$$\left(\frac{\partial}{\partial t}+\mathbf{v}\cdot\nabla\right)f_W(\mathbf{p},\mathbf{x},t)=\frac{1}{i\hbar}\int\frac{d^3\mathbf{s}d^3\mathbf{p}'}{h^3}e^{i(\mathbf{p}-\mathbf{p}')\cdot\mathbf{s}/\hbar}\times\left[V\left(\mathbf{x}+\frac{\mathbf{s}}{2}\right)-V\left(\mathbf{x}-\frac{\mathbf{s}}{2}\right)\right]f_W(\mathbf{p}',\mathbf{x},t), \tag{3.22}$$

which can be rewritten as

$$\left(\frac{\partial}{\partial t}+\mathbf{v}\cdot\nabla\right)f_W(\mathbf{x},\mathbf{p},t)=\int d^3\mathbf{p}'W(\mathbf{x},\mathbf{p}')f_W(\mathbf{x},\mathbf{p}+\mathbf{p}',t) \tag{3.23}$$

where

$$W(\mathbf{x},\mathbf{p}')=\int d^3\mathbf{s}\,\sin\left(\frac{\mathbf{p}'\cdot\mathbf{s}}{\hbar}\right)\left[V\left(\mathbf{x}+\frac{\mathbf{s}}{2}\right)-V\left(\mathbf{x}-\frac{\mathbf{s}}{2}\right)\right]. \tag{3.24}$$

We can now use the expansion

$$V\left(\mathbf{x}\pm\frac{\mathbf{s}}{2}\right)=V(\mathbf{x})\pm\frac{\mathbf{s}}{2}\cdot\nabla V(\mathbf{x})\pm\ldots=e^{\pm\frac{1}{2}\mathbf{s}\cdot\nabla}V(\mathbf{x}). \tag{3.25}$$

In fact, the operation in (3.25) is the traditional notation for a displacement operator in quantum mechanics [21, 30]. Noting that the factor **s** becomes the gradient in momentum, we find that

$$\begin{aligned}W(\mathbf{x},\mathbf{p}-\mathbf{p}')&=\frac{2}{\hbar}\int\frac{d^3\mathbf{s}}{\hbar^3}e^{i(\mathbf{p}-\mathbf{p}')\cdot\mathbf{s}/\hbar}\sin\left(\frac{\hbar}{2}\nabla_{\mathbf{p}'}\cdot\nabla_{\mathbf{x}}\right)V(\mathbf{x})\\&=\frac{2}{\hbar}\sin\left(\frac{\hbar}{2}\nabla_{\mathbf{p}'}\cdot\nabla_{\mathbf{x}}\right)V(\mathbf{x})\delta(\mathbf{p}-\mathbf{p}').\end{aligned} \tag{3.26}$$

We note that, in (3.26), the momentum gradients operate on the Wigner function, while the position gradients operate on the potential. In addition, we should note that all of the even powers of the derivatives cancel between the two terms, and this also may be seen from (3.24). Hence, if the potential is of no more than quadratic behavior in the position, we achieve the fact that (3.24) becomes

$$\left(\frac{\partial}{\partial t} + \mathbf{v} \cdot \nabla\right) f_W(\mathbf{p}, \mathbf{x}, t) = \nabla V \cdot \frac{\partial f_W(\mathbf{p}, \mathbf{x}, t)}{\partial \mathbf{p}} = -e\mathbf{E} \cdot \frac{\partial f_W(\mathbf{p}, \mathbf{x}, t)}{\partial \mathbf{p}}, \tag{3.27}$$

which is just the collisionless Boltzmann equation. Thus, we have recovered the classical behavior when the potential is of no more than quadratic behavior. On the other hand, if we keep the full form of (3.26), the sine function can be expanded to give

$$\left(\frac{\partial}{\partial t} + \mathbf{v} \cdot \nabla\right) f_W(\mathbf{p}, \mathbf{x}, t) = \sum_{n\ odd} \frac{(i\hbar)^{n-1}}{2^{n-1} n!} (\nabla_\mathbf{p} \cdot \nabla_\mathbf{x})^n V(\mathbf{x}) f_W(\mathbf{p}, \mathbf{x}, t), \tag{3.28}$$

where, again, we must keep in mind the convention that the position gradient operates only upon the potential. This is sometimes referred to as the star product convention.

3.3 Generalizing the Wigner function

Further insight into the structure of the Wigner function may be obtained by introducing a generalization with the use of the Weyl operator [16]. To ease the reading, we pursue this in one dimension, but the generalization to the full three dimensions does not add any great difficulty. We begin by defining a characteristic function derived from the expectation value of an exponential operator with respect to the wave function. This is given as

$$C(\tau, \vartheta) = \int_{-\infty}^{\infty} dx\, \psi^*(x) e^{i(\tau\hat{p} + \vartheta\hat{x})} \psi(x), \tag{3.29}$$

where the $\hat{p}$ and $\hat{x}$ in the exponential are normal quantum-mechanical operators. In addition to discussing the one-dimensional problem, we will take care to distinguish between the quantum operators and the classical variables in this section. That is, the $\hat{p}$ is a differential operator which works on the second wave function. The τ and the θ are classical transform variables for position and momentum, respectively, and the exponent is properly dimensionless (any needed factors of $\hbar$ are incorporated in the latter variables). Then, the Wigner distribution can be obtained from this characteristic function as

$$f_W(x, p) = \frac{1}{4\pi^2} \int_{-\infty}^{\infty} d\tau \int_{-\infty}^{\infty} d\vartheta\, C(\tau, \vartheta) e^{-i(\tau p + \vartheta x)}, \tag{3.30}$$

where p and x are now the c-numbers associated with the Wigner distribution. To see how the normal Wigner function arises from this approach, we can rewrite the exponential in (3.29) using the Baker–Hausdorf formula [17] as

$$e^{i(\tau\hat{p} + \vartheta\hat{x})} = e^{i\tau\hat{p}/2} e^{i\vartheta\hat{x}} e^{i\tau\hat{p}/2}. \tag{3.31}$$

Using this result in (3.29) leads to

$$\begin{aligned} C(\tau, \vartheta) &= \int_{-\infty}^{\infty} dx [e^{-i\tau\hat{p}/2}\psi^*(x)] e^{i\vartheta\hat{x}} e^{i\tau\hat{p}/2}\psi(x) \\ &= \int_{-\infty}^{\infty} dx \psi^*\left(x - \frac{\tau\hbar}{2}\right) e^{i\vartheta\hat{x}} \psi\left(x + \frac{\tau\hbar}{2}\right). \end{aligned} \tag{3.32}$$

Now, we can put this back into (3.30) to yield

$$\begin{aligned} f_W(x, p) &= \frac{1}{4\pi^2}\int_{-\infty}^{\infty} d\tau \int_{-\infty}^{\infty} d\vartheta e^{-i\tau p} \int_{-\infty}^{\infty} dx' \psi^*\left(x' - \frac{\hbar\tau}{2}\right) e^{i\vartheta(x-x')} \psi\left(x' + \frac{\hbar\tau}{2}\right) \\ &= \frac{1}{2\pi}\int_{-\infty}^{\infty} d\tau \int_{-\infty}^{\infty} dx' \psi^*\left(x' - \frac{\hbar\tau}{2}\right) e^{-i\tau p} \psi\left(x' + \frac{\hbar\tau}{2}\right) \delta(x - x') \\ &= \frac{1}{h}\int_{-\infty}^{\infty} ds \psi^*\left(x' + \frac{s}{2}\right) e^{isp/\hbar} \psi\left(x' - \frac{\hbar\tau}{2}\right), \end{aligned} \tag{3.33}$$

where we have made the change of variables $s = -\hbar\tau$. Clearly, this last form is the one-dimensional version of (3.2).

The use of the characteristic function to arrive at the Wigner function allows a more general approach, which we will return to later. Now, however, we want to show that the characteristic function can be generalized to a wider usage [18]. It can be used in connection with an observable, but does so in connection with a specific state $|\psi\rangle$, written in Dirac notation. That is, the characteristic function of an observable A, with respect to this state, may be expressed as

$$C_A(\xi) = \langle \psi \mid e^{i\xi A} \mid \psi \rangle, \tag{3.34}$$

where ξ is a real parameter. If the operator A has an eigenvalue spectrum, which may be defined via the equation

$$A \mid k \rangle = A_k \mid k \rangle, \tag{3.35}$$

then the characteristic function can be evaluated in this representation as

$$\begin{aligned} C_A(\xi) &= \sum_k \sum_{k'} \langle \psi | k \rangle \langle k \mid e^{i\xi A} \mid k' \rangle \langle k' | \psi \rangle \\ &= \sum_k \sum_{k'} \langle \psi | k \rangle e^{i\xi A_k} \langle k' | \psi \rangle = \sum_k e^{i\xi A_k} \mid \psi_k \mid^2. \end{aligned} \tag{3.36}$$

An equivalent version for continuous eigenvalues may be obtained. Here, we see that the characteristic function for an operator A is the Fourier transform of the probability distribution for the eigenvalues of that operator.

While the above approach suggests other types of distributions based upon replacing the operator A with those in the Weyl operator appearing in (3.29), such an approach is hindered by the non-commutative nature of most conjugate operators. To illustrate this point, let us assume a characteristic function for two non-commuting operators A and B to be [19]

$$C_{AB}(\xi_1, \xi_2) = \langle \psi \mid e^{i(\xi_1 A + \xi_2 B)} \mid \psi \rangle. \tag{3.37}$$

Let us assume that each of the two operators has a valid basis set which are formed from the natural eigenvalues of these operators, so that these basis sets may be defined via

$$\begin{aligned} A \mid k \rangle &= A_k \mid k \rangle \\ A \mid k' \rangle &= B_{k'} \mid k' \rangle. \end{aligned} \tag{3.38}$$

Now, we also want to impose a special relationship between the two operators. That is, we require that the two commutators satisfy the relations $[A, [A, B]] = [B, [A, B]] = 0$, and we must remember that each commutator yields a c-number, so that we can utilize the identity [17]

$$e^{i(\xi_1 A + \xi_2 B)} = e^{i\xi_1 A} e^{i\xi_2 B} e^{-\xi_1 \xi_2 [A, B]/2}. \tag{3.39}$$

Inserting this relation into (3.36), and using the various basis sets, we have

$$\begin{aligned} C_{AB}(\xi_1, \xi_2) &= e^{-\xi_1\xi_2[A, B]/2} \sum_{k,k'} \langle \psi | k \rangle \langle k \mid e^{i(\xi_1 A - \xi_2 B)} \mid k' \rangle \langle k' | \psi \rangle \\ &= e^{-\xi_1\xi_2[A, B]/2} \sum_{k,k'} \langle \psi | k \rangle e^{i(\xi_1 A_k, -\xi_2 B_k)} \langle k | k' \rangle \langle k' | \psi \rangle. \end{aligned} \tag{3.40}$$

Here, the last exponential in (3.39) has been brought to the front, as the commutator produces a c-number as noted above. Note that, since the two basis sets do not have a mutual orthogonality relation, the central expectation is not a delta function. Now, we can define the set of expectations as a generalized phase space distribution:

$$\begin{aligned} f(A_k, B_k) &= \langle \psi | k \rangle \langle k | k' \rangle \langle k' | \psi \rangle \\ &= \frac{1}{(2\pi)^2} \int d\xi_1 \int d\xi_2 e^{\xi_1\xi_2[A, B]/2} C_{AB}(\xi_1, \xi_2) e^{-1(\xi_1 A_k + \xi_2 B_{k'})}. \end{aligned} \tag{3.41}$$

The actual form which the result takes is dependent upon the manner in which the last exponential in (3.37) is expanded. The present result depends upon the approximations and assumptions detailed above. The first line of (3.41) differs from the previous definition of the Wigner function due to the presence of the first exponential. If we take A to be x and B to be p, then this last expression can be shown to reduce to [19]

$$f(x, p) = \frac{1}{h} \int_{-\infty}^{\infty} ds\, \psi^*(x) \psi(x - s) e^{ips/\hbar} = \frac{1}{\sqrt{h}} \psi^*(x) e^{ipx/\hbar} \varphi(p). \tag{3.42}$$

Here, the Fourier transform of one wave function has given us the momentum wave function in (3.42). This remains a bilinear wave function in phase space, but it is not the Wigner function proper. In fact, it is an alternative phase space function. We will return to discuss other alternative phase space functions in the next section.

3.4 Other phase space approaches

While the Wigner function has a great many advantages, it is not the only phase space function that has been used in quantum transport. There have been several alternative formulations, most of which depend upon the Weyl exponential form [16, 20], which we have already used several times above. This form is given as

$$e^{i(\xi_1\hat{x}+\xi_2\hat{p})} = \frac{1}{2}(e^{i\xi_1\hat{x}}e^{i\xi_2\hat{p}} + e^{i\xi_2\hat{p}}e^{i\xi_1\hat{x}}). \tag{3.43}$$

As discussed above, it is well known that the ordering of operators is quite important in quantum mechanics, as different ordering produces different expectation values [21]. It is well known that non-commutating operators must be carefully ordered to assure the desired result, as is seen from just px and xp. But, how should one deal with a term such as $(\xi_1 x)^2(\xi_2 p)^3$ that arises in the expansion of (3.43). There is no single preferred ordering for these operators, such as *normal ordering* (where each operator is written in terms of its creation and annihilation operators, with all of the latter moved to the left-hand side of the expression) used in Green's functions [22]. Lacking any recognized protocol for choosing an ordering, many different ones have appeared in the literature, each of which has led to its own version of a phase space distribution function [20, 23].

Mehta [24] has given a standard review of the Wigner function, in which he explored 'standard ordering', where he retains only the first term on the right-hand side of (3.43). By standard ordering, we mean that if one expands the exponential term into a power series, all powers of x preceed all powers of p, where x and p are the position and momentum. This ordering occurs whether we are referring to the c-numbers that appear in the Wigner function or the operators that appear in the quantum functions. With this exponential term, he finds the characteristic function to be

$$C_S(\xi_1, \xi_2) = \int dx\, \psi^\dagger(x)e^{i\xi_1\hat{x}}e^{i\xi_2\hat{p}}\psi(x) = \int dx\, \psi^\dagger(x)e^{i\xi_1\hat{x}}\psi(x + \hbar\xi_2). \tag{3.44}$$

If we now use (3.30) to compute the phase space distribution function from the characteristic function, we find

$$\begin{aligned} f_S(x, p) &= \frac{1}{4\pi^2}\iint d\xi_1\, d\xi_2\, C_S(\xi_1, \xi_2)e^{-i(\xi_1 x+\xi_2 p)} \\ &= \frac{1}{4\pi^2}\iint d\xi_1\, d\xi_2 \int dq\, \psi^\dagger(q)e^{i\xi_1 q}\psi(q + \hbar\xi_2)e^{-i(\xi_1 x+\xi_2 p)} \\ &= \frac{1}{2\pi}\int d\xi_2 \int dq\, \psi^\dagger(q)e^{-i\xi_2 p}\psi(q + \hbar\xi_2)\delta(q - x) \\ &= \frac{1}{2\pi}\int d\xi_2\, \psi^\dagger(x)e^{-i\xi_2 p}\psi(x + \hbar\xi_2). \end{aligned} \tag{3.45}$$

This form differs from (3.42) by a factor of $\hbar$, which can be recovered by a change of the integration variable. Here, we have not specifically required this distribution function to preserve normalization of the wave function (we have not included the factor of 2 that appears in (3.42), and this would indicate that the normalization may not be carried through properly). In fact, Lee [20] suggests that normalization is not maintained with this distribution function, as will be the case with most others that do not use the complete form of the Weyl operator on the left of (3.43).

Kirkwood used anti-standard ordering [24], in which just the last term of (3.43) is retained, and considered the many-body statistics of the distribution. By anti-standard ordering, we mean that if one expands the exponential term into a power series, all powers of p precede all powers of q, where q and p are the position and momentum. The single-particle form differed little from the Wigner function, or from the above form. We can see this, as the characteristic function becomes

$$C_{AS}(\xi_1, \xi_2) = \int dx\, \psi^{\dagger}(x) e^{i\xi_2 \hat{p}} e^{i\xi_1 \hat{x}} \psi(x) = \int dx\, \psi^{\dagger}(x - \hbar\xi_2) e^{i\xi_1 \hat{x}} \psi(x). \tag{3.46}$$

If we now use (3.30) to compute the phase space distribution function from the characteristic function, we find

$$\begin{aligned} f_{AS}(x, p) &= \frac{1}{4\pi^2} \iint d\xi_1\, d\xi_2 C_{AS}(\xi_1, \xi_2) e^{-i(\xi_1 x + \xi_2 p)} \\ &= \frac{1}{4\pi^2} \iint d\xi_1 d\xi_2 \int dq\, \psi^{\dagger}(q - \hbar\xi_2) e^{i\xi_1 q} \psi(q) e^{-i(\xi_1 x + \xi_2 p)} \\ &= \frac{1}{2\pi} \int d\xi_2 \int dq\, \psi^{\dagger}(q - \hbar\xi_2) e^{-i\xi_2 p} \psi(q) \delta(q - x) \\ &= \frac{1}{2\pi} \int d\xi_2\, \psi^{\dagger}(x - \hbar\xi_2) e^{-i\xi_2 p} \psi(x). \end{aligned} \tag{3.47}$$

Again, the result bears significant similarity to (3.42) except that it is the complex conjugate wave function that is shifted by the parameter ξ_2. Now, to compare the three versions, we complete the Fourier transforms to yield the three versions of the phase space functions as

$$\begin{aligned} f(x, p) &= \frac{1}{\sqrt{h}} \psi^{\dagger}(x) e^{ipx/\hbar} \varphi(p) \\ f_S(x, p) &= \frac{1}{\sqrt{h}} \psi^{\dagger}(x) e^{ipx/\hbar} \varphi^*(p) \\ f_{AS}(x, p) &= \frac{1}{\sqrt{h}} \varphi^{\dagger}(p) e^{-ipx/\hbar} \psi(x). \end{aligned} \tag{3.48}$$

Mehta has also examined the use of normal-ordered products of the quantum operators through the introduction of creation and annihilation operators [24]. A somewhat similar discussion of the use of these operators and the connection with classical statistical mechanics for quantum optics has been discussed as well [26, 27]. For this purpose, these creation and annihilation operators may be defined as [24]

$$a = \frac{1}{\sqrt{2h}}(x + ip), \quad a^{\dagger} = \frac{1}{\sqrt{2h}}(x - ip), \tag{3.49}$$

where, as before, q and p are the quantum-mechanical operators. Within some constants, these operators are precisely those used for the harmonic oscillator or for angular momentum raising and lowering operators. Here, the adjoint (second) term is the creation operator, while the first form is the annihilation operator. If we use the full form of the left-hand side of (3.43), and expand the exponential, then the various terms in the exponential can be rearranged into two exponentials, each of which contains one of the two operators of (3.49). Specifically, Mehta [24] has shown that (3.43) can be rewritten as

$$e^{i(\xi_2 p+\xi_1 q)} = e^{ia^\dagger(\xi_1+i\xi_2)\sqrt{\hbar/2}}e^{ia(\xi_1-i\xi_2)\sqrt{\hbar/2}}. \tag{3.50}$$

The phrase 'normal ordering' implies that all the creation operators ($a^\dagger$) are kept to the left of all the annihilation operators. When this expansion is used to generate the characteristic function, the resulting distribution function is given as

$$f_N(x, p) = \exp\left[-\frac{\hbar}{4}\left(\frac{\partial^2}{\partial p^2} + \frac{\partial^2}{\partial x^2}\right)\right]f_W(x, p). \tag{3.51}$$

Thus, this normal-ordered distribution is an infinite series of corrections to the Wigner function itself. It is not readily apparent under which circumstances one may prefer this latter distribution. Nevertheless, it may fit well into a situation where the full expansion of the potential via (3.28) is also included in the treatment.

Finally, there is a further phase space function that has been quite useful in optics and also in quantum chaos [28]. This is the so-called Husimi function [29]. Husimi refers to our coherent wave packet (3.18) as a Heisenberg wave packet, and points out that these are not mutually orthogonal. This means that a group of Gaussian wave packets of this form can not be used as an orthonormal basis set, as any two are not orthogonal to one another. This is a result of them both being positive-definite functions. But it is important to note that one is not required to use orthonormal basis sets in quantum mechanics. When the basis set is not orthonormal, the mathematics becomes more complex but not impossible. Hence under certain conditions the set of Gaussians with different initial points (for example on each of the grid points of a discretization of the Schrödinger equation) can form a complete, normalized, almost orthogonal basis set (Husimi attributes this to work of von Neumann [30]). Generally, the required conditions are a discretization of phase space according to which the coherent wave packets are spaced according to

$$\begin{aligned} \Delta k &= \frac{\sqrt{\pi}}{\sigma} \\ \Delta x &= \sqrt{\pi}\sigma, \end{aligned} \tag{3.52}$$

where σ is the standard deviation of the wave packet, as indicated in (3.18). The prescription given by Husimi is to first smooth the density matrix, the kernel of the integral in (3.2) via

$$\bar{\rho}(x') = \frac{1}{\sqrt{2\pi}\sigma}\int d\xi\psi^*\left(\xi - \frac{x'}{2}\right)\psi\left(\xi + \frac{x'}{2}\right)e^{-(\xi-x')^2/2\sigma^2}, \tag{3.53}$$

which provides a Gaussian smoothing of the density matrix. Then, this smoothed density matrix is then transformed using a coherent wave packet as

$$f_H(x, k) = \int \bar{\rho}(x') e^{-\frac{x'^2}{8\sigma^2} + ikx'} dx'. \tag{3.54}$$

When this procedure is done with the sample coherent wave packet introduced in (3.18), we find the Husimi function is

$$f_H(x, k) = e^{-\frac{x^2}{4\sigma^2} - \sigma^2(k-k_0)^2}, \tag{3.55}$$

which has some differences from the previous result (3.20). To begin, the Husimi function is broader in position space, but these small differences mask some fundamental differences between the Wigner and Husimi functions. It is well known that the Wigner function can be negative, but these regions disappear when it is averaged over a region the size of an uncertainty box in phase space. On the other hand, the Husimi function is already averaged via the procedure (3.53), and is therefore a positive-definite and a semi-classical distribution. In fact, Takahashi has pointed out that the Husimi distribution is a coarse graining of the Wigner function where the coarse graining is accomplished with a Gaussian smoothing [31]. Moreover, the Husimi function does not produce the correct probability functions [32]. That is, when the momentum is integrated out of the Husimi function, the resulting probability distribution in real space is broader than that from the Wigner function. Correspondingly, when the Husimi function is integrated over real space, the resulting momentum probability function is broader than that resulting from the Wigner function. Both of these results arise from the coarse graining that is inherent in the Husimi function, and is apparent when (3.55) is compared with (3.20). It should be remarked that a different prescription for obtaining the Husimi function has been given in the literature. In this form, we find the Husimi function from [27, 33–35]

$$f_H = \left| \int dx' \psi(x', k_0) \varphi_T(x' - x, k) \right|^2, \tag{3.56}$$

where $\psi(x', k_0)$ is our specific wave function (3.18) with which we want to obtain the Husimi function. The second wave function φ_T is the transforming (or smoothing) function by which the wave function is modified. This is a generalized coherent state written in terms of the two coordinates and a general wave vector k, which can be the exponential term in (3.54). This form gives the same result (3.55) for our initial coherent state. This particular formulation is actually just a reworking of the integrals in (3.53) and (3.54), and has found wide application in quantum optics. In addition to the optical applications, as we remarked above, the Husimi function has found a great deal of usage in the arena of quantum chaos [28], where the natural simulation space is a Poincaré section of classical phase space. Since the quantum simulation can be done for the same (equivalent) system to yield the wave function, the projection of this onto the classical Poincaré section gives direct information on the classical-to-quantum crossover. The use of (3.56) gives a quick transformation from the wave function to the Husimi distribution for this purpose, and can be done quite easily numerically [36].

3.5 Wigner–Weyl transforms

One of the most important aspects of either classical mechanics or quantum mechanics is the determination of the expectation values of various operators. For example, when we have a range of states described by the one-particle distribution, we want to know the expectation, or average, value of variables like position and momentum. These are determined taking the appropriate moments of the variable with the distribution function. Similarly, in quantum mechanics, we are interested in the expectation of an operator $\hat{A}(x)$, which is typically evaluated with the probability function as

$$\langle A \rangle = \int \psi^{\dagger}(\mathbf{x})\hat{A}(\mathbf{x})\psi(\mathbf{x})\, d^3\mathbf{x}. \tag{3.57}$$

(If we were using the density matrix, the operator would be a function of two positions, and the two wave functions would correspond to the two different positions. We will generally use less explicit forms for the operator, so as not to confuse the reader.) We write the quantum operator with the hat over the top. The question that we address here is just how this expectation value is taken in the Wigner formulation. The variables in (3.57) are quantum-mechanical variables, which are operators. On the other hand, the Wigner function (3.2) is a function of classical variables after the transformation occurs. That is, the formulation of the Wigner function changes the quantum operators into classical variables, as we have discussed in the above sections. Weyl pointed out that to compute the expectation of the operator, we first have to transform the quantum operator into the classical equivalent and this is done with the same prescription (3.2) [16]. Hence,

$$A(\mathbf{x}, \mathbf{p}) = \int d^3\mathbf{y}\psi^{\dagger}\left(\mathbf{x} + \frac{\mathbf{y}}{2}\right)\hat{A}\psi\left(\mathbf{x} - \frac{\mathbf{y}}{2}\right)e^{-i\mathbf{p}\cdot\mathbf{y}/\hbar}. \tag{3.58}$$

As we mentioned, the Weyl transform converts the operator form into a function of the classical phase space variables. Then, the classical expectation value is given by the normal form [37]

$$\langle A \rangle = \int d^3\mathbf{x} \int d^3\mathbf{p}\, A(\mathbf{x}, \mathbf{p}) f_W(\mathbf{x}, \mathbf{p}). \tag{3.59}$$

In the following, we wish to show that this is correct.

Let us begin by considering the classical expectation value of two such operators, each of which is given by the Weyl transformation of (3.58). We write the expectation of the product in analogy with (3.59)

$$\langle AB \rangle = \int d^3\mathbf{x} \int d^3\mathbf{p}\, A(\mathbf{x}, \mathbf{p}) B(\mathbf{x}, \mathbf{p}). \tag{3.60}$$

To begin, we introduce the expression (3.58) for each of the two operators, as

$$\begin{aligned}
\langle AB\rangle &= \int d^3\mathbf{x}\int d^3\mathbf{p}\int d^3\mathbf{y}\int d^3\mathbf{y}'\psi^\dagger\left(\mathbf{x}+\frac{\mathbf{y}}{2}\right)\hat{A}\psi\left(\mathbf{x}-\frac{\mathbf{y}}{2}\right)\\
&\quad\times\psi^\dagger\left(\mathbf{x}+\frac{\mathbf{y}'}{2}\right)\hat{B}\psi\left(\mathbf{x}-\frac{\mathbf{y}'}{2}\right)e^{-i\mathbf{p}\cdot(\mathbf{y}+\mathbf{y}')/\hbar}\\
&= h^3\int d^3\mathbf{x}\int d^3\mathbf{y}\int d^3\mathbf{y}'\psi^\dagger\left(\mathbf{x}+\frac{\mathbf{y}}{2}\right)\hat{A}\psi\left(\mathbf{x}-\frac{\mathbf{y}}{2}\right)\\
&\quad\times\psi^\dagger\left(\mathbf{x}+\frac{\mathbf{y}'}{2}\right)\hat{B}\psi\left(\mathbf{x}-\frac{\mathbf{y}'}{2}\right)\delta(\mathbf{y}+\mathbf{y}')\\
&= h^3\int d^3\mathbf{x}\int d^3\mathbf{y}\psi^\dagger\left(\mathbf{x}+\frac{\mathbf{y}}{2}\right)\hat{A}\psi\left(\mathbf{x}-\frac{\mathbf{y}}{2}\right)\psi^\dagger\left(\mathbf{x}-\frac{\mathbf{y}}{2}\right)\hat{B}\psi\left(\mathbf{x}+\frac{\mathbf{y}}{2}\right).
\end{aligned}\tag{3.61}$$

From the first line to the second, we used the properties of the integration over momentum to yield a delta function. Then, we integrated over this delta function to give the final line of the equation. To proceed, we will introduce a change of variables $\mathbf{u} = \mathbf{x} - \mathbf{y}/2$, and $\mathbf{v} = \mathbf{x} + \mathbf{y}/2$. Then, we will transition to Dirac notation. With these changes, (3.61) becomes

$$\begin{aligned}
\langle AB\rangle &= h^3\int d^3\mathbf{v}\int d^3\mathbf{u}\psi^\dagger(\mathbf{v})\hat{A}\psi(\mathbf{u})\psi^\dagger(\mathbf{u})\hat{B}\psi(\mathbf{v})\\
&= h^3\int d^3\mathbf{v}\int d^3\mathbf{u}\langle\mathbf{v}\mid\hat{A}\mid\mathbf{u}\rangle\langle\mathbf{u}\mid\hat{B}\mid\mathbf{v}\rangle = h^3\mathrm{Tr}[\hat{A}\hat{B}].
\end{aligned}\tag{3.62}$$

In the last term on the right, if we replace, for example, the B operator by the density matrix, we recover the normal method of finding the expectation value for the operator A. If we replace h^3B by the Wigner function, then we recover (3.59).

To illustrate some of these concepts, let us consider the simple harmonic oscillator, which we treat in one dimension. The oscillator is described by the Hamiltonian

$$\hat{H} = \frac{\hat{p}^2}{2m} + \frac{m\omega^2\hat{x}^2}{2},\tag{3.63}$$

where m is the mass of the oscillator particle and ω is a measure of the restoring force. It is convenient to introduce creation and annihilation operators a and $a^\dagger$, respectively, for which the Hamiltonian can be written as

$$\hat{H} = \hbar\omega\left(a^\dagger a + \frac{1}{2}\right) = \hbar\omega\left(n + \frac{1}{2}\right),\tag{3.64}$$

where n is the energy level index given by the product of operators in the first term on the right-hand side. The transformations between position and momentum and the operators is given as

$$\begin{aligned}
a &= \sqrt{\frac{m\omega}{2\hbar}}\left(\hat{x} + i\frac{\hat{p}}{m\omega}\right), \quad a^\dagger = \sqrt{\frac{m\omega}{2\hbar}}\left(\hat{x} - i\frac{\hat{p}}{m\omega}\right),\\
\hat{x} &= \sqrt{\frac{\hbar}{2m\omega}}(a + a^\dagger), \quad \hat{p} = \frac{1}{i}\sqrt{\frac{m\omega\hbar}{2}}(a - a^\dagger).
\end{aligned}\tag{3.65}$$

The ground state wave function is given as

$$\varphi_0 = \left(\frac{m\omega}{\pi\hbar}\right)^{1/4} \exp\left(-\frac{m\omega x^2}{2\hbar}\right). \tag{3.66}$$

With this wave function, we can determine the Wigner function from the ground state from (3.2) to be

$$f_{W,0}(x, p) = \frac{2}{h} \exp\left(-\frac{p^2}{m\omega\hbar} - \frac{m\omega x^2}{\hbar}\right). \tag{3.67}$$

Using this Wigner function, we then can find the energy of the ground state of the harmonic oscillator to be

$$\langle H \rangle = \int dx \int dp\, f_{W,0}(x, p)\left(\frac{p^2}{2m} + \frac{m\omega^2 x^2}{2}\right) = \frac{1}{2}\hbar\omega, \tag{3.68}$$

which agrees with (3.64).

As we have discussed earlier, the harmonic oscillator is largely a purely classical object in terms of its motion in phase space. The quantization has to be put in by hand from the quantum-mechanical solution; it does arise in the equation of motion for the Wigner function. If, however, either the adjoint equation is used to get the initial time Wigner function or the actual harmonic oscillator wave functions are used to form the Wigner function, then the proper physics will result. We show this with the general temporal motion of the harmonic oscillator, which may be found to be

$$\begin{aligned} x &= x_0 \cos(\omega t) + \frac{p_0}{m\omega}\sin(\omega t) \\ p &= p_0 \cos(\omega t) - m\omega x_0 \sin(\omega t), \end{aligned} \tag{3.69}$$

which is nothing more than a time-varying coordinate rotation in phase space. Hence, if we have an initial condition at $t = 0$ of $f_W(x_0, p_0, 0)$, then the time-varying function becomes

$$f_W(x, p, t) = f_W\left(x\cos(\omega t) - \frac{p}{m\omega}\sin(\omega t), p\cos(\omega t) + m\omega\sin(\omega t), 0\right). \tag{3.70}$$

Hence, one sees that all the discussion of the harmonic oscillator, either in terms of the operators or in terms of the Hermite polynomials, really only relates to the nature of the states that are found, and not with the time evolution. This latter is a property of the evolution for the Wigner function. Of course, we have followed only this simple problem which is mainly classical. But it is characteristic of a great many such problems. Once we find the eigenstate wave function, we can transform it with the Weyl transform to reach the Wigner function, and then the temporal motion can be established quite easily. It should be noted here that the use of the operators is common, particularly as these operators can be applied to a wide class of potentials which may be addressed through the use of supersymmetric quantum mechanics [38].

In closing this section, we need to remember that different forms of the Wigner function can be obtained by different variations of the Weyl operator (3.43). We have used the usual variant in our discussion of the Weyl transforms and the method of obtaining the expectation values. One should determine, when using a different ordering of the operators, whether or not this leads to some change in the expectation value. For product operators, it is known that different ordering gives different expectation values, and each case needs to be carefully investigated.

3.6 The hydrodynamic equations

The development of a transport model based upon a parameterized version of the distribution function is more than a century old. The idea is that there are certain constants of the motion in classical physics. If we consider, for example, the simple Maxwellian distribution function used often in classical physics, it is described by an exponential function of the energy, normalized by the thermal energy $k_B T$. In transport, where we are interested in the constants of motion, we can think about the argument of the exponential being a description of the entropy production of the transport process [39, 40]. In semiconductors, the usual constants of the motion are the density, the linear momentum, and the energy. These are typically handled by the introduction of a proper normalization constant, a drift velocity, and an effective temperature to replace the lattice temperature, respectively. These are then evaluated by a set of equations, often referred to as the hydrodynamic equations, that are derived from moments of the equation of motion of the distribution. The latter would be the Boltzmann equation in classical physics, and (3.28) for the Wigner function, albeit with the proper collisional terms added. This approach was added to nonequilibrium transport in semiconductors by Fröhlich and Paranjape in the mid-twentieth century [41]. These authors argued that it was a valid approach when the electron density was sufficiently high to ensure that interelectronic collisions dominated the energy and momentum transfer and forced the distribution into a shifted Maxwellian form. These collisions provide a very fast timescale that dominates the distribution function and forces it into a quasi-equilibrium form as described by Bogoliubov [42]. In this quasi-equilibrium form, the distribution function is shifted in momentum space by the drift velocity and has a greater spread in energy due to the enhanced electron temperature.

A great deal of resemblance between the equation of motion (3.28) for the Wigner distribution function and the classical equivalent Boltzmann equation exists, as we have already pointed out. But it is also useful to extract the hydrodynamic equations for this Wigner equation of motion. For this, we will deal primarily with the streaming terms given in (3.28), leaving the collision term for later considerations. We will eventually deal here with the relaxation time approximation for the scattering term of the Wigner equation of motion. The moment equations are obtained by multiplying the equation of motion by an arbitrary function of momentum $\varphi(\mathbf{p})$ and then integrating over the momentum. Using (3.28), we then find

$$\begin{aligned}\frac{\partial\langle\varphi(\mathbf{p})\rangle}{\partial t}+\frac{1}{m}\nabla\cdot\langle\varphi(\mathbf{p})\mathbf{p}\rangle=\sum_{s\ odd}\frac{(i\hbar)^{s-1}}{2^{s-1}s!}\frac{\partial^s V(\mathbf{x})}{\partial\mathbf{x}^s}\cdot\int d^3\mathbf{p}\,\varphi(\mathbf{p})\frac{\partial^s f_W}{\partial\mathbf{p}^s}\\+\left\langle\varphi(\mathbf{p})\frac{\partial f_W}{\partial t}\bigg|_{coll.}\right\rangle,\end{aligned}\tag{3.71}$$

where

$$\langle\varphi(\mathbf{p})\rangle=\int d^3\mathbf{p}\,\varphi(\mathbf{p})f_W(\mathbf{x},\mathbf{p},t).\tag{3.72}$$

The last equation defines the expectation value for particular function of momentum.

Let us first begin by taking the function $\varphi(\mathbf{p})$ as just the c-number 1. This produces the simple

$$\frac{\partial n(\mathbf{x})}{\partial t}+\frac{1}{m}\nabla\cdot(n(\mathbf{x})\langle\mathbf{p}\rangle)=\frac{\partial n(\mathbf{x})}{\partial t}+\frac{1}{m}\nabla\cdot(n(\mathbf{x})\mathbf{v}_d)=0,\tag{3.73}$$

which is just a form of the continuity equation. We note here that the single value of momentum that exists in the second term of (3.70) corresponds exactly to the quantum-mechanical current, and this term just takes the average of that quantity. In the case where only a single conduction band valley is considered, the scattering conserves the local density, although nonlocal scattering could change this with a broadening effect. This does not occur with the relaxation time approximation. Note that the $s = 0$ term does not exist in the sum on the right-hand side of (3.71). The $s = 1$ term yields an asymmetric argument of the integral and this also vanishes. There are no higher-order terms, as one can do integration by parts (in momentum) which moves the derivative to the included function of momentum. As this is constant, the higher-order terms vanish.

Now, let us turn to the next possible term, which occurs when we let $\varphi(\mathbf{p}) = \mathbf{p}$. This leads us to the equation

$$\frac{\partial(n\langle\mathbf{p}\rangle)}{\partial t}+\frac{1}{m^*}\nabla\cdot(n\langle\mathbf{pp}\rangle)-\frac{\partial V(\mathbf{x})}{\partial\mathbf{x}}\cdot\int\mathbf{p}\frac{\partial f_W}{\partial\mathbf{p}}d^3\mathbf{p}=\left\langle\mathbf{p}\frac{\partial f_W}{\partial t}\bigg|_{coll}\right\rangle.\tag{3.74}$$

While there are higher-order terms in the potential term, they can be eliminated by integration by parts, so that we are left with only the $s = 1$ term. That is, when one integrates the momentum integral by parts, only the first derivative of the momentum does not vanish, and so only the single term is found. Thus, the potential term can be taken to be the voltage and then the derivatives give the electric field, so that this term becomes $e\mathbf{E}n$. If we introduce the relaxation time approximation as

$$-\frac{f_W(\mathbf{x},\mathbf{p})-f_{W0}(\mathbf{x},\mathbf{p})}{\tau_m},\tag{3.75}$$

where the second distribution is the equilibrium value, then the relaxation term becomes just

$$-\frac{n\langle \mathbf{p}\rangle}{\tau_m} = -\frac{nm^*\mathbf{v}_d}{\tau_m}. \tag{3.76}$$

The second term on the left-hand side of (3.74) is more problematic. The average produces a second-rank tensor, with the diagonal elements being proportional to the energy of the distribution. This is a major point of coupling between this equation and the next higher order. Even if we keep just the diagonal elements, as is common for the classical drifted Maxwellian, even these terms go beyond the classical energy, and we show this by expanding as [43]

$$\langle p^2\rangle = \left\langle -\frac{\hbar^2}{4}\left(\psi^*\frac{d^2\psi}{dx^2} - 2\frac{d\psi^*}{dx}\frac{d\psi}{dx} + \frac{d^2\psi^*}{dx^2}\psi\right)\right\rangle. \tag{3.77}$$

If we now assume that we can write the wave function as

$$\psi(x) = A(x)e^{iS(x)/\hbar}, \tag{3.78}$$

then (3.77) becomes

$$\langle \mathbf{p}^2\rangle = (m\mathbf{v}_d)^2 n(\mathbf{x}) - \frac{\hbar^2}{4}n(\mathbf{x})\frac{d^2(\ln n(\mathbf{x}))}{d\mathbf{x}^2}, \tag{3.79}$$

where the last term is another form of the Wigner potential, which serves as a correction to the semi-classical potential. We will discuss this further in the next chapter. Finally, if we fold in the derivatives of the density from (3.73) with the approximations found above, we find the new form of the time derivative of the drift velocity to be

$$\frac{\partial \mathbf{v}_d}{\partial t} = \frac{e\mathbf{E}}{m^*} - \frac{\mathbf{v}_d}{\tau_m} + \frac{\mathbf{v}_d}{n}\nabla\cdot(n\mathbf{v}_d) - \frac{1}{nm^*}\nabla\cdot(nE) + \frac{\hbar^2}{4m^*}\frac{d^2(\ln n)}{dx^2}. \tag{3.80}$$

Only the last term differs from what would be found if we had used the classical Boltzmann equation. So, it is somewhat natural that the Wigner equation approach brings in the Wigner potential as a quantum correction.

We need to do one more moment to account for the continuity of energy in the system, since the relaxation time τ_m is often an energy dependent quantity, and the energy is the third of our normal constants of the motion. For this purpose, we let $\varphi(\mathbf{p}) = \mathrm{p}^2/2m^*$. Again, we multiply the basic transport equation with this quantity and integrate over the momentum. This leads us to the equation

$$\frac{1}{2m^*}\frac{\partial(n\langle p^2\rangle)}{\partial t} + \frac{1}{2m^{*2}}\nabla\cdot(n\langle p^2\mathbf{p}\rangle) + \frac{n}{m^*}\nabla V\cdot\langle \mathbf{p}\rangle = \left\langle \frac{p^2}{2m^*}\frac{\partial f_w}{\partial t}\bigg|_{coll}\right\rangle. \tag{3.81}$$

Again, the last term on the left-hand side results from the fact that only the $s = 1$ term arises from the summation in (3.68). In fact, it would be at the next higher-order equation where the $s = 2$ term first appears. As before, the integral in this summation is integrated by parts (twice), which leaves just this simple term. The second term on the left-hand side can be evaluated just as was done in the momentum equation. The term is written out in terms of the wave function, and using (3.77), we arrive at

$$\langle p^2 \mathbf{p} \rangle = n(\mathbf{x})(mv_d)^2 m\mathbf{v}_d - \frac{n(\mathbf{x})\hbar^2}{4}[3m\mathbf{v}_d \nabla^2(\ln n(\mathbf{x})) + \nabla^2(m\mathbf{v}_d)]. \tag{3.82}$$

As before, the terms involving an explicit dependence upon Planck's constant are quantum corrections to the potential that appears in the equation. One further note is that, in the generalized expansions of a parameterized distribution that we used in the last chapter, the leading term involves the time derivative

$$\frac{1}{2m^*}\frac{\partial(n\langle p^2 \rangle)}{\partial t} = \frac{\partial}{\partial t}\left[n\left(\frac{3k_B T_e}{2} + \frac{1}{2}m^* v_d^2\right)\right]. \tag{3.83}$$

Not only does this term fold (3.73) into the present equation, the derivative of the drift term also folds the complete (3.81) into the present equation. This will lead to a very long and complicated final equation for the time variation of the electron temperature. For brevity, we will omit this final expansion of (3.81), leaving it for the reader to explore.

The accuracy that is typically found when using the hydrodynamic equations to study semiconductor transport ranges from very good to exceedingly poor. The former results are found when the carrier density is relatively high and the assumptions about the form of the distribution function hold. The latter results arise when these assumptions just do not hold, such as the case in which the scattering is dominated by the polar optical modes and the density is relatively low. This form of scattering, like impurity scattering, is very anisotropic and does not lead to a nice Maxwellian form for the distribution function. Rather the distribution function becomes strongly peaked in the electric field direction, even in the quantum case.

References

[1] Wigner E P 1932 *Phys. Rev.* **40** 749
[2] Moyal J E 1949 *Proc. Cambridge Phil. Soc.* **45** 99
[3] Levinson I B 1969 *Zh. Eksp. Teor. Fiz.* **57** 660
1970 *Sov. Phys. JETP* **30** 362
[4] Nedjalkov M, Selberherr S, Ferry D K, Vasileska D, Dollfus P, Querlioz D, Dimov I and Schwaha P 2013 *Ann. Phys.* **328** 220
[5] Dias N C and Prata J N 2004 *Ann. Phys.* **313** 110
[6] Ravaioli U, Osman M A, Pötz W, Kluksdahl N and Ferry D K 1985 *Physica* B **134** 36
[7] Frensley W R 1987 *Phys. Rev.* B **36** 1570
[8] Kluksdahl N C, Kriman A M, Ferry D K and Ringhofer C 1989 *Phys. Rev.* B **39** 7720
[9] Smith T B 1978 *J. Phys.* A **11** 2179
[10] Janusis A, Streklas A and Vlachos K 1981 *Physica* A **107** 587

[11] Royer A 1991 *Phys. Rev.* A **43** 44
[12] Carruthers P and Zachariasen F 1983 *Rev. Mod. Phys.* **55** 245
[13] Ferrari G, Giacobbi N, Bordone P, Bertoni A and Jacoboni C 2004 *Semicond. Sci. Technol.* **19** 8254
[14] Bordone P, Bertoni A and Jacoboni C 2002 *Physica* B **314** 123
[15] Remler E A 1975 *Ann. Phys.* **95** 455
[16] Weyl H 1931 *The Theory of Groups and Quantum Mechanics* (New York: Dover)
[17] Messiah A 1961 *Quantum Mechanics* vol 1 (New York: Wiley) p 442
[18] Grubin H L, Ferry D K, Iafrate G J and Barker J R 1982 *VLSI Microelectronics: Microstructure Science* vol 3 ed N Einspruch (New York: Academic) 197–300
[19] Iafrate G J, Grubin H L and Ferry D K 1982 *Phys. Lett.* A **87** 145
[20] Lee H-W 1995 *Phys. Rep.* **259** 147
[21] Ferry D K 2001 *Quantum Mechanics* 2nd edn (Bristol: Institute of Physics)
[22] Fetter A L and Walecka J D 1971 *Quantum Theory of Many-Particle Systems* (New York: McGraw-Hill)
[23] Fan H-Y 2003 *Commun. Theor. Phys.* **40** 409
[24] Mehta H 1964 *J. Math. Phys.* **5** 677
[25] Kirkwood J G 1933 *Phys. Rev.* **44** 31
[26] Glauber R J 1963 *Phys. Rev.* **131** 2766
[27] Sundarshan E C G 1963 *Phys. Rev. Lett.* **10** 277
[28] Gutzwiller M C 1990 *Chaos in Classical and Quantum Mechanics* (New York: Springer)
[29] Husimi K 1940 *Proc. Phys.-Math. Soc. Jpn.* **22** 264
[30] von Neumann J 1955 *Mathematical Foundations of Quantum Mechanics* trans. R T Beyer (Princeton, NJ: Princeton Univ. Press)
[31] Takahashi K 1986 *J. Phys. Soc. Jpn.* **55** 762
[32] Ballentine L E 1998 *Quantum Mechanics: A Modern Development* (Singapore: World Scientific)
[33] Luna-Acosta G A, Na K, Reichl L E and Krokhin A 1996 *Phys. Rev.* B **53** 3271
[34] Bäcker A, Fürstberger S and Schubert R 2004 *Phys. Rev.* E **70** 036204
[35] Weingartner B, Rotter S and Burgdörfer J 2005 *Phys. Rev.* B **72** 115342
[36] Brunner R, Meisels R, Kuchar F, Akis R, Ferry D K and Bird J P 2007 *Phys. Rev. Lett.* **98** 204101
[37] Case W B 2008 *Am. J. Phys.* **76** 937
[38] Cooper F, Khare A and Sukhatme U 1995 *Phys. Rep.* **251** 267
[39] Zubarev D N 1974 *Nonequilibrium Statistical Thermodynamics*trans. from Russian by P J Shephard (New York: Consultants Bureau)
[40] Degond P and Ringhofer C 2003 *J. Stat. Phys.* **112** 587
[41] Fröhlich H and Paranjape V V 1956 *Proc. Phys. Soc. London* B **69** 21
[42] Bogoliubov N N 1962 *Studies in Statistical Mechanics* ed J de Boer and G E Uhlenbeck (Amsterdam: North-Holland)
[43] Iafrate G J, Grubin H L and Ferry D K 1981 *J. Phys. Coll.* **C7** 307

IOP Publishing

The Wigner Function in Science and Technology

David K Ferry and Mihail Nedjalkov

Chapter 4

Effective potentials

In each of the previous three chapters, we have introduced and discussed briefly the idea of an effective potential. Now, it is fair to ask just why we need to have such a concept. For one thing, approaches to quantum mechanics by both Bohm and his predecessors, as well as by Wigner, have shown that such an effective potential arises naturally. But why is this so? In some regard, the idea relates to the size of the particle, or electron, in which we are interested. In classical physics, the electron is incredibly small, being something of the order of 10^{-24} m or so. But in quantum mechanics, the electron is described by a wave packet. In the atom, the wave packet is comparable to the size of the electron orbit around the atom, but when removed from the atom, this wave packet can be substantially larger. In some cases, this wave packet may well be a Gaussian as described in (3.18). The quantum wave packet is much larger than the classical size, if for no other reason that it cannot be smaller in position and momentum space than dictated by the uncertainty relation. Hence, it is reasonable to ask just how large this wave packet will be. Again, it is important to understand that the size matters. If we have a potential barrier, a classical particle can sit next to the barrier and nothing of importance arises. But, if we try to put the centroid of the Gaussian next to the barrier, then we must address the issue of wave function penetration into the barrier. For sure, the barrier is such that the wave function does not penetrate far into it. But this in turn leads to the inability of the centroid of the Gaussian to actually be next to the barrier. It is repelled from the barrier, and this leads to quantization if we try to use two barriers to confine the electron. This becomes even more important if we turn our attention to nano-scale semiconductor devices.

Electron transport in nano-scale semiconductor devices has become very important as CMOS architecture is scaled toward 10–20 nm characteristic lengths. In this regime, the transport is dominated by quantum effects that arise throughout the active region. Several approaches to simulation of semiconductor devices have appeared in which the transport is handled quantum mechanically [1]. In these small structures, one must begin to worry about the effective size of the carriers themselves. This size

doi:10.1088/978-0-7503-1671-2ch4

can be several nanometers. Then one begins to understand the problem, and to ask how the electrons can fit into the nano-scale device. Is the number of electrons going to be limited by their size? This is an important question. In the next section, some arguments for various sizes will be considered for electrons in semiconductors. In particular, in the quasi-two-dimensional electron gas in the channel of a MOSFET, it will be argued that the effective size of the electron packet is only λ_F/π, where λ_F is the Fermi wavelength, a value providing an almost minimum uncertainty packet. This size also reflects the 'squeezing' of the packet in two dimensions as the carrier density is increased (and results in a greater extent in the third dimension). The case of a non-degenerate semiconductor will also be considered.

The importance of the size of the electron wave packet becomes clearer when we talk about the role of the potential in the total energy of the system. Typically, this is given by a term which is the product of the potential at a position and the local density at that position. Since this then is an integral over the spatial coordinates, it becomes feasible to transfer the form of the wave packet from a highly localized electron to the potential itself, which produces an effective smoothing of the barrier. From this smoothing, we see that the potential has been modified by an effective potential. Now, the nature of these effective potentials has been discussed significantly since the beginnings of quantum mechanics. The point is, however, that consideration of the size of the wave packet leads us directly into the subject of effective potentials. In subsequent sections, we will consider a variety of such effective potentials.

4.1 Size of the electron

One of the problems that we have to face, as discussed above, is that the simple wave packet treatment considers an *isolated* packet for a single electron. In actual fact, the totality of the electron gas composes a range of plane waves, or momentum states, or wave packets, depending upon whatever description is selected. In order to describe the packets in real space, one must account for the contributions to the wave packet from all occupied plane wave states [2]. That is, the states that exist in momentum space are the Fourier components of the real space wave packets. If we want to estimate the size of this wave packet, we must utilize all Fourier components, not just a select few. At low temperature in a two-dimensional semiconductor system, all states up to the Fermi energy are occupied, and the Fermi wave vector is defined by the interface (surface) carrier density n_s as

$$k_F = \sqrt{2\pi n_s}. \tag{4.1}$$

This means that, in the momentum representation, all states up to this value are occupied, or

$$\phi_m(k) = \frac{\sqrt{2\pi}}{k_F} u_0(k_F - k), \tag{4.2}$$

for any state below the Fermi energy, where u_0 is the Heavyside function, $u_0(x) = 1$ for $x \geqslant 0$, and zero otherwise. From this momentum space representation, we can now define a wave packet (centered at the origin) by taking the Fourier transform of (4.2), which leads to

$$\psi(\mathbf{r}) = \frac{1}{r}\sqrt{\frac{2}{\pi}}J_1(k_F r), \tag{4.3}$$

where J_1 is the Bessel function of the first kind and first order. It must be remembered that **k** and **r** are really two-dimensional vectors in this calculation, and the scalar values shown are the magnitude of the radial component. These wave functions may now be manipulated to show that they possess an uncertainty of

$$\Delta p = \frac{\hbar k_F}{\sqrt{2}}, \quad \Delta r = \frac{1}{k_F}. \tag{4.4}$$

Hence, the spatial extent of the real space wave packet can be estimated as the full-width at half-maximum value, or twice the uncertainty in position, which leads to

$$\delta r = 2\Delta r = \frac{2}{k_F} = \frac{\lambda_F}{\pi}. \tag{4.5}$$

The result (4.5) is the central result for the degenerate two-dimensional electron gas, and tells us that the spatial extent of the wave packet is quite small and related to the de Broglie wavelength of the electron at the Fermi surface (where transport takes place).

One reassuring feature of the value (4.5) for the size of the electron wave packet is that it is reduced as the electron density is increased. An increase of the density leads to an increase of k_F, which reduces the value in (4.5). On the other hand, the total volume of space occupied by the electron is relatively constant, so that the reduction in the two-dimensional plane must be accompanied by an expansion in the normal plane, which means the electrons extend to a higher level in the confinement potential of this third dimension. This means a higher Fermi energy, which is consistent with the increased density. A similar 'squeezing' of the wave functions in two dimensions has been discussed by Kubo *et al* [3] for the wave functions of carriers in a magnetic field. In this case, an increase in the magnetic field leads to a reduction in the magnetic length (the radius of the lowest Landau level), and this causes an elongation in the direction normal to the two-dimensional plane. This also leads to the interesting result that the amount of area occupied by each electron in real space is approximately one-half of that given by the reciprocal of the electron density. We return to the effect of the other carriers on the wave packet below.

For non-degenerate semiconductors, the distribution of allowed momentum states is defined with the Maxwell–Boltzmann distribution. This brings the temperature into the problem. As the temperature is increased, higher momentum states become occupied as the distribution spreads under the influence of the temperature. A wider momentum space distribution means a tighter distribution in real space for the electron wave packet. As previously, we consider the momentum space distribution as a description of the occupied plane wave states which contribute to the electron wave packet. The normalized momentum space distribution may then be defined to be

$$\varphi(k) = 2\left(\frac{\lambda_D}{2}\right)^{3/4} \exp\left(-\frac{\lambda_D^2 k^2}{4\pi}\right), \tag{4.6}$$

where

$$\lambda_D = \sqrt{\frac{2\pi\hbar^2}{m^* k_B T}} \tag{4.7}$$

is the thermal de Broglie wavelength [4]. We can now Fourier transform this to obtain the real space wave packet:

$$\psi(r) = \frac{\pi}{(\sqrt{2}\lambda_D)^{3/2}} \exp\left[-\frac{\pi r^2}{\lambda_D^2}\right]. \tag{4.8}$$

We then find that the effective size of the electron packet is given by

$$\delta r \sim \sqrt{\frac{3}{8}}\lambda_D \sim 0.61\lambda_D. \tag{4.9}$$

As expected, the size of the electron's wave packet is inversely proportional to the temperature, through the thermal de Broglie wavelength.

One interesting aspect of this last result is that there is no density dependence in the thermal wavelength, and hence in the electron wave packet size. While this may seem strange at first sight, it is quite natural, as the non-degenerate limit is one of a dilute electron gas. When the density becomes sufficiently large that this limit is no longer appropriate, then the distribution changes character to that of the Fermi–Dirac distribution. In the latter situation, the density dependence reappears through the importance of the Fermi wavelength, rather than the thermal wavelength, since the Fermi wavelength (and Fermi energy) depend explicitly upon the carrier density. However, this should not be construed to mean that the other electrons have no effect on the size of the packet on the non-degenerate thermal wavelength, as we have not considered the explicit electron–electron interaction in this discussion.

We can estimate the above sizes with some simple structures. For the two-dimensional electron gas in a Si MOSFET, or for that in a GaAs HEMT, the typical size is about 8 nm (this depends only upon the density and not the electron mass, and the density has been taken to be 10^{12} cm^{-2}). On the other hand, the thermal de Broglie wavelength at room temperature in Si is about 4.3 nm, and this will be reduced at the higher electron temperatures expected in active devices when bias is applied. As critical dimensions in modern day devices (of either type) have become of the order of a very few tens of nanometers, these sizes are certainly in the range that they must be taken into account. Further, one must also consider that the carrier density and the carrier temperature are not homogeneous quantities. Rather, these vary with position within the device. The results obtained here suggest that the effective 'size' of the electron is given either by its temperature (in a non-degenerate situation) or its density (in a degenerate situation). Consequently, this leads to the situation in actual devices that *the effective wave packet size for the electron actually changes with position* throughout the device! Moreover, the interaction of this wave packet with a scattering 'center', such as, for example, an impurity, is also a nonlocal event—some parts of the wave packet are closer to the impurity than others. This, in fact, is part of the need to consider off-shell corrections in the treatment of quantum transport. With barriers, the leading portion of the packet arrives, and begins interacting with the barrier, well before the trailing part of the packet [5–7].

The above considerations mean that the size of the electron wave packet is a result of the interactions with its environment. This environment includes not only the confining potentials, but also the effective confinement provided by the repulsive forces of the other particles, whether electrons or impurities. The shape and size of the wave packet is a balance between these environmental forces and the self-force provided by the diffusive nature of the Schrödinger equation. This latter is often evaluated from the shape of the wave packet by, for example, a generalized quantum potential [8]. Since the environment changes, through changes in the confining potentials and the local carrier densities, the shape and size of the wave packet are quantities that vary throughout a real device. Moreover, the interaction with the distributed image charge near barriers means that the shape of the packet may well be significantly deformed at these points. The upshot of this is that any simulations of quantum effects must be carried out with the full many-electron Hamiltonian and the real environment must appear through self-consistent potentials. Moreover, the contacts, and importantly the transition regions where decoherence is expected to occur, become real parts of the device and must be considered with the entire quantum-mechanical problem, as discussed in chapter 2. The nonlocal nature of quantum mechanics provides further complications, and approaches such as those of the Wigner function offer some attraction. However, the clear point is that it will no longer be adequate to consider the electron as a point particle in these future device simulations, and new approaches to kinetic pictures for transport are needed, unless the effective potential can provide some relief from this need.

We remarked in the beginning of this chapter that the concept of the size of the electron could be mapped into an effective potential. How does this occur? We arrive at this by noting that the potential in an inhomogeneous system enters the Hamiltonian as [9]

$$H_V = \int d\mathbf{r}\; V(\mathbf{r})n(\mathbf{r}), \tag{4.10}$$

Using the approximation in which a Gaussian wave packet is adopted for each single electron now leads to

$$\begin{aligned} V &= \int d\mathbf{r} V(\mathbf{r}) \sum_i n_i(\mathbf{r}) \\ &= \int d\mathbf{r} V(\mathbf{r}) \sum_i \int d\mathbf{r}' \exp\left(-\frac{|\mathbf{r}-\mathbf{r}'|^2}{\alpha^2}\right)\delta(\mathbf{r}'-\mathbf{r}_i) \\ &= \sum_i \int d\mathbf{r}\delta(\mathbf{r}-\mathbf{r}_i) \int d\mathbf{r}' V(\mathbf{r}')\exp\left(-\frac{|\mathbf{r}-\mathbf{r}'|^2}{\alpha^2}\right), \end{aligned} \tag{4.11}$$

where the summation over i is a summation over the individual electrons themselves. The last form has been achieved by interchanging the primed and unprimed variables, and rewriting the integrals. The term in the primed integration is now the *effective potential*, and the finite size of the electron has been moved to a smoothing of the real potential.

As with our above discussion, we are interested in how this plays out in a modern semiconductor device. The traditional approach to quantization in the MOSFET begins with the assumption of a triangular potential well. The sharp barrier at $x = 0$ is between the Si conduction band and the SiO_2 conduction band—a barrier usually taken to be $V_0 = 3.2$ eV. Within the Si, the conduction band rises linearly in response to an effective electric field of [10]

$$E_{eff} = \frac{e}{\varepsilon}\left[N_A w_B + \frac{n_s}{2}\right], \tag{4.12}$$

where N_A and w_B are the background doping of the channel region (acceptors in an n-channel device) and the depletion depth arising from the gate voltage at inversion of the surface, respectively, and n_s is the inversion density. The triangular barrier then leads to an effective potential given by [11]

$$\begin{aligned} W(x) = \frac{V_0}{2}\left[1 - erf\left(\frac{x}{\sqrt{2}\lambda_D}\right)\right] + \frac{E_{eff}X}{2}\left[1 + erf\left(\frac{x}{\sqrt{2}\lambda_D}\right)\right] \\ + \frac{E_{eff}\lambda_D}{\sqrt{2}\pi}\exp\left(-\frac{x^2}{2\lambda_D^2}\right), \end{aligned} \tag{4.13}$$

where the various parameters have the meanings defined above, and x is the direction normal to the interface. The form (4.13) has two distinct advantages in that it provides the initial observables of quantization in the channel. We sketch the real and effective potentials in figure 4.1. The solid black curve is the potential determined by the oxide offset V_o and the electric field of (4.12). The dashed curve is the result (4.13). First, it may be observed that the minimum of the potential, and therefore the peak of the classical density, has been moved away from the interface by an amount Δx. This accounts for the quantum wave packet being near zero at the interface and having a peak some distance from the interface. Second, it may be seen that the minimum of the effective potential lies at a higher energy than the real minimum at the interface. This accounts for the quantization in the channel, where the minimum energy of the lowest sub-band lies above the actual zero of the conduction band energy. These modifications of the potential lead to observable changes in the characteristic curves of the MOSFET [12, 13].

4.2 The Bohm potential

In treating the quantum transport of carriers through small semiconductor devices, it is often desired to find an approach that gives the important aspects of the

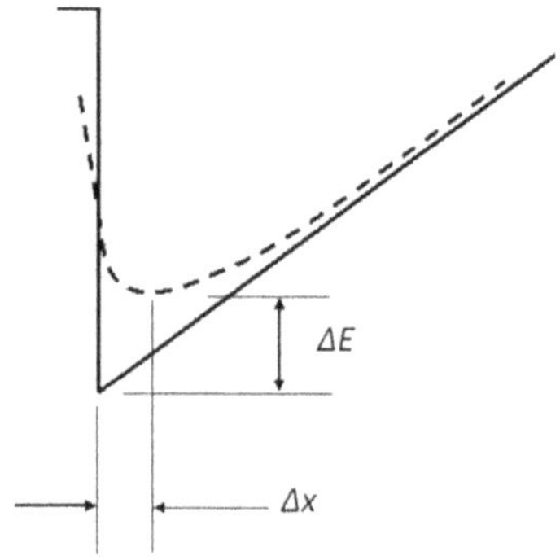

Figure 4.1. The triangular potential (solid curve) formed by the oxide band offset and the electric field within the semiconductor is compared with the effective potential (dashed curve) of (4.13). The effective potential causes the density to lie away from the interface by Δx and the minimum of the conduction band to be raised by ΔE due to quantum confinement.

quantum effects without all the mathematical detail that is necessary for some of the more fundamental approaches. Of course, the scattering matrix approach of section 2.2 allows this, but one would like to have something closer to the ubiquitous ensemble Monte Carlo approach, which has been used for years in the simulation of semiconductor devices. As we will find in this book, there are many approaches that rely upon an effective mass solution of the Schrödinger equation coupled to a Monte Carlo introduction of scattering. The problem with this approach is that one needs to have the particle trajectories to fully employ the Monte Carlo procedure, and these do not come from the normal Schrödinger equation. In chapter 1, however, we introduced the effective potential of Madelung and Bohm [14, 15]. Here, the Bohm potential provides an addition to the total energy, and provides a non-classical force which guides the wave functions in a self-consistent manner. This can provide a basis to use a particle representation, where the particles move through the presence of both the classical and the quantum forces [16]. In chapter 1, we gave a brief overview of the derivation of the Bohm potential. Here, we want to give the full derivation in order to discuss possible applications.

It is important to note that the Madelung–Bohm approach is developed from the Schrödinger equation itself, so that it is another interpretation of the wave function theory for quantum mechanics. We begin by defining the wave function to consist of a real amplitude A and real phase $S/\hbar$, where S is the action (integral of the energy over time). Thus, we write the wave function as

$$\psi(\mathbf{x}, t) = A(\mathbf{x}, t)e^{iS(\mathbf{x}, t)/\hbar}. \tag{4.14}$$

Now, we insert this wave function into the time-dependent Schrödinger equation

$$-\frac{\hbar^2}{2m}\nabla^2\psi(\mathbf{x}, t) + V(\mathbf{x})\psi(\mathbf{x}, t) = i\hbar\frac{\partial\psi(\mathbf{x}, t)}{\partial t}. \tag{4.15}$$

This leads to the complex equation

$$\begin{aligned} -A\frac{\partial S}{\partial t} + i\hbar\frac{\partial A}{\partial t} = \frac{A}{2m^*}(\nabla S)^2 - \frac{i\hbar A}{2m^*}\nabla^2 S \\ - \frac{i\hbar}{m^*}\nabla S \cdot \nabla A - \frac{\hbar^2}{2m^*}\nabla^2 A + V(\mathbf{x})A. \end{aligned} \tag{4.16}$$

For this equation to be valid, it is required that both the real parts and the imaginary parts balance separately, so that this is really two equations, which are given as

$$\frac{\partial S}{\partial t} + \frac{1}{2m^*}(\nabla S)^2 + V - \frac{\hbar^2}{2m^* A}\nabla^2 A = 0 \tag{4.17}$$

and

$$\frac{\partial A}{\partial t} + \frac{A}{2m^*}\nabla^2 S + \frac{1}{m^*}\nabla S \cdot \nabla A = 0. \tag{4.18}$$

Equation (4.18) can be rearranged by multiplying by A, which leads to

$$\frac{\partial A^2}{\partial t} + \nabla \cdot \left(\frac{A^2}{m^*}\nabla S\right) = 0. \tag{4.19}$$

The factor A^2 is obviously the magnitude squared of the wave function itself and therefore relates to the probability density provided by the wave function. Dimensionally, the quantity in the parentheses is the product of the probability and the velocity of the wave function, so that

$$\mathbf{p} = \nabla S, \quad \mathbf{v} = \frac{1}{m^*}\nabla S. \tag{4.20}$$

The values defined in (4.20) are explicit quantum-mechanical operators, which in this case are differential operators as introduced by Schrödinger and Heisenberg. This latter relation is important in connecting the quantum-mechanical operators to the usual definitions of the action, which is the time integral of the energy, as discussed above.

There is another constraint on the momentum and action that is not apparent from this present situation. There is a requirement for the quantum system that

$$\oint_C \mathbf{p} \cdot d\mathbf{l} = 2\pi\hbar n, \tag{4.21}$$

where n is an integer and C is a closed contour. This is just a quantization condition, and normally the contour is an extremal orbit that the closed trajectory may take on what is called the invariant torus of the system. But this contour can be almost any closed path. Now this integral appears in many guises. For example, the Einstein, Brillouin [17], and Keller [18] form is often called EBK quantization. In a more modern version, the right-hand side of (4.21) is modified, such as by the addition of a factor of 1/2 in the Wentzel–Kramer–Brillouin (WKB) approximation, and more generally it is written as

$$\oint_C \mathbf{p} \cdot d\mathbf{l} = 2\pi\hbar\left(n + \frac{1}{4}\beta\right), \tag{4.22}$$

where β is the Morse or Maslov index [19, 20]. In general, this index relates to the number of *turning points* that the trajectory makes. But all this index really does is to shift the energy levels that arise from this quantization. If we now apply Stoke's theorem, we can rewrite (4.21) as

$$\frac{1}{m^*}\int_W (\nabla \times \mathbf{v}) \cdot \hat{\mathbf{n}} dW = 2\pi n, \tag{4.23}$$

where W is the surface area enclosed by the contour C and $\hat{\mathbf{n}}$ is the surface normal unit vector. The quantity in parentheses in (4.23) is the vorticity of the hydrodynamic 'fluid' and this equation requires it to be quantized. In the integral (4.21), there is also the important concept that passing around the contour may introduce an additional phase to the wave function, often called the Berry phase or geometric phase [21]. Examples of this arise in, for example, the Aharonov–Bohm effect [22], in which conductance through a ring, when a magnetic field is threaded through the ring, exhibits oscillatory behavior.

Now, let us return to equation (4.17). In this equation, there is only one term that involves Planck's constant and this would vanish if we let this constant go to zero. We have already remarked that the action is the time integral of the energy, so that the first term in (4.17) is obviously the energy, while the third term is the potential energy. From (4.20), we recognize the second term as the kinetic energy, so that (4.17) can be recognized as just the condition of conservation of energy. But, the last term is new and has not been recognized as an energy, which it must be. This last term is usually referred to as the Bohm potential, given as

$$Q = -\frac{\hbar^2}{2m^* A}\nabla^2 A, \tag{4.24}$$

and is a true quantum potential. This quantum potential is important to the trajectory approach, as we can define the acceleration of the 'particle' trajectory from the total potential as [16]

$$\frac{d\mathbf{v}}{dt} = -\frac{1}{m}\nabla[V(\mathbf{x}, t) + Q(\mathbf{x}, t)]. \tag{4.25}$$

This directly allows us to develop a time-resolved approach to electron transport using the quantum trajectories [23]. Once the Schrödinger equation is solved at each instant of time, the quantum potential can be computed along with the self-consistent potential energy (subject to any charge movement or rearrangement), and then the directional velocities of the various trajectories determined. These velocities now allow one to create the motion of the trajectories during the next time increment. This is a consequence of the Bohm theory being a wave AND particle

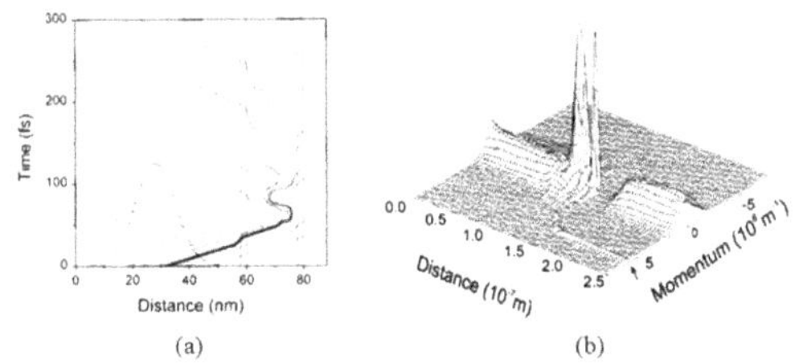

Figure 4.2. (a) Bohm trajectories for a Gaussian wave packet tunneling through a double-barrier resonant-tunneling diode. Details are given in the text. (b) Phase space distribution for such a diode with full density under a bias of 0.39 V. Reprinted with permission from Oriols *et al* [25]. Copyright 1998 American Institute of Physics, Publishing LLC.

approach. First, the wave mechanics are solved via the Schrödinger equation, and THEN the particle trajectories are determined. These resulting trajectories of the particles are real, and may be observed in experiments, as discussed below.

This approach has been applied to a study of a resonant-tunneling diode by Oriols *et al* [24]. In figure 4.2(a), we illustrate the motion of a Gaussian wave packet, indicated at the lower left of the figure, through a double-barrier, resonant-tunneling diode [23]. Here, the barriers are each 2 nm thick and have an energy height of 0.3 eV. The enclosed quantum well is 18 nm thick. It is assumed that the material is GaAs and the barriers are AlGaAs. The Gaussian wave packet is a coherent wave packet, such as used earlier in (3.18), with a momentum corresponding to a central energy of 0.16 eV and has a spatial dispersion of 10 nm. No bias is considered as being applied to the structure. In figure 4.2(b), the resulting phase space distribution function for the full density of the tunneling device with a bias of 0.39 V. Note the tunneling ridge in the lower right (arrow) that corresponds to the ballistically tunneling electrons. It can be shown that the Bohm trajectory approach agrees very well with use of the Schrödinger equation directly [23] and with use of the Wigner phase space distribution function [25].

The use of Bohm trajectories has also been applied to semiconductor nanostructures such as quantum wires and quantum point contacts. In figure 4.3, the results for a simple asymmetric quantum point contact are shown, with the trajectories overlaid on the quantum potential itself [26]. In the top panel, the Schrödinger equation was solved assuming hard walls for the boundaries and without considering a self-consistent rearrangement of the charge in response to the barrier potential. This was then used to find the trajectories and the quantum potential, both of which are shown in the top panel. Note that here, as well as in figure 4.2, none of the trajectories cross, which is required since the velocity field is uniquely determined, and only one trajectory can pass through any point [8, 27], as indicated in (4.25). One can also see that the trajectories are bent around the constriction, a consequence of electron diffraction through the slit. In the lower panel of the figure, we now consider results in which the electron density is computed self-consistently with the potential. This would lead to a self-energy that causes more modes to propagate, so the Fermi energy has been lowered to maintain the same number of propagating modes. Now, the trajectories are laid over the sum of the quantum and self-consistent potentials. Use of

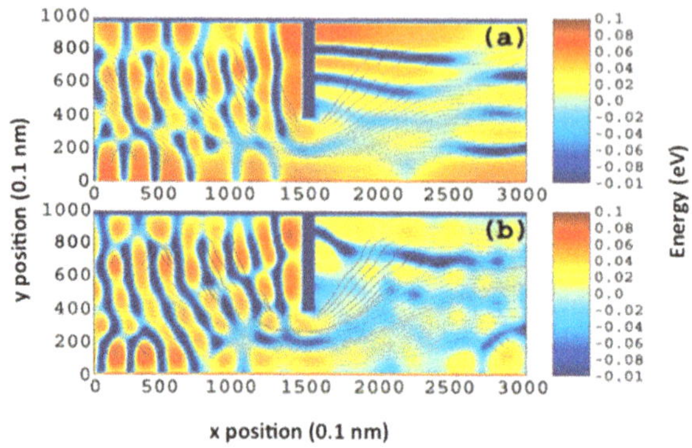

Figure 4.3. Bohm trajectories for a quantum point contact in GaAs. (a) The background shading corresponds to the quantum potential. The constriction is the vertical blue bar at $x = 150$ nm. The electrostatic potential is taken to be zero. (b) Now, the electrostatic potential and the density are computed self-consistently to account for the effect of the barrier on the density. The background shading is now the sum of the electrostatic potential and the quantum potential. Reprinted from Schifren *et al* [27], copyright 2000 with permission from Elsevier.

the self-consistent potential leads to variations of the individual trajectories between the two panels.

4.3 Bohm and the two-slit experiment

It is also important to realize that the Bohm interpretation and the trajectories are relevant to the interpretation of quantum mechanics itself. From the earliest days, there was the consideration of the particle–wave either/or situation. But with the guiding wave theory it is an 'and' situation. Nowhere is this better depicted than in the two-slit experiment. The two-slit experiment has been one of the most confusing problems in quantum mechanics since the earliest days. As we saw in chapter 1, Taylor had investigated diffraction effects with very weak light as early as 1909 [28]. Even when the light was sufficiently weak that only a single photon at a time was in the experiment, the diffraction corresponded to the wave picture and the arrival at the photographic plate corresponded to the particle picture. The two-slit experiment has been demonstrated with electrons in an electron microscope under conditions in which only a single electron was in the machine at any one time [29]. Whether for photons or for electrons, the actual fringe that is built up by the particles arriving at the detector clearly shows the effects of their wave nature as they passed through the diffraction region.

The experiment is relatively simple, and one can pose it clearly. A single particle arrives at a screen with two slits in it. We take the normal direction to the screen to be the z direction, so that the particle can be said to have a momentum $p_z = \hbar k_0$. After passing through one of the two slits it arrives at a second screen, where its position is recorded, for example by a photographic plate in the experiment of Taylor. In the usual interpretation, we describe the particle by a wave that arrives at the screen as $\psi \sim \exp(ik_0 z)$. When this wave passes through the two slits, it undergoes both interference and diffraction and develops a complicated spatial variation that illustrates the interference at the second screen. At this second screen,

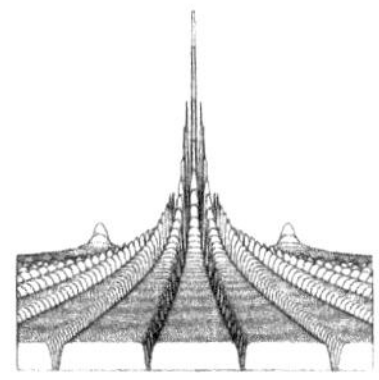

Figure 4.4. The quantum potential for the two-slit experiment, as seen from the screen upon which the interference is found. The two slits are located between the three peaks at the rear. The peaks are regions where the wave function is small and the particles are not allowed to appear. Figure is taken from Springer Il Nuovo Cimento, 'Quantum Interference and the Quantum Potential', vol. 52B (1979) pp 15–28, C Philippidis, C Dewdney, and B J Hiley, with permission of Springer.

the wave has a probability in the lateral direction as, for example, $P(x) = |\psi(x)|^2$. If the experiment is repeated many times, the interference fringe is gradually built up on the photographic plate. The question has always been how a single particle going through a single slit can generate a diffraction pattern.

The causal interpretation that Bohm put forward clarifies this picture considerably [13]. The wave function used is the same as that in the usual interpretation. But in the causal picture, this wave function is regarded as a real field; that is, it is a mathematical representation of a field and provides force upon the particle. The wave function generates this force in agreement with the Schrödinger equation through the quantum potential (4.24). It is the quantum force which leads to the fact that the particles illustrate the interference fringes when they arrive at the detector. If it happens that there are places where the wave function is zero, this leads to an infinitely large force which prevents the particles from approaching these places. Moreover, if one of the slits is closed, the diffraction of the wave function changes. This in turn dramatically changes the quantum potential, so that the particle is now allowed to reach places that it had no access to when both slits were open. Hence, the opening or closing of a slit affects the particle indirectly, and only through the changes produced in the quantum potential by such an action. Thus it is the actual ψ-field resulting from the detailed state of the two slits that generates the forces that cause the particle motion corresponding to that particular state. Here, there is no strange mystery about the quantum behavior of such a system.

The role of the quantum potential was investigated by Philippidis, Dewdney, and Hiley [30], where they calculated the quantum potential for the two-slit experiment. In figure 4.4, we show the quantum potential they determined from the wave function in the region between the two slits and the screen upon which the interference was developed. It is clear that the quantum potential can give rise to very nonlocal behavior. There are three clear peaks in the quantum potential, with the major peak being located between the two slits. The slits themselves lie between these three peaks. The peaks are regions where the wave function is small and the particles cannot effectively reach. Clearly, there are troughs in the potential, and the idea is that the particles will be focused into these troughs by the quantum potential. In figure 4.5, the corresponding particle trajectories are plotted. The trajectories correspond to different initial conditions within the two slits. One can clearly see

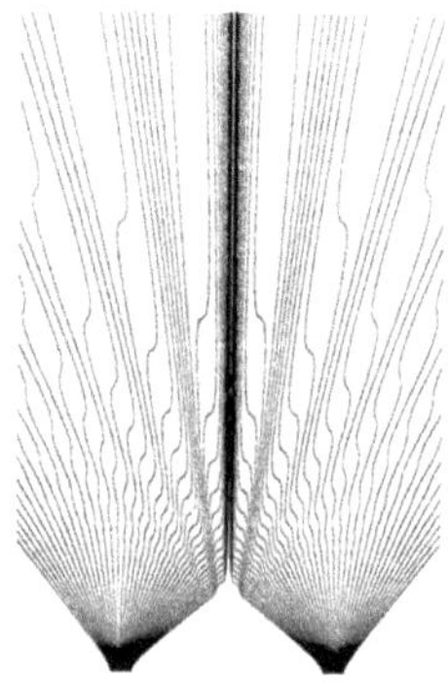

Figure 4.5. The trajectories of particles that are initiated within the two slits (at the bottom of the figure). It can be seen how the quantum potential focuses these particles in a manner to create the interference fringes on the screen (top). Figure is taken from Springer Il Nuovo Cimento, 'Quantum Interference and the Quantum Potential', vol. 52B (1979) pp 15–28, C Philippidis, C Dewdney, and B J Hiley, with permission of Springer.

how the particles are focused into beams which create the interference fringes on the screen. Clearly, if a particle is excited out of one of the troughs, it moves over the intervening potential into another trough. There are few particles where the potential is high, and a great many particles where the potential is low in the troughs.

It has always been assumed that it would be impossible to determine which slit a single particle passes through. Yet recent experiments have shown that this may be possible. Menzel *et al* [31] have used an optical parametric down conversion process to generate entangled photon pairs, and to obtain which-slit information on the signal photon by a coincidence measurement of the two photons. They observe in a different coincidence measurement interference fringes in the signal photon. This surprising result is contrary to the principle of complementarity. The authors have suggested that this result is possible because of the transverse mode structure which produces a superposition of two intensity maxima representing two macroscopically distinguishable wave vectors for the signal photon. Hence it appears to be not exactly a single photon, but a superposition.

Perhaps a more relevant experiment was carried out by Kocsis *et al* [32]. In this experiment, single photons generated from a semiconductor quantum dot laser are passed through a double slit apparatus, which of course creates the interference fringes. But through a set of measurements of the photon momentum, which was post-selected according to the results of the measurements of photon position as they passed a series of planes, they claim to be able to determine which-slit information. The single-photon nature of the laser is confirmed by correlation measurements. The ensemble of photons creates the interference fringes in the detector array, as a strong measurement. Hence, they are able to obtain a measure of both the photon position and momentum, and these are averaged over a great many photons. They are then able to reconstruct the photon trajectories, and they find that these photons seem to follow typical Bohmian trajectories, as in figure 4.5. In spite of whether or not there is some confusion or ambiguity about which slit the particle actually passed through, these experiments confirm that the photons are following individual trajectories that are affected by the two slits and the resulting quantum potential. They thus have particle properties, but when the ensemble of photons arrive at the final observation screen they build up the interference pattern, in complete agreement with Taylor's experiments more than a century earlier. As a result, it is hard not to draw the conclusion that the Bohmian dynamics, and the corresponding interpretation of the physics, are a valid description of the effects. Hence, it is absolutely clear that the quantum potential is an essential part of the solutions to quantum mechanics.

4.4 The Wigner potential

Let us now turn to the full development of the potential corrections found by Wigner [33]. So far, we have used only a Wigner function for a single particle state or for a single basis state. If we have occupation of a number of basis states, then the total Wigner function can be expanded in terms of a series of Wigner functions for each of the basis states as

$$f_W(\mathbf{x}, \mathbf{p}, t) = \sum_i w_i f_{W,i}(\mathbf{x}, \mathbf{p}, t), \tag{4.26}$$

where $f_{W,I}$ is the Wigner distribution function for state ψ_i. This now describes a mixture corresponding to the density matrix for such a mixture. Here, w_i is the probability that the particular state is occupied. If the system is in equilibrium at a temperature $T = 1/k_B\beta$, where k_B is the Boltzmann constant, the relative probability of a particular state being occupied is given by the eigen-energy E_i of that state through the function $\exp(-\beta E_i)$. That is, this exponential function gives us the coefficient w_i for the mixture when it is in thermodynamic equilibrium. In another way, we can write the total Wigner function as

$$f_W(\mathbf{x}, \mathbf{p}) = \frac{1}{h^3}\sum_i \int d^3\mathbf{y}\psi_i^*\left(\mathbf{x} + \frac{\mathbf{y}}{2}\right)e^{-\beta E_i}\psi_i\left(\mathbf{x} - \frac{\mathbf{y}}{2}\right)e^{i\mathbf{p}\cdot\mathbf{y}/\hbar}. \tag{4.27}$$

The term

$$\sum_i \psi_i^*(\mathbf{x})g(H)\psi_i(\mathbf{z}) \tag{4.28}$$

is clearly the positional matrix element of the operator $g(H)$, where H is the Hamiltonian operator from the Schrödinger equation. So, under the integral sign in (4.27), we clearly have the matrix element of the exponential operator $\exp(-\beta H)$, transformed by the last exponential into momentum. Thus, instead of transforming this exponential, we could have used the Weyl approach to transform H first and then taken the exponential with the transformed version. By doing this latter transformation of H, Wigner arrives at the operator [33]

$$H = e^{i\mathbf{x}\cdot\mathbf{p}/\hbar}\left(-\frac{\hbar^2}{2m}\nabla^2 + V(\mathbf{x})\right)e^{-i\mathbf{x}\cdot\mathbf{p}/\hbar}, \tag{4.29}$$

which is equal to

$$\tilde{H} = E_c + \left(i\hbar \frac{\mathbf{p} \cdot \nabla}{m} - \frac{\hbar^2}{2m} \nabla^2 \right) \tag{4.30}$$

where

$$E_c = \frac{\mathbf{p}^2}{2m} + V(\mathbf{x}) \tag{4.31}$$

may be interpreted as the classical energy. One can then replace (4.27) by the integral of (4.30) over the difference in the two coordinates in (4.27), using the Barker–Campbell–Hausdorf expansion of the exponential operator. The last term in (4.30) gives then an expansion of the potential in the integral. To lowest order, one reclaims the classical result for the probability as $\exp(-\beta E_c)$. Higher approximations are found by using this operator, and the next order in the expansion gives

$$e^{-\beta E_c} \frac{\hbar^2}{m} \left(-\frac{\beta^2}{12} \nabla^2 E_c + \frac{\beta^3}{24} (\nabla E_c)^2 \right). \tag{4.32}$$

The factors 1/12 and 1/24 come from the first two commutators in the above exponential expansion. Since only the potential is an explicit function of position, this leads to the result given in chapter 1 as (1.19). This then leads to a modified potential of the form

$$V(\mathbf{x}) = V_c(\mathbf{x}) - \frac{\hbar^2}{12m} \nabla^2 E_c + \frac{\hbar^2}{24m} (\nabla E_c)^2. \tag{4.33}$$

We have left the classical energy in the last two terms for convenience. In chapter 1, it was demonstrated that the second and third terms can be combined to lead to a term that differs from the Bohm potential by only a numerical factor. Here, we pursue a slightly different combination, by noting that the spatially varying density is related to the probability as

$$E_c(\mathbf{x}) \simeq -\frac{1}{\beta} \ln(n(\mathbf{x})). \tag{4.34}$$

This form can be used to replace the classical energy in (4.33) and yield the Wigner correction to the potential as [34]

$$V_W(\mathbf{x}) = -\frac{\hbar^2}{8m} \nabla^2 \ln(n(\mathbf{x})). \tag{4.35}$$

This form has been termed the *density gradient potential* as well as the Wigner potential and is another approach to the effective potential. This effective potential has been applied to a wide variety of semiconductor devices [35–38]. In figure 4.6, we show one such quantum correction to the potential, obtained in an n^+–n–n^+ structure through simulation with moments of the density matrix. The central 60 nm is doped at 10^{15} cm^{-3} and the exterior regions are doped an order of

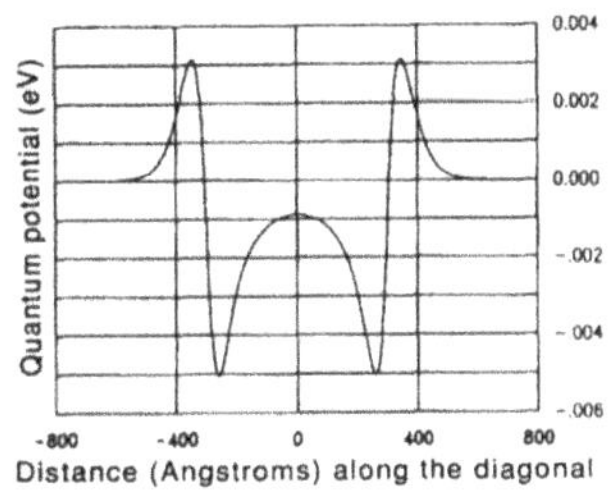

Figure 4.6. The quantum potential in the central region for an n^+–n–n^+ structure, where the central region of 60 nm is doped to 10^{15} cm^{-3} and the end regions are doped to 10^{16} cm^{-3}. The quantum potential adds to the normal band bending in such a structure. Figure is taken from Springer Solid-State Electronics, 'Transport via the Liouville equation and Moments of Quantum Distribution Functions', vol. 36 (1993) pp 1697–1709, H L Grubin, T R Govindan, J P Kreskovsky, and M A Stroscio, with permission of Springer.

magnitude higher, in GaAs [35]. Normal band bending, due to the variation in doping is augmented by the quantum potential, shown in the figure, to affect the carrier density itself, as well as the depletion and accumulation layers near the doping interface.

While it is easy to talk about the various contributions to the band bending and local potential, it is sometimes not easy to understand if one is not really familiar with semiconductors. So, let us try to explain what the local physics is for figure 4.6. First, it is important to understand that semiconductors are rather different than metals, and it is possible to control the density at each point in the device. It is easily shown that, in the absence of any current flow in the semiconductor, the Fermi energy must be constant with position. Hence, as one varies the impurity doping, the conduction band edge moves relative to the Fermi energy. For non-degenerate semiconductors, which is the condition for the above doping, we may express the local carrier density as

$$n(\mathbf{x}) = N_C \exp\left[-\frac{E_c(\mathbf{x}) - E_F}{k_B T}\right], \tag{4.36}$$

where N_C is the effective density of states in the conduction band for GaAs (= 4.7×10^{17} cm^{-3} at room temperature), $E_c(\mathbf{x})$ is the local conduction band energy, and E_F is the Fermi energy. Thus, for a doping of 10^{16} cm^{-3}, the conduction band edge is ~100 meV above the Fermi energy. For a doping of 10^{15} cm^{-3}, the conduction band edge is ~160 meV above the Fermi energy. This means that there is a potential step of 60 meV between the two doping regions. Such a potential step requires a charge dipole to exist at the interface between the two regions. This dipole arises from the diffusion of electrons from the heavily doped region into the lightly doped region. This leaves uncompensated positively charged donors in the heavily doped region and excess electrons in the lightly doped region, and this constitutes the dipole charge necessary to support the potential discontinuity. This diffusion of electrons means that we have to solve the local potential self-consistently with Poisson's equation, and this smoothes the potential step classically, as shown in figure 4.7.

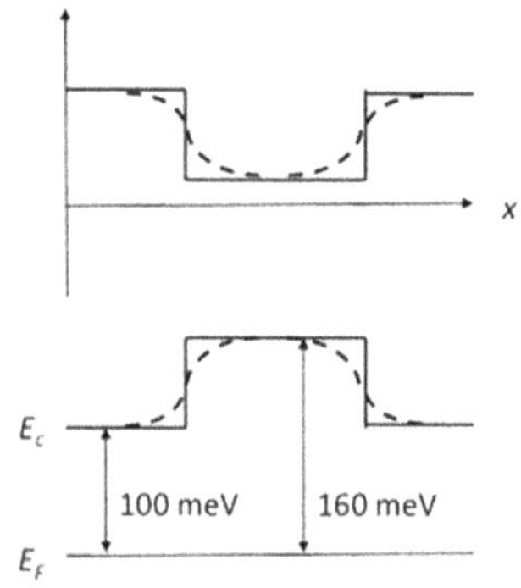

Figure 4.7. The upper panel shows the doping profile (solid curve) and the self-consistent electron density (dashed curve) that arises from diffusion of the electrons across the interface between the two doping levels. The lower panel depicts the band alignment (solid curve) and the resulting self-consistent conduction band edge (dashed curve).

What we see in figure 4.6 is the quantum potential that arises from the resulting carrier density after diffusion into the lightly doped regions. Since the structure has mirror symmetry (between the two highly doped regions), the induced quantum potential reflects this mirror symmetry. The peak in the potential on either side lies outside the metallurgical boundary of the lightly doped region, while the two negative peaks lie just inside the lightly doped region. The peaks in the quantum potential are trying to raise the potential over the self-consistent potential of figure 4.7 (outside the lightly doped region). On the other hand, the negative excursions of the quantum potential are trying to lower the self-consistent potential. That is, the quantum effects are attempting to make the potential transition more spread out than indicated in figure 4.7 (it is smoothing the potential in the same manner as in figure 4.1). This implies that the differential form of the Wigner potential is providing almost the same effect as the integral smoothing of the effective potential discussed in regard to figure 4.1. The conclusion is that quantum mechanics abhors potential discontinuities and would prefer to reduce spatial quantization if at all possible. It is certainly possible to impose sharp potentials on a quantum system, but the quantum system would prefer to rid itself of such sharp potentials. We have to remember that spatial confinement and quantization require the addition of energy to the system. A hard-wall quantum well has levels which go as n^2, where n is the level index, while the soft-wall harmonic oscillator potential has levels which vary as n. The tendency of the quantum system to prefer soft-wall potentials is just a matter of seeking a lower energy state for a given electron density.

4.5 Feynman and effective potentials

So far, we have seen both effective potentials that are smoothing functions and additive functions. These, of course, are integral and differential forms for correcting the classical potential. Each approach is one method of incorporating a quantum correction, which is separate from the Hamiltonian, into the total potential that goes beyond the classical potential. Still another approach was put forward by

Feynman [39]. Feynman was concerned about the evaluation of path integrals in his approach to quantum mechanics, and sought a way to impose quantization onto the paths. He first decided that he would introduce the mean position corresponding to one of the paths as

$$\bar{x} = \frac{1}{\beta\hbar}\int_0^{\beta\hbar} x(u)du, \tag{4.37}$$

where $x(u)$ is a path variable for the position. In essence, this is smoothing over a thermal spread to get an effective value. Then, he points out that all paths need to have the same initial and final points, so that it can be suggested that corrections to the classical potential will be of second order (in the spatial derivative) or higher, and this leads him to suggest that the correction to the free energy will be of the order of

$$1 - \frac{\beta^2\hbar^2}{24m}\frac{\partial^2 V(\bar{x})}{\partial \bar{x}^2}. \tag{4.38}$$

We note that this has definite similarities to the Wigner potential (4.33). He goes on to point out that this may be transferred into a discussion of the corrected value for the Helmholtz free energy F', and shows that these terms lead to a new value for the partition function:

$$e^{-\beta F'} = \sqrt{\frac{m}{2\pi\hbar^2\beta}}\int e^{-\beta U(\bar{x})}d\bar{x}, \tag{4.39}$$

where

$$U(\bar{x}) = \sqrt{\frac{12m}{2\pi\hbar^2\beta}}\int_{-\infty}^{\infty} V_c(\bar{x} + y)e^{-6y^2m/\beta\hbar^2}dy. \tag{4.40}$$

Hence the correction arrives by a smoothing of the potential by a Gaussian smoothing function, as we discussed in section 4.1, as well as smoothing over the path itself.

Feynman points out the conclusion that an approximate value for the free energy F' can be achieved in a classical manner, and we then can get a good estimate of the quantum corrections by using the smoothed classical potential described by (4.40). However, this approach is basically computing a local linear average of the actual potential when we compute the mean position in (4.37). One can see that this becomes inapplicable when the local potential might not have well-defined derivatives or might not be integrable. These lead to problems with the simple hard-wall potential. It should be noted, however, these problems are avoided even for the hard-wall potential if we skip the path integral discussion and begin with the free energy description of (4.39). In fact, the case for a finite height of the hard wall is just the problem discussed in figure 4.1.

Many people have extended the Feynman approach to the case of bound particles which appear in the presence of quantum confinement [40–42] and to the case of particles at interfaces [43, 44]. These approaches use the fact that the most likely trajectory in the path integral no longer follows the classical path when the electron is bound inside a potential well. The introduction of the effective potential and its effective Hamiltonian is closely connected to the return to a phase space description. This can be done at present only for Hamiltonians containing a *kinetic* energy quadratic in the momenta and a coordinate-only dependence in the potential energy. That is, it is clear that some modifications will have to be made when non-parabolic energy bands, or a magnetic field, are present. However, the Gaussian approximation is well established as the method for incorporating the purely quantum fluctuations around the resulting path.

Let us first consider the bound states. The key new ingredient for bound states (such as in the potential well at the interface of a MOSFET) is the need to determine variationally the dominant path and hence the 'correct' value for the smoothing parameter α in (4.11). Here, we will use λ for the characteristic range of the smoothing potential, to differentiate it from the earlier treatment. For the case in which the bound states are well defined in the potential, both Feynman and Kleinert [41] and Cuccoli *et al* [45] find

$$\lambda^2 = \frac{\hbar^2}{4m^* k_B T}\left[\frac{\coth(f)}{f} - \frac{1}{f^2}\right], \quad f = \frac{\hbar\omega}{2k_B T}, \tag{4.41}$$

where $\hbar\omega$ is the spacing of the sub-bands. If we take the high-temperature limit, as appropriate for semiconductor devices at room temperature, we can expand for small f and arrive at

$$\lambda^2 = \frac{\hbar^2}{12m^* k_B T} \tag{4.42}$$

to leading order. In Si, this gives a value for the smoothing distance of 0.52 nm normal to the oxide interface, which is much smaller than the values discussed above. For transport along the channel, a different mass appears, and this gives a value of 1.14 nm. Nevertheless, these values differ little from those discussed earlier in this chapter.

At this point, we would like to pursue a somewhat more formal approach in an attempt to bring these various descriptions of the effective potential together in a common approach. The natural place for starting a discussion is in terms of a statistical ensemble that is just the density matrix. We would like to develop from this basis a more formal method of obtaining the effective potential that modifies the classical density matrix with the onset of quantum phenomena. Our beginning point is the adjoint equation. The time evolution of the density matrix is determined by the time evolution of the two wave functions with which the density matrix is formed. We can directly form the proper products with

$$\begin{aligned}\frac{\partial\rho}{\partial t} &= \frac{\partial\psi^\dagger(\mathbf{r})}{\partial t}\psi(\mathbf{r}') + \psi^\dagger(\mathbf{r})\frac{\partial\psi(\mathbf{r}')}{\partial t} \\ &= -\frac{i}{\hbar}[H, \rho(\mathbf{r}, \mathbf{r}')].\end{aligned} \tag{4.43}$$

The last line introduces the Liouville equation, which is the equation of motion for the density matrix. The last form, however, is somewhat different in that a superoperator has been introduced. The Hamiltonian operates on different wave functions in the two terms of the commutator that follows. This means that the derivatives operate on specific wave functions, and the explicit form of (4.43) that is given by (2.69) becomes

$$i\hbar\frac{\partial\rho}{\partial t} = \left[-\frac{\hbar^2}{2m}\left(\frac{\partial^2}{\partial\mathbf{r}'^2} - \frac{\partial^2}{\partial\mathbf{r}^2}\right) + V(\mathbf{r}') - V(\mathbf{r})\right]\rho(\mathbf{r}', \mathbf{r}). \tag{4.44}$$

Now, in fact, we want to work with the adjoint equation, in which the commutator relation is replaced by the anti-commutator relation, and which leads to

$$i\hbar\frac{\partial\rho}{\partial t} = \left[-\frac{\hbar^2}{2m}\left(\frac{\partial^2}{\partial\mathbf{r}'^2} + \frac{\partial^2}{\partial\mathbf{r}^2}\right) + V(\mathbf{r}') + V(\mathbf{r})\right]\rho(\mathbf{r}', \mathbf{r}). \tag{4.45}$$

In dealing with thermodynamic quantities, it is better to use results arising from the thermodynamic, or temperature, Matsubara Green's functions. To get to these, we make the change of variables $t \to -i\hbar\beta$, $\beta = 1/k_BT$, so that the last equation becomes

$$-\frac{\partial\rho}{\partial\beta} = \left[-\frac{\hbar^2}{2m}\left(\frac{\partial^2}{\partial\mathbf{r}'^2} + \frac{\partial^2}{\partial\mathbf{r}^2}\right) + V(\mathbf{r}') + V(\mathbf{r})\right]\rho(\mathbf{r}', \mathbf{r}). \tag{4.46}$$

We introduce the Wigner coordinate transformations as

$$\begin{aligned} \mathbf{x} &= \frac{\mathbf{r} + \mathbf{r}'}{2}, \quad & \mathbf{s} &= \mathbf{r} - \mathbf{r}' \\ \mathbf{r} &= \mathbf{x} + \frac{\mathbf{r}}{2}, \quad & \mathbf{r}' &= \mathbf{x} - \frac{\mathbf{s}}{2}. \end{aligned} \tag{4.47}$$

With these variable changes, we can rewrite (4.46) in the form

$$\frac{\partial\rho}{\partial\beta} = \left[\frac{\hbar^2}{8m}\nabla^2 + \frac{\hbar^2}{2m}\frac{\partial^2}{\partial\mathbf{s}^2}\right]\rho - W_s(\mathbf{w}, \mathbf{s})\rho, \tag{4.48}$$

where

$$W_S(\mathbf{x}, \mathbf{s}) = \cos\left(\frac{1}{2}\mathbf{s}\cdot\nabla\right)V(\mathbf{x}). \tag{4.49}$$

The last term is the addition of the two Wigner shifted potential terms, and differs from the integrand of (3.6) by the sign change. While only odd values of the derivatives appeared in (3.6), only even values appear in (4.49), which is appropriate to the adjoint equation. Hence, this equation will give rise to quantization effects, a result with often eludes the Wigner equation of motion. The result (4.49) provides a nonlocal average of the two potentials, and it is in this spirit that effective potentials arise in the first place.

One clear approach to solving (4.48) is to use the function W_S as an integrating factor. Before doing this, however, we wish to introduce the defining equation for F as a differential operator:

$$F(\mathbf{x},\,\mathbf{s})\rho = \frac{\hbar^2}{2m}\left[\frac{1}{4}\nabla^2 + \frac{\partial^2}{\partial \mathbf{s}^2}\right]\rho, \tag{4.50}$$

so that using the integrating factor in (4.48) allows us to write the solution for the density matrix in the functional form:

$$\rho(\mathbf{x},\,\mathbf{s}) = \exp\left(-\beta W_S(\mathbf{x},\,\mathbf{s}) + \int_\beta F(\mathbf{x},\,\mathbf{s})d\beta'\right)\rho_0 \tag{4.51}$$

This form is clearly similar to that introduced by Feynmann in (4.38) above and the terms in the exponent should be related to the overall free energy of the system. If we assume that the constant term at the end of (4.51) is the classical value of this energy, then our exponential is the changes that are desired in this classical result.

To proceed, it is convenient to divide the function F into its 'potential' parts and its 'dynamic' parts, through the defined separation

$$F(\mathbf{x},\,\mathbf{s}) = -Q(\mathbf{x},\,\mathbf{s}) + S(\mathbf{x},\,\mathbf{s}). \tag{4.52}$$

Let us deal with the second term first, as we assert that this term is the dynamic portion of the function. In general, the dynamic terms are related to the integral invariants of the motion, and may be written in terms of a sum over these quantities as [46]

$$\int_\beta F(\mathbf{x},\,\mathbf{s})d\beta' = -\sum_{i=0}^{N}\zeta_i P_i, \tag{4.53}$$

where ζ_i is the integral invariant and P_i is the conjugate quantum-mechanical operator. The most common application is to let ζ_0 be β so that P_0 is H_0, the bare Hamiltonian (minus the external potential) used to solve the Schrödinger equation. Other terms would involve, for example, a term in momentum so that the conserved momentum is one of the integral invariants and so on. Here, however, it is the off-diagonal difference variable $\mathbf{s}$ that transforms, through a Wigner–Weyl transform (to be discussed in the next chapter) into the equivalent momentum, so that the form should be somewhat different in the current application. Nevertheless, it is easy to formulate the form that will be useful here, since it must satisfy certain limitations that follow the derivatives with respect to this difference variable. Hence, we are guided by the knowledge that follows from the moment equations in section 3.6. Thus, we will let the integration over the function S take the form

$$\int_\beta S(\mathbf{x},\,\mathbf{s})d\beta' = J(\mathbf{x},\,\mathbf{s}) = \frac{i}{\hbar}\mathbf{p}_d\cdot\mathbf{s} - \frac{m}{2\beta}\left(\frac{\mathbf{s}}{\hbar}\right)^2 - \frac{1}{2}\ln(\beta). \tag{4.54}$$

This follows from considerations of the drifted Maxwellian discussed in section 3.6, and from the form of the Schrödinger equation when separated into its magnitude and phase as done above in section 4.2, so that the second term on the right is effectively the kinetic energy of the system. The last term is a normalization, which leads to adding $-d/2\beta$ to S, where d is the dimensionality of the system. After differentiating with respect to β, we note that the second term is equivalent to adding a harmonic oscillator potential, which is normalized by the temperature, to the transverse coordinate. This is reassuring in that its form is reminiscent of that used by Feynman and Kleinert [41]. While the drift momentum term is usual in its form, it does not involve β and will therefore vanish in S. While this satisfies the general requirements of (4.53), it does not allow for the replication that we expect to find in solving the differential equation for the quantum potential, but we only need to incorporate the approximations that will be required in this task. So, the present form that is now achieved for (4.52) may be written as

$$-Q(\mathbf{x}, \mathbf{s}) + S(\mathbf{x}, \mathbf{s}) = -Q(\mathbf{x}, \mathbf{s}) + \frac{m\mathbf{s}^2}{2\hbar^2\beta^2} - \frac{3}{2\beta} = \frac{\hbar^2}{2m}\left[\frac{1}{4}\nabla^2 + \frac{\partial^2}{\partial \mathbf{s}^2}\right]. \tag{4.55}$$

The last expression is just a restatement of (4.50) and (4.52). If we now insert (4.50), using (4.52) and the first two terms in (4.54), we can write the above equation as

$$\begin{aligned}[-Q(\mathbf{x}, \mathbf{s}) + S(\mathbf{x}, \mathbf{s})]\rho = {} & \frac{\hbar^2}{8m}\nabla^2\rho - \frac{\hbar^2\beta}{2m}\frac{\partial^2(W_s + Q)}{\partial \mathbf{s}^2} + \frac{\hbar^2}{2m}\frac{\partial^2 J}{\partial \mathbf{s}^2} \\ & + \frac{\hbar^2}{2m}\left[-\beta\frac{\partial(W_s + Q)}{\partial \mathbf{s}} + \frac{\partial J}{\partial \mathbf{s}}\right].\end{aligned} \tag{4.56}$$

Using the values of S in (4.54), this simplifies to

$$\begin{aligned}-Q(\mathbf{x}, \mathbf{s}) = {} & \frac{\hbar^2}{8m}\nabla^2\rho - \frac{\hbar^2\beta}{2m}\frac{\partial^2(W_s + Q)}{\partial \mathbf{s}^2} + \frac{\hbar^2}{2m}\left[\beta\frac{\partial(W_S + Q)}{\partial \mathbf{s}}\right]^2 \\ & - \frac{\hbar^2\beta}{2m}\frac{\partial(W_s + Q)}{\partial \mathbf{s}} \cdot \frac{\partial^2 J}{\partial \mathbf{s}^2}.\end{aligned} \tag{4.57}$$

In this form, the term J is given in (4.54).

In general, the term involving J can be taken to be small and thus ignored. Also, consistency means that the derivatives of both W and Q with respect to $\mathbf{s}$ must be small and vanish in the limit of $\mathbf{s}$ going to zero. This is the limit in which the density matrix becomes diagonal and basically becomes classical. This means that we can ignore, for practical purposes, the nonlinear quadratic term. Now, we also have to restore a few terms that have been omitted in this last form, and when this is done we find the resulting differential equation for the sum of the applied and quantum potentials to be

$$-\frac{\hbar^2\beta}{2m}\frac{\partial^2(W_S + Q)}{\partial \mathbf{s}^2} + \mathbf{s} \cdot \frac{\partial(W_S + Q)}{\partial \mathbf{s}} + (W_S + Q) = -\frac{\hbar^2}{8m\rho}\nabla^2\rho + W_S. \tag{4.58}$$

Hence it is now clear that the total effective potential includes the quantum potential defined earlier as a driving force as well as a driving force from the applied potential. To simplify the form, let us scale this displacement by the thermal wavelength as

$$\mathbf{s}' = \sqrt{\frac{2m}{\hbar^2\beta}}\mathbf{s}, \tag{4.59}$$

so that our differential equation is now

$$-\frac{\partial^2(W_S + Q)}{\partial \mathbf{s}'^2} + \mathbf{s}' \cdot \frac{\partial(W_S + Q)}{\partial \mathbf{s}'} + (W_S + Q) = -\frac{\hbar^2}{8m\rho}\nabla^2\rho + W_S. \tag{4.60}$$

This latter equation can be solved through the use of a Green's function, which itself is found from the adjoint equation [47]

$$\frac{\partial^2 G}{\partial \mathbf{s}^2} + \frac{\partial(\mathbf{s}G)}{\partial \mathbf{s}} - G = -\delta(\mathbf{s} - \mathbf{s}') \tag{4.61}$$

under the restrictions that both the Green's function and its derivatives vanish at infinity. Then, by the normal methods, the Green's function may be found to be [48]

$$G(\mathbf{s}, \mathbf{s}') = \frac{1}{4\pi \mid \mathbf{s} - \mathbf{s}'|} e^{-|\mathbf{s}-\mathbf{s}'|^2/2}, \tag{4.62}$$

and we can now find the quantum potential from

$$Q(\mathbf{x}, \mathbf{s}) = \frac{1}{4\pi}\int \frac{d^3\mathbf{s}'}{|\mathbf{s} - \mathbf{s}'|}\left[-\frac{\hbar^2}{8m\rho}\frac{\partial^2\rho}{\partial \mathbf{s}'^2} + W_S(\mathbf{w}, \mathbf{s}')\right]e^{-|\mathbf{s}-\mathbf{s}'|^2/2} - W_S(\mathbf{w}, \mathbf{s}). \tag{4.63}$$

Now, we remember that Q is the correction to the potential, so that the sum of Q and W gives the new total effective potential. Thus, there are two parts to this effective potential. The first is the smoothed correction that arises from the quantum potential while the second is the smoothing of the potential itself.

The final result tells us that this more extensive consideration yields not much new to the arguments presented in the preceding parts of this chapter. That is, in the end, there is not much to choose between one form of the effective potential or another different form. The fact that we need to evaluate just what the smoothing parameter, or some coefficient as in the case of the Bohm and Wigner potentials, just means that it is not at all clear just what form of the effective potential is preferred for a given situation, and much more work is necessary in this area to clarify any cloud of confusion.

References

[1] Ferry D K and Grubin H L 1996 *Solid State Physics* vol 49 (New York: Academic) pp 283–448

[2] Ferry D K 1995 *Quantum Mechanics* (Bristol: Institute of Physics) p 11

[3] Kubo R, Miyake S J and Hashitsume N 1965 *Solid State Physics* (New York: Academic) p 269

[4] Fetter A L and Walecka J D 1971 *Quantum Theory of Many-Particle Systems* (New York: McGraw-Hill) p 277
[5] Dewdney C and Hiley B J 1982 *Found. Phys.* **12** 27
[6] Kluksdahl N, Pötz W, Ravaioli U and Ferry D K 1987 *Superlatt. Microstruc.* **3** 41
[7] Barker J R, Roy S and Babiker S 1992 *Science and Technology of Mesoscopic Structures* (Berlin: Springer) p 213
[8] Bohm D and Hiley B J 1993 *The Undivided Universe* (London: Routledge)
[9] Kadanoff L P and Baym G 1962 *Quantum Statistical Mechanics* (Reading, MA: Benjamin) ch 6
[10] Ando T, Fowler A B and Stern F 1982 *Rev. Mod. Phys.* **54** 437
[11] Ferry D K 2000 *Superlatt. Microstruc.* **28** 419
[12] Vasileska D and Ferry D K 1999 *Nanotechnology* **10** 192
[13] Vasileska D, Akis R, Knezevic I, Milicic S N, Ahmed S S and Ferry D K 2001 *Microelectron. Eng.* **63** 233
[14] Madelung E 1926 *Z. Phys.* **40** 322
[15] Bohm D 1952 *Phys. Rev.* **85** 166
[16] Kennard E H 1928 *Phys. Rev.* **31** 876
[17] Brillouin L 1926 *J. Phys. Radium* **7** 353
[18] Keller J B 1958 *Ann. Phys.* **4** 180
[19] Stockmann H-J 1999 *Quantum Chaos: An Introduction* (Cambridge: Cambridge Univ. Press)
[20] Miller W H 1975 *J. Chem. Phys.* **63** 995
[21] Berry M V 1984 *Proc. R. Soc. Lond.* A **392** 45
[22] Ferry D K, Goodnick S M and Bird J P 2009 *Transport in Nanostructures* 2nd edn (Cambridge, UK: Cambridge Univ. Press)
[23] Albareda G, Marian D, Benali A, Yaro S, Zanghi N and Oriols X 2013 *J. Comput. Electron.* **12** 405
[24] Oriols X, Garcia-Garcia J J, Martin F, Suñé J, Gonález T, Mateos J and Pardo D 1998 *Appl. Phys. Lett.* **72** 806
[25] Colomés E, Zhan Z and Oriols X 2015 *J. Comput. Electron.* **14** 894
[26] Shifren L, Akis R and Ferry D K 2000 *Phys. Lett.* A **274** 75
[27] Holland P R 1997 *The Quantum Theory of Motion* (Cambridge: Cambridge Univ. Press)
[28] Taylor G I 1909 *Proc. Camb. Phil. Soc.* **15** 114
[29] Tonomura A, Endo J, Matsuda T, Kawasaki T and Ezawa H 1989 *Am. J. Phys.* **57** 117
[30] Philippidis C, Dewdney C and Hiley B J 1979 *Il Nuovo Cim.* **52B** 15
[31] Menzel R, Puhlmann D, Heuer A and Schleich W P 2012 *Proc. Nat. Acad. Sci.* **109** 9314
[32] Kocsis S, Braverman B, Ravets S, Stevens M J, Mirin R P, Krister Shalm L and Steinberg A M 2011 *Science* **332** 1170
[33] Wigner E 1932 *Phys. Rev.* **40** 749
[34] Iafrate G J, Grubin H L and Ferry D K 1981 *J. Phys.* **41-C7** 307
[35] Zhou J-R and Ferry D K 1993 *IEEE Trans. Electron Dev.* **40** 421
[36] Grubin H L, Govindam T R, Kreskovsky J P and Stroscio M A 1993 *Sol.-State Electron.* **36** 1697
[37] Grubin H L, Kreskovsky J P, Govindam T R and Ferry D K 1994 *Semicond. Sci. Technol.* **9** 855
[38] Kosina H, Langer E and Selberherr S 1995 *Microelectron. J.* **26** 217

[39] Feynman R P and Hibbs A R 1965 *Quantum Mechanics and Path Integrals* (New York: McGraw-Hill) section 10.3
[40] Giachetti R and Tognetti V 1986 *Phys. Rev.* B **33** 7647
[41] Feynman R P and Kleinert H 1986 *Phys. Rev.* A **34** 5080
[42] Kleinert H 1986 *Phys. Lett.* A **118** 267
Kleinert H 1986 *Phys. Lett.* B **118** 324
[43] Kriman A and Ferry D K 1989 *Phys. Lett.* A **285** 217
[44] Kriman A M, Zhou J R and Ferry D K 1989 *Phys. Lett.* A **138** 8
[45] Cuccoli A, Macchi A, Neumann M, Tognetti V and Vaia R 1992 *Phys. Rev.* B **45** 2088
[46] Zubarev D N 1970 *Nonequilibrium Statistical Thermodynamics* (New York: Consultants Bureau)
[47] Morse P M and Feshbach H 1953 *Methods of Theoretical Physics* (New York: McGraw-Hill) ch 7
[48] Ferry D K and Zhou J-R 1993 *Phys. Rev.* B **48** 7944

IOP Publishing

David K Ferry and Mihail Nedjalkov

Chapter 5

Numerical solutions

In the previous chapters, we have developed the Wigner function and rationalized the importance of it to physical systems. As we have tried to emphasize, the Wigner function offers many advantages for modeling of quantum devices and systems, and for investigating coherent properties of these systems [1]. The Wigner function is a phase space description, which connects naturally to classical studies using the Boltzmann description. In addition, it is thought that the scattering is in many cases a local phenomena, although there is serious debate about this point, which we return to in chapter 7. Because the Wigner function is a phase space description, it is conceptually possible to use the correspondence principle to examine in detail just where quantum corrections to the classical picture are found in the device. Moreover, as we shall see, the phase space description permits an easier handling of the boundary conditions, as it allows a separation of the problem into incoming and outgoing parts, exactly as is handled in the Landauer description of section 2.1. Still another advantage is that the Wigner function is usually a purely real function, which simplifies both its calculation and the interpretation of the results. Just as one does in the classical situation when the results from the Boltzmann transport equation are found, the results from the Wigner function evolution are easily coupled to solutions of the Poisson equation to obtain self-consistent results. This allows one to examine the role of quantum effects in the redistribution of charge in the device.

As we shall see in later chapters, Wigner functions have been widely used in the field of quantum optics, as well as other fields of physics, as well. The quantum phase space distribution is readily usable to describe the coherence of optical fields and to describe the polarization of these fields [2]. It has even been investigated for use in elementary pattern recognition [3]. From the earliest interest in quantum effects in electronic devices, there has been interest in the Wigner function approach [1, 4–6]. Much of this builds upon the very early work on the Wigner formalism [7, 8].

doi:10.1088/978-0-7503-1671-2ch5

There are several major concerns in the simulation and modeling of electronic devices, as well as in many other areas in which the Wigner function is utilized. First and foremost for numerical simulations are the stability and convergence of the numerical techniques. Because a transport system is, by necessity, an open system, the nature of the boundary conditions is of importance, especially as these may well dominate questions of stability and convergence. Self-consistency between solutions of the Wigner equation of motion and solutions of Poisson's equation is also of utmost concern. Finally, the role of collisions is exceedingly important. In this chapter, we will severely limit the latter to a simple approximation, so as not to cloud the numerical questions with the details of the scattering processes. We leave detailed discussion of the scattering to chapter 7.

The most commonly studied electronic device has been the resonant tunneling diode (RTD), if for no other reason than that this device is clearly a quantum-mechanical structure. Moreover, it has been fabricated and built quite some years ago [9, 10], and has been shown to be a good source for terahertz radiation [10]. Consequently, most illustrations of the Wigner function in this chapter will be based upon such a device. In short, a RTD is composed of a quantum well placed between two potential barriers. The bound state in the well provides 'resonant' tunneling through the structure, which in turn provides a filter on the electronic states which pass through. As a result, the device exhibits negative differential conductance, which can be used for a two-terminal oscillator [11]. The prototypical system is fabricated in the GaAs–AlGaAs heterostructure system via molecular-beam epitaxy, a process which gives exceedingly abrupt interfaces between one layer and the next. Usually, thick GaAs layers are grown on either side of the RTD structure, while the barriers and wells are often of the order of 5 nm thickness each.

5.1 The initial state

A serious consideration for numerical simulations of the Wigner function is the initial state. Although quantum-mechanical effects can enter through the potential function in the Wigner equation of motion, a quadratic potential reduces to the classical Boltzmann equation, and the source of quantum effects has vanished. However, a quadratic potential is known to produce quantization of the harmonic oscillator, which is the prototype quadratic potential. Hence, using the Wigner equation of motion is inadequate to produce the quantization required for an appropriate initial (or later) condition. In spite of this, one might suspect that one could use (3.4), with the time derivative term set equal to zero, to find a correct quantum-mechanical steady-state solution that could serve as the initial condition. The fallacy in this procedure is actually quite subtle, as using (3.4) with just the boundary conditions is ill-defined. The correct quantum-mechanical boundary conditions presuppose complete knowledge of the state of the system, and how this state appears at the boundaries. Yet, this state is a function of the internal potential. Such knowledge of the state of the quantum system and the boundaries implies knowing this without resorting to the Wigner equation of motion.

In order to include the quantum corrections to all orders, one of two approaches can be utilized. The first is to extend the computational domain a distance sufficiently far from any quantum structure that the system will be in a near-classical state. Then, a classical distribution may be assumed at the boundaries. It has been claimed that quantum corrections 'heal' over several thermal de Broglie wavelengths [12], but this assumes that long-range coherence, such as entanglement, does not exist in the system. Even so, this length could be more than 100 nm in a material such as GaAs. The second approach to incorporate quantum effects is to solve the adjoint equation to determine the equilibrium initial condition for the Wigner function. The adjoint equation has been given as (4.46), which is independent of time in this formulation. Solving this equation will give an initial condition which includes quantum corrections to all orders, and will assure that solutions of the Wigner equation of motion (3.4) will also include all the quantum corrections. But, there is also a numerical problem with this second approach. The problem arises from the need to solve the adjoint equation on the same discretized space that will be used for solving the Wigner equation of motion. As a result, the questions of stability and convergence are also present in finding the initial condition. Any errors, whether to discretization or other sources, will then be propagated through the full temporal solutions of (3.4). While the solution of the adjoint equation is the preferred approach, it may be limited by this discretization.

There remains another approach, which could be called the 'third way'. In this approach, the Landauer approach is used to build up an equilibrium density matrix for the complete quantum system. This density matrix is then transformed to produce the Wigner function for the initial state at $t = 0$. What we mean by using the Landauer approach is to construct a set of scattering states, e.g. waves initialized from one boundary, which are then incident upon, and reflected from, the quantum structure using the Lippmann–Schwinger equation [12]. In principle, one wants to know the density matrix throughout space, and to determine it self-consistently with the electrostatic corrections to the heterostructure potential, as in the RTD structure. Since the presence of the heterostructure affects the local electron density in the vicinity, this necessitates the self-consistent approach. To carry out this task with the Landauer method, we need to have both orthonormality and completeness for the basis set of scattering states. Then, the density matrix is populated by that sum of the scattering states which gives the correct distribution function at a great distance from the quantum structure.

The usual method of applying scattering states is often considered to be heuristic. The normal current-conservation arguments give the correct probability that a state will be occupied, but they do not provide the needed normalization for these states nor do they provide the overlap integral corrections to assure orthogonality [13]. On the other hand, we have shown that the scattering states are, by construction, solutions of the Lippmann–Schwinger equation, and so obey the same orthonormality relations as any set of unperturbed states [12]. An orthogonal, normalized basis is obtained when scattering states are normalized so that the incident part of the wave function equals the incident part of a free state wave function with the same momentum. This provides that even degenerate states incident from opposite directions are orthogonal. Since the scattering states are eigenstates of the Hamiltonian, the Boltzmann factor is diagonal in this basis. As a result, simple expressions for the density matrix, and any other quantities of interest, are found from the scattering-state basis.

If the potential approaches a constant as $x \to \pm\infty$, i.e. $V(x) = V^-$ for $x < x^-$ and $V(x) = V^+$ for $x > x^+$, where the parameters x^- and x^+ are the left and right edges of the tunneling region, the basis states are plane waves. In the situation in which the device is in equilibrium, we may take the case that $V^- = V^+ = 0$. Hence, for states which are incident from the left-hand side of the barriers, we may say that, for $k > 0$ (that is propagating from the left to the right toward the barriers),

$$\psi_k(x) = \frac{1}{\sqrt{2\pi}}[e^{ikx} + r(k)e^{-ikx}], \tag{5.1}$$

for $x < x^-$, and

$$\psi_k(x) = \frac{1}{\sqrt{2\pi}}t(k)e^{ikx}, \tag{5.2}$$

for $x > x^+$. Here, the quantities $r(k)$ and $t(k)$ are the reflection and transmission coefficients, respectively, for the tunneling barriers. In a similar manner, the states incident from the right may be defined as

$$\psi_k(x) = \frac{1}{\sqrt{2\pi}}[e^{-ikx} + r(k)e^{ikx}], \tag{5.3}$$

for $x > x^+$, and

$$\psi_k(x) = \frac{1}{\sqrt{2\pi}}t(x)e^{-ikx}, \tag{5.4}$$

for $x < x^-$. Because the tunneling barrier is a passive element, the transport is linear, and therefore the reflection and transmission coefficients are symmetrical, although they depend explicitly upon the momentum.

In some cases, even though the potential at the ends of the device are zero, there may be cases in which a potential well appears at each end of the tunneling barriers. This is the situation if the regions adjacent to the barriers are doped to a lower level than that of the contact regions. This was the case for figure 4.6, and this leads to values of the local potential that are < 0. Consequently, there may be bound states in these regions, which must be determined from solving the Schrödinger equation directly. These bound states can also then propagate through the tunnel barriers. Such bound states may be described by an eigen-energy E_n and a wave function $\psi_n(x)$. The density matrix is then populated by both the bound states and the propagating states described by (5.1)–(5.4). As a result, the density matrix then becomes

$$\rho(x, x') = \frac{1}{Z}\left[\sum_n \psi_n(x)\psi_n^*(x')f(E_n) + \int_{-\infty}^{\infty} dk\ \psi_k(x)\psi_k^*(x')f(E_k)\right], \tag{5.5}$$

where Z is the partition function and $f(E)$ is the Fermi–Dirac distribution function in the equilibrium regions at the boundaries of the device (or the Boltzmann distribution in the non-degenerate situation). Here, the equilibrium partition function provides the normalization of the various waves and is given by

$$Z^{-1} = 2\beta\sqrt{\pi}e^{\beta V}\lim_{x\to-\infty}\rho(x, x). \tag{5.6}$$

Here, the parameter $\beta = 1/k_{\mathrm{B}}T$.

We may initially start with an unnormalized set of basis functions, and compute the values for the reflection and transmission coefficients for each function using translation matrices, or analytically solve the Schrödinger equation directly. These states are then normalized with scattering theory, in which wave functions in the presence of a scatterer are compared to those in a reference space. These are related through the Lippmann–Schwinger equation. A useful consequence of this equation is that the scattering states satisfy precisely the same orthonormality relations as unperturbed states [12]. Each state contributes to the density matrix according to the thermal equilibrium distribution function, with the partition function evaluated at the ends of the device, where thermal equilibrium is established. We may evaluate the density matrix analytically or on a discrete lattice. Using the lattice may introduce errors as discussed above, so the analytical solution is desired. Then, this initial density matrix may be used to determine the initial Wigner distribution by direct transformation, as discussed in chapter 3.

As a first example of this approach, we consider the case of an infinite barrier existing in a half space. We take $V(x) = 0$ for $x > 0$, and infinite for $x < 0$. There can only be right-incident waves in this situation, and these are clearly simple sine waves, as may be found in any introductory book on quantum mechanics. These sine waves are restricted to the region $x > 0$. For this potential, the density matrix may be computed exactly as

$$\rho(x, x') = \rho_+\left\{\exp\left[-\frac{1}{4}(x - x')^2\right] - \exp\left[-\frac{1}{4}(x + x')^2\right]\right\}, \tag{5.7}$$

where both x and x' have been normalized to the thermal de Broglie wave length

$$\lambda_T = \hbar\sqrt{\frac{\beta}{2m}}. \tag{5.8}$$

The corresponding momentum is then normalized by $\hbar/\lambda_T$. Equation (5.7) is valid for the regime in which $x,x' > 0$, otherwise the density matrix is zero. This immediately allows us to specify the density as

$$\rho(x) = \rho(x, x) = \rho_+(1 - e^{-x^2}), \quad \text{for} \quad x > 0. \tag{5.9}$$

We notice here that the density approaches its limiting value exponentially fast, but near the barrier the density approaches zero in a quadratic fashion. It is important to note that this is *not* a depletion layer in the normal sense. It is *not* caused by band bending. Rather, the decreased density at the barrier is a result of quantum

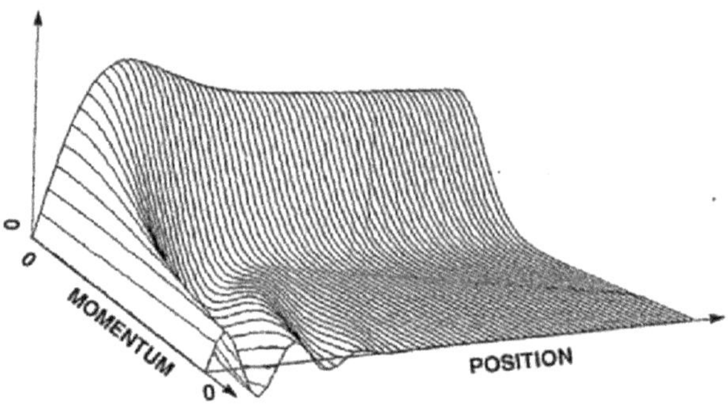

Figure 5.1. The Wigner function for a barrier of infinite height in the region $x < 0$. Reprinted with permission from A M Kriman, N C Kluksdahl, and D K Ferry, *Phys. Rev.* B **36** 5953 (1987). Copyright 1987 by the American Physical Society.

repulsion. The infinite barrier in the left-half plane requires the wave function to vanish at the barrier, and the peak in the sine wave away from the barrier depends upon the specific value of the momentum wave number k. Waves with larger momentum will have a peak closer to the barrier, as may be seen in figure 5.1, which displays the Wigner function. Since high-momentum states have a lower population, the oscillations may be seen at the edges of the Wigner function. The reduction in density extends a finite distance away from the barrier, and this distance is of the order of the thermal de Broglie wave length. In order to preserve charge neutrality, there is a peak in the density as well, as the carrier density repelled from the barrier accumulates a short distance from it, as may be seen in the figure.

From the density matrix (5.7), we can now compute the Wigner function analytically. This is found to be

$$f_W(x, p) = \frac{\rho_+ \lambda_T}{2\pi\hbar}\theta(x)\left[f_W^{(1)}(x, p) - f_W^{(2)}(x, p)\right], \tag{5.10a}$$

where

$$f_W^{(1)}(x, p) = \sqrt{\pi}e^{-p^2}[erf(ip - x) - erf(ip + x)] \tag{5.10b}$$

$$f_W^{(2)}(x, p) = -4xe^{-x^2}\frac{\sin(2px)}{2px}. \tag{5.10c}$$

It should be noted here that these remain the normalized position and momentum so that no dimensionality problem exists, for example in the error functions of (5.10*b*). The error functions themselves are given by the standard definition

$$erf(u) = \frac{2}{\sqrt{\pi}}\int_0^u e^{-t^2}dt. \tag{5.11}$$

Near the barrier, the Wigner distribution exhibits a number of non-classical features. One such feature, clearly observable in figure 5.1, is the negative regions

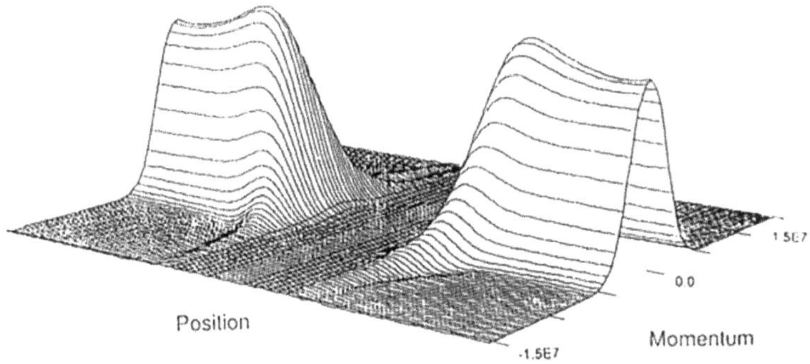

Figure 5.2. The self-consistent Wigner distribution for an RTD structure, determined by the scattering-state basis discussed in the text. The units on momentum are nm^{-1}. Reprinted from 'Intrinsic bistability in the resonant tunneling diode', N Kluksdahl, A M Kriman, D K Ferry, and C Ringhofer, *Superlatt. Microstruc.* **5** 297 (1989), with permission of Elsevier.

that result from the oscillatory behavior of the sine waves. These negative regions do not contradict the idea that the Wigner function is a probability distribution. As discussed in chapter 3, these negative excursions exist over an area of phase space of the order of $\hbar$, and any smoothing in either position or momentum will eliminate these excursions. To actually investigate the behavior in such a small region would require a measurement that violates the Heisenberg uncertainty principle. We have already discussed the depletion that appears near the barrier, and the corresponding peak in density a short distance from the barrier. These latter two effects are of course coupled.

In a finite barrier, penetration of the barrier by the wave functions will occur. This will reduce somewhat the depletion that is observed as well as the density peak. But these two effects are not large, as one must remember that the tunneling coefficient is exponentially small. This is especially true in the case of the resonant tunneling diode where there are two barriers closely spaced to one another. In the prototype simulation that will be used here, the two barriers are assumed to be AlGaAs, while the remainder of the material is GaAs. The composition of the alloy is such that the barriers in the conduction band are 0.3 eV high and are 5 nm each in thickness. The well between the two barriers is also taken to be 5 nm in thickness. On the outside of the actual RTD structure, there is a 5 nm undoped region, while the remainder of the device is doped to 10^{18} cm^{-3}, which provides a degenerate density in the leads. In figure 5.2, we display the equilibrium Wigner distribution function for this device. The depletion next to the barriers and the peaks in the density away from the barriers that were found for the infinite barrier may also be seen in this plot. The position of the two barriers is indicated by the set of transverse lines between the two density peaks. While this is a resonant tunneling structure, the basic properties of the Wigner function away from the barriers is quite similar to that of the infinite barrier case. In this figure, the units on momentum are nm^{-1}.

5.2 Numerical techniques

It is easy to understand that the accuracy of a numerical solution to a partial differential equation depends to a large extent upon the choice of the discretization of phase space, any numerical approximations, and the boundary conditions. For example, the Wigner equation of motion (3.27) involves first-order derivatives in time, position, and momentum, although higher-order derivatives in momentum can also arise. The first problem is the discretization of position and momentum space, which we discuss in one dimension for clarity. It is clear that the points raised here will extend to higher dimensionality without difficulty.

The Wigner function that is found as the initial condition by the approach of the previous section is then mapped onto the grid points that are established by the discretization scheme. Let us consider that the spatial region that is to be used to describe our device extends from $x = 0$ to $x = L$. This length is divided into a set of grid points with a spacing $\Delta x = a$. This latter parameter defines a mesh in real space that is sufficiently small to fully characterize the barriers and wells in the system. This grid size is often taken to be approximately the lattice spacing in, for example, GaAs, or about 0.25 nm. But this adoption of a grid introduces other considerations. For example, it represents a lattice with lattice spacing a. Corresponding to this lattice is a reciprocal lattice upon which the momentum variable is evaluated. The range of the reciprocal lattice is generally described by the Brillouin zone, for which the principal range of momentum lies in the range $-\hbar\pi/a \leqslant p \leqslant \hbar\pi/a$. Because the Brillouin zone is periodic, the discrete Wigner function will be periodic in momentum with a period of $2\hbar\pi/a$. So, this dimension in momentum space is discretized with a convenient grid in momentum space. Nevertheless, one is constrained to consider momentum values that lie within the first Brillouin zone to avoid numerical artifacts that arise from this momentum periodicity. Usually, the choice on the range of momentum is that defined by the first Brillouin zone so that one half of the values are momentum values directed in the positive x direction and the other half are momentum values directed in the negative x direction.

There is an additional constraint that arises from the statistical mechanics of the problem. This comes from the fact that the number of points in momentum space is dictated by the grid size in real space. In condensed matter physics, we are familiar with the fact that the finite length of a material sample determines the number of points in momentum space via the fact that the solid is composed of primitive unit cells. As a result, the number of unit cells in the solid is determined by the lattice constant. In our one-dimensional lattice above, the number of cells is $N = L/a$. A subtle point is that the density of states in momentum space is determined by the number of unit cells per unit volume. Each lattice point in momentum space contains a 'volume' (in one dimension) of $2\pi/L$. Consequently, the Brillouin zone contains exactly N values of momentum, and this constrains our grid size in momentum space[1]. In a metal, where each unit cell

[1] Those familiar with Fourier transforms will note the similarity. The number of Fourier points is defined by the grid size and the overall length. While I can place the Fourier frequencies wherever desired, and concentrate on a narrow frequency range, I cannot increase the resolution over that determined by the length.

usually contains a single atom, there are thus $2N$ states, when we consider spin. Since each metal atom usually has a single bonding electron, this means that the N electrons for the metal atoms fill half the band. On the other hand, in a semiconductor where there are typically two atoms per unit cell, there are enough electrons to completely fill the band, and the semiconductor is normally a narrow gap insulator. As a result, the number of electrons in, for instance, the conduction band is set by the Fermi level and the density of dopant atoms, that provide extra electrons (or holes). We will see below that this subtle density of states requirement affects our choice of a representation for spatial derivatives. Here, however, it signifies that our optimum grid has exactly N points in both the spatial and momentum directions. Certainly, one is free to use a larger value of the momentum grid size, but in doing so it is important to remember how the statistical physics enters the problem.

In the above, the initial Wigner distribution is found at each and every point in the phase space for the device under consideration. It is possible to ease this restriction by defining a 'region of interest' [14]. This is supplemented by requiring also the knowledge of the device on the boundaries at all times less than the observation time. The idea is that Wigner paths can be used to connect the two regions, and these paths will open the way to Monte Carlo approaches, which we discuss in the next chapter. It has also been suggested that one can narrow the momentum domain in phase space, since there are regions where we do not need complete accuracy for the Wigner function [15].

Still the above gridding and the approaches of section 5.1 and discussions below tend to still use the basic finite difference scheme for solving the Poisson equation. One problem with this is that the electric field exists on an intermediate spatial grid that is interpolated between the normal grid points. Computation with the nonlocal potential (or even with the local field) is the major difficulty in the self-consistent solution we discuss in this section, as the Wigner potential extends over the entire space of interest, even if the potential itself does not. Ringhofer has suggested that it may be more efficient to use a spectral method in which Fourier transformation of the potential is introduced [16]. The idea of spectral methods has come back in recent years as a technique for solution [17–20]. Ringhofer has suggested using a collocation method that introduces an analytic fit to the potential solution within each grid cell [21]. This allows the field to be computed analytically rather than having to interpolate [22]. A somewhat different approach was suggested earlier by Barker and Murray, in which the nonlocal potential was addressed by expanding a set of classical distribution functions for different fields, and using this set as a basis set for the Wigner function [23].

The general approach to solving the Wigner equation of motion has been discussed by many through the years. Several important considerations have been suggested [24–27]. Cai *et al* have discussed a variation of the moment method discussed in chapter 3 to close the Wigner equation of motion [28], while others have suggested a maximum entropy method to better derive the various moment equations [29, 30]. Several groups have discussed extending the above discussion to the case of spatially dependent effective mass, such as in a heterostructure [31–34], and to the case of inhomogeneous electric fields [35]. There has also been discussion of extending the treatment to relativistic energy structures [36, 37] and non-parabolic energy bands [38, 39]. The role of spin and the exclusion principle have also been considered [40–42]. The approach has been extended to multiband [43, 44] and two-dimensional transport, as well [41, 45, 46]. Further considerations on self-consistency and fluctuations in the transport have also been considered [47, 48]. There is also discussion of the Wigner potential itself and approaching it as a scattering process [49, 50], a treatment we will return to below and in chapter 7.

5.2.1 Stability and convergence

Perhaps the most common numerical technique for solving discretized equations in time is with explicit differencing. The stability of an explicit approach requires that the error in the discretized equation remains bounded, and doesn't grow with time. By a Fourier analysis of the growth of error in typical explicit schemes, a stability criterion has been found by Courant, Friedrichs, and Lewy (CFL) [51]. In this criterion, the time step is basically required to be sufficiently small that a component of the solutions propagates sufficiently slow that it remains within one grid cell during the time step, or

$$\frac{\Delta x}{\Delta t} = \frac{a}{\Delta t} \geqslant v_{\max} \rightarrow \Delta t \leqslant \frac{a}{v_{\max}}. \tag{5.12}$$

Here, $v_{\max}$ is the maximum velocity in the function being simulated. Clearly, in a system where the momentum is continuous, the choice of the maximum velocity is somewhat arbitrary. For example, in the RTD problem, the barriers are typically 0.3 eV high, and this energy corresponds to a group velocity in GaAs of 1.2×10^8 cm s^{-1}. If we use this value twice to set the time step, the result is a time step of 0.2 fs. It has been suggested that one can avoid the temporal integration by directly seeking the steady-state solution via a conjugate gradient approach for the nonlinear system of equations when the Wigner equation of motion is coupled to Poisson's equation [52]. But this is trading one complicated approach for another, and the temporal approach avoids the possible attraction to a non-global minimum in the solution space. Kim has suggested that a nonlinear mesh could be used to ease the computational burden [53].

The time derivative in (3.27) is usually stepped by the Lax–Wendroff explicit time differencing approach [54, 55]. This approach is considered to provide greater stability as it retains second-order terms in the expansion of $f(t + \Delta t)$. Yet, there are many concerns that have to be faced in the present situation due to the fact that (3.27) involves only first derivatives in space and momentum. Normally, the Lax–Wendroff scheme introduces a second-order spatial difference term into the equation of motion, and that is lacking here. The Lax–Wendroff approach works well with the Schrödinger equation, primarily because it involves second derivatives in position. Now, one might try to get at the second-order approach for our first derivative in position through the central difference

$$\frac{\partial f}{\partial x} \simeq \frac{f(x + a) - f(x - a)}{2a}, \tag{5.13}$$

but this leads to disastrous physics. By introducing the factor $2a$, one is doubling the unit cell in the grid. This is equivalent to introducing a superlattice of twice the size. Such a superlattice produces a Brillouin zone that is only one half the normal value. In this reduced zone, the regions $-\hbar\pi/a \leqslant p \leqslant -\hbar\pi/2a$ and $\hbar\pi/2a \leqslant p \leqslant \hbar\pi/a$ are folded back into the new zone $-\hbar\pi/2a \leqslant p \leqslant \hbar\ \pi/2a$, a process known as zone-folding that is common in studies of superlattices. This doubles the number of states at a given value of momentum and creates the fermion doubling problem. As a result, this central difference creates unphysical statistical mechanics in the problem, and has to be avoided. If this were not bad enough, (5.13) also creates intractable problems at the boundaries where one or the other of the two values in the numerator (on the right-hand side) do not exist. We then find that this approach is intractable, and we must turn to first-order differencing, but with a twist.

Each point in the position and momentum grid has a characteristic direction in which information propagates. This direction is associated with the direction of the momentum. For positive momentum, propagation is in the positive x direction. For negative momentum, propagation is in the negative x direction. That is, for any given point on the grid, the characteristic direction is defined by the local direction of the momentum. In the discretized Wigner function on this grid, there is a *slice* of the mesh over which the momentum is constant. Each of these slices can be considered as an equation, and the various slices are coupled by the potential term in (3.27). Each slice of the mesh is characterized by the same characteristic velocity or momentum. For positive momentum, information flows from the boundary at $x = 0$, and moves toward the boundary at $x = L$, where it leaves the simulation domain. For negative momentum, the characteristic direction and the roles of the boundaries are reversed. Recognition of these characteristic directions solves our inherent problem of first-order derivatives. That is, we adopt the scheme of first-order upwind differencing [54] to propagate the Wigner function to the outgoing boundary along the characteristic direction. Then, we evaluate the derivatives by

$$\frac{\partial f}{\partial x} = \begin{cases} \dfrac{f(x) - f(x-a)}{a} & p(x) > 0, \\ \dfrac{f(x+a) - f(x)}{a} & p(x) < 0. \end{cases} \tag{5.14}$$

Again, the spatial and temporal time steps are still chosen to satisfy the CFL condition (5.12).

There is still one further consideration that we have to make in regards to the stability and convergence of the solution, and this relates to the derivatives in (5.14). Quantum mechanically, the gradient that appears in (5.14) is related to the momentum operator, so that the left-hand side of the equation is proportional to the momentum times the distribution function. This implies that the right-hand side should also be proportional to the momentum and the distribution function. However, in analogy with quantum displacement operators, we introduce exponential factors into the right-hand side of the equation to give

$$\frac{\partial f}{\partial x} = \begin{cases} f(x)\dfrac{e^{-ipa/2\hbar}}{a}\left(e^{ipa/2\hbar} - e^{-ipa/2\hbar}\right) & p > 0 \\ f(x)\dfrac{e^{ipa/2\hbar}}{a}\left(e^{ipa/2\hbar} - e^{-ipa/2\hbar}\right) & p < 0 \end{cases}. \tag{5.15}$$

Thus, the coefficient of the distribution function is proportional to

$$\frac{2}{a}\sin\left(\pm\frac{pa}{2\hbar}\right). \tag{5.16}$$

This highlights an important point of using a discretized grid. It is discussed most with quantum simulations, but it appears equally often in classical simulations. For small values of p, this term is proportional to p as expected. However, as one approaches the edge of the Brillouin zone in momentum space, the sine function becomes 1, and is not proportional to p. As a result, the energy spectrum deviates from the classically expected spectrum due to the introduction of the periodic behavior of the momentum. One must actually limit the range of p and the energy range of interest to, approximately, the central one half of the Brillouin zone ($-\hbar\pi/4a<p<\hbar\pi/4a$). This is well-known in the numerical solution of the discretized Schrödinger equation [56], but is often overlooked in studies of the Wigner function.

5.2.2 The boundary/contact

The boundaries and contact regions create their own special problems that greatly affect the performance of any device with which we deal. In order to specify how these problems arise, let us consider the generic model often used in conjunction with the Landauer formula. This is shown schematically in figure 5.3. We consider the bias is applied to the right reservoir, so that the left Fermi energy can be used as the reference level for the applied potential (and indeed for the energies throughout the structure). The right contact now emits carriers into the constriction with energies up to the local Fermi level plus the applied bias, $E_F + eV$ (note that the energy eV will be negative for a positive voltage). The left contact emits electrons into the constriction with energies only up to E_F. In the following discussion, we will assume that the applied voltage is quite small, although this also is not a stringent requirement. The question to be asked is just where in this structure is thermodynamic equilibrium found. Usually, it only exists at the far left and at the far right edges. That is, even though we specify a contact region, equilibrium usually does not exist through much of this contact region. So, boundaries and contacts are distinct entities. We can illustrate this by considering a uniform electric field through the central region, which we refer to as the device.

In figure 5.4, we illustrate the variation of the potential from one end of the active region to the other. This is an approximation. We assume the voltage drop across the active region must correspond to a constant electric field, which can be quite low, as

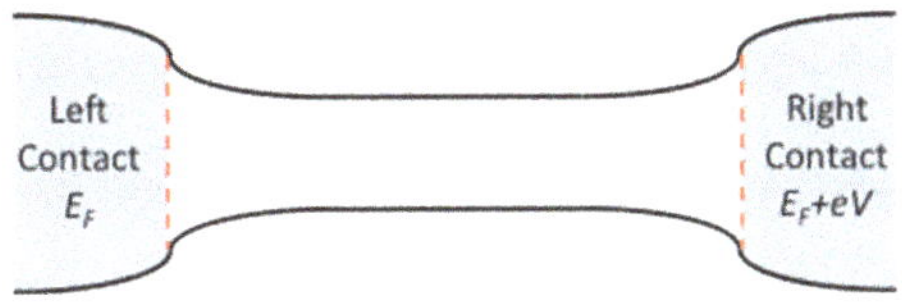

Figure 5.3. A generic device. One can treat the region between the two contacts as a quantum wire, a ballistic constriction, or an active and resistive region. Bias is assumed to be applied to the right reservoir, so that the left reservoir provides the reference level for the energy.

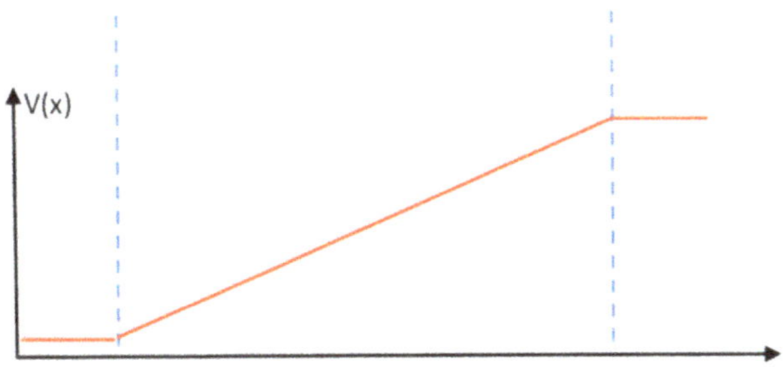

Figure 5.4. Linear potential rise is shown under the assumption of a constant electric field. The blue dashed lines correspond to the dashed red lines in figure 5.3.

we will see. In reality, the voltage is not dropped smoothly across the device. Notice particularly the discontinuity in the electric field at the edge of each contact. These discontinuities mean that charge must accumulate at the contacts in order to support the variations in the voltage. Such a discontinuity in the electric field requires that a sheet of excess charge be present at this discontinuity. In fact, the charge at the left must be negative charge, while the charge at the right must be positive charge. So, we must have a depletion of electrons at the left and an accumulation of electrons at the right. If electrons are the major charge carrier, then the left-hand reservoir is the cathode through which the electrons enter the structure. The depletion of charge at the transition point contributes to the contact resistance.

There is still an additional problem with the potential as drawn in this figure. The linear potential drop leads to a sizable electric field, which will accelerate the electrons. If the ballistic electron is injected from the left contact, it will gain energy from the electric field. That is, the ballistic electron travels at constant energy, and the potential energy at the cathode is converted to kinetic energy as it travels. By the time it reaches the anode at the right-hand side, it will have gained kinetic energy given by eV. A more important aspect is introduced by Kirchoff's current law. As the electrons are accelerated by the electric field, their increase in velocity must be accompanied by a decrease in carrier density if the current is to be conserved as required by the continuity equation. Since the cross-sectional area doesn't change, this requires the product of density and velocity to be a constant value. As the velocity rises, the density must decrease. But in a semiconductor, the density is usually set by the doping. As the number of carriers drops, this creates an additional space charge in the channel, which in turn requires that the potential be nonlinear, i.e. the potential drop cannot be a linear one. The result of this argument is that we

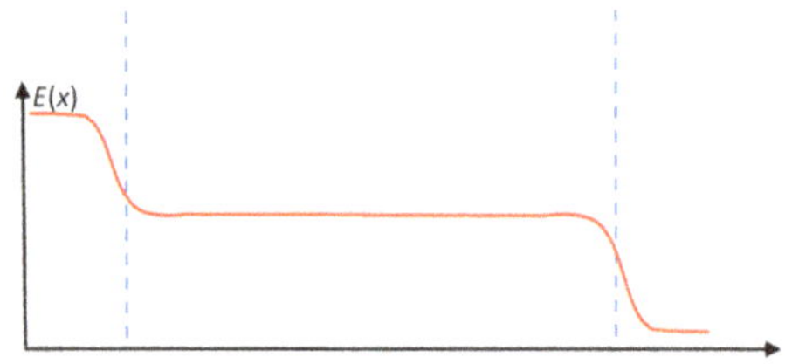

Figure 5.5. The potential energy distribution within the ballistic device. The vertical dashed lines correspond to the red dashed lines in figure 5.3.

cannot have the linear potential shown in figure 5.3 if the carriers are moving via ballistic transport. The linear potential drop only can occur if there is sufficient scattering to assure that the carriers move with a near-equilibrium energy. Our conclusion, then, can only be that the electric field must essentially be near zero in the channel if the carriers are to move via ballistic transport.

As a result of the above discussion, the potential drop must divide between the cathode and the anode transition regions, as shown in figure 5.5 for a ballistic device. While we have drawn the two voltage drops as nearly the same, there is no real requirement that this be the case. In fact, it is usually assumed in the transport world that most of the discontinuity is related to the cathode. But now, ballistic transport can occur through the constriction without the carriers gaining excess energy from the applied bias. At the same time, we note that the potential drops in the transition regions now require that dipole charge densities exist at each transistion. The potential drop can only occur through the existence of such dipole charge densities. Originally, in figure 5.3, the dipole was split between the two transition regions. Now, each transition region has its own dipole charge density, and this leads to the contact resistance.

If we were to consider the Landauer formula, these results give rise to an important modification. The transmission from one mode to another is defined in the reservoirs, which means to the far left of the cathode transition region and to the far right of the anode transition region; that is, these modes are defined in the equilibrium regions. What if we actually do measurements in the channel itself? If these are two-terminal measurements, in which the same contacts are used to source the current and to measure the voltage, we will obtain an answer, but the voltage drop has to occur in these contacts, which have merely replaced the normal ones. On the other hand, if we use four-terminal measurements, then we will find that the Landauer formula becomes modified in order to account for the contact regions with their inherent dipoles.

The dipole charge at the transition between the contact region and the channel gives rise to the contact resistance. It is an important consideration in the treatment of the device itself. This resistance characterizes the transition region between the pure contact and the channel and appears regardless of whether the transport is described by the mobility or is ballistic. In some sense, this resistance describes a barrier between the source and the channel [57], and this barrier is incorporated within the contact resistance. Because of this contact resistance, it is clear that one must solve the Poisson equation in the structure, even in the case of ballistic transport. One cannot evaluate correctly the charge in the two dipoles with this self-consistent calculation. This also confirms that, just because a contact region exists, it cannot be considered as being in equilibrium. This reinforces the view that the boundary is not equivalent to the contact. We may regard the contact as the source of the current into the active channel. But the boundary is a position whose properties determine the Wigner function that describes the particles, or current, that enter the overall structure. It is a significant mistake to confuse these two quantities.

We can see that the regions which we have designated as contacts include some model of the interface between the active region and any external circuit. The charge dipoles are part of this interface, but the external circuit must, at a minimum, be reflected within the boundary conditions on the Wigner function. The conditions throughout the contact must be consistent with physical reality. If a current is flowing through the device, current continuity requires that an identical current flows through the external circuit. Current continuity is required from Maxwell's equations. In particular, we use

$$\nabla \times \mathbf{H} = \mathbf{J} + \frac{\partial \mathbf{D}}{\partial t}, \tag{5.17}$$

where $\mathbf{H}$ is the magnetic field intensity, $\mathbf{J}$ is the conduction current density, and $\mathbf{D}$ is the electric flux density. The last term on the right accounts for displacement current and can *never* be ignored, particularly in time-varying devices. To clarify this, we take the divergence of this equation, noting that the divergence of the curl of any vector is zero. We find

$$\nabla \cdot \mathbf{J} + \frac{\partial}{\partial t}(\nabla \cdot \mathbf{D}) = \nabla \cdot \mathbf{J} + \frac{\partial \rho}{\partial t} = 0. \tag{5.18}$$

This is the continuity equation that guides our device considerations. In the time-invariant situation, it requires that the current be continuous in the device. On the other hand, with a time-varying situation, the displacement current leads to temporal variations in the charge, and these have to be considered in current continuity. All of this means that the appropriate boundary condition must reflect the current in the external circuit and must then characterize a distribution function that represents current flow. In near-equilibrium situations, this can be a shifted thermal distribution, such as a 'drifted Maxwellian' or the equivalent form of the Fermi–Dirac distribution. Of course, it is usually not easy to determine the shift *a priori*. Thus, this drift current-induced shift must be part of the self-consistent simulation.

As we discussed above, we have left and right boundaries, at $x = 0$ and $x = L$, respectively. Mathematical constraints only allow us to specify a single boundary. However, the discretized Wigner field may be thought of as a coupled set of systems, each of which is a slice in momentum space. Each of these slices has a boundary from which electrons enter the slice, and they leave the slice from the opposite boundary. The entering contact is constrained by the current, which must be in the boundary condition. This is accomplished with the thermal-drifted distribution, with the amount of this shift determined self-consistently. The distribution at the

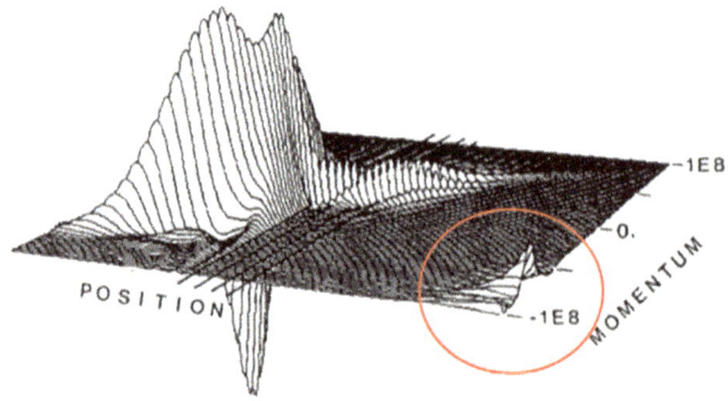

Figure 5.6. Tunneling of a Gaussian-derived Wigner function through a resonant tunneling diode structure. The Wigner function is depicted 50 fs into the transit. The units of momentum are cm^{-1}. The numerically induced artificial reflections appear within the red circle. Adapted from [61].

boundary is then matched to the corresponding momentum slice of the model. Because each momentum slice has only a single boundary condition, the entering boundary, the mathematical constraints are satisfied. Further discussion of the problems of contacts have appeared and may be found at [58–60].

5.2.3 Artificial reflections

Another concern is the need to remove all carriers that leave through the exit boundary. In the so-called 'ideal' contact, the carriers are extracted without reflection. However, simulations with Wigner functions seem to not have ideal contacts. Studies with Gaussian-derived Wigner functions have shown that the momentum nonlocality in the equation of motion couples the incoming and outgoing boundaries and serves as a source of artificial reflections [61]. Such phenomena have been observed in other well-posed numerical systems [62], and are thought to arise strictly from the coupling at the boundaries. (Having said this, these effects perhaps also could arise from the sharpness of the potential, and the nonlocal effect this has through the domain. More work is needed in this area.) Such artificial reflections mar the concept of a rational boundary condition. We can illustrate these artificial reflections by simulating a Gaussian-derived Wigner function tunneling through a double potential barrier, as arises in a resonant tunneling diode [61]. This is depicted in figure 5.6. Clearly evident in the lower-right corner of the figure (within the red circle) is the effect of the spurious numerical reflection that results from the small, but non-zero, coupling of the distribution in this region to the outgoing wave. While small, this spurious reflection will grow rapidly despite the fact that the systesm is well-posed mathematically. This must be addressed with a corrected boundary condition on the outgoing waves.

In order to remove the spurious numerical reflection, we must couple part of the outgoing wave back into the incoming wave during each time step operation. For this, we rewrite (3.5) and (3.6) as

$$\frac{\partial f_W(x, p, t)}{\partial x} = a(p)\frac{\partial f_W(x, p, t)}{\partial t} + \frac{1}{h}\int dp' M(x, p') f_W(x, p + p', t), \qquad (5.19)$$

where $a(p)$ is the inverse of the velocity and $M(x,p')$ is $W(x,p')/v(p)$. With this form of the equation, it has been suggested that the boundary condition for the proper removal of the spurious reflection can be created [62, 63]. At the left edge, $x = 0$, where the outgoing wave is for $p < 0$, the incoming wave is subject to the condition

$$\frac{\partial f_W(0, p, t)}{\partial t} + \frac{1}{h}\int dp' \frac{M(0, p')}{a(p) - a(p')} f_W(0, p + p', t) = 0. \tag{5.20}$$

Similarly, the boundary condition for incoming particles at $x = L$, where $p > 0$ is given by the temporal adjustment

$$\frac{\partial f_W(L, p, t)}{\partial t} + \frac{1}{h}\int dp' \frac{M(L, p')}{a(p) - a(p')} f_W(L, p + p', t) = 0. \tag{5.21}$$

These two equations constitute an absorbing boundary condition for the Wigner equation of motion that removes the artificial reflections from the outgoing boundary. The reflected waves are absorbed at least to second order. These boundary conditions have been used successfully in full device simulations [6].

In subsequent work, Arnold has recast these boundary conditions into an analytical framework from which he has demonstrated the well-posedness of these equations, as well as showing that they could be considered to be members of a hierarchy of possible boundary conditions [64]. In addition, Arnold and coworkers have suggested an improvement over the normal Lax–Wendroff procedure that incorporates an operator-splitting approach [65–67]. If W_n and W_{n+1} are the approximations to the Wigner distribution function $f_W(t)$, where $t = t_0 + n\Delta t$ and Δt is the time step, and we have two operators A and B, then one first solves the iteration (that is constrained within the range $t_n \leqslant t \leqslant t_{n+1}$)

$$\frac{\partial u}{\partial t} = Au, \quad u(t_n) = W_n, \quad W_{n+1/2} = u(t_{n+1}), \tag{5.22}$$

where u is an arbitrary intermediate function, whose initial condition at time t_n is set by the Wigner function at that time. The Lax–Wendroff approach can be used in this intermediate iterative step as well as in the following one. The second part of the split-operator approach then takes the form

$$\frac{\partial u}{\partial t} = Bu, \quad u(t_n) = W_{n+1/2}, \quad W_{n+1} = u(t_{n+1}). \tag{5.23}$$

This has obvious similarities to a predictor-corrector approach to time splitting, except that the operators A and B arise from the spatial gradient and the potential terms in (3.5) and (3.6).

5.2.4 Spectral methods

In section 5.2.1, we introduced spectral methods initially. Here, however, we want to discuss them in some detail as an approach to highlight the fact that the nonlocal potential creates significant problems in solving the Wigner equation of motion

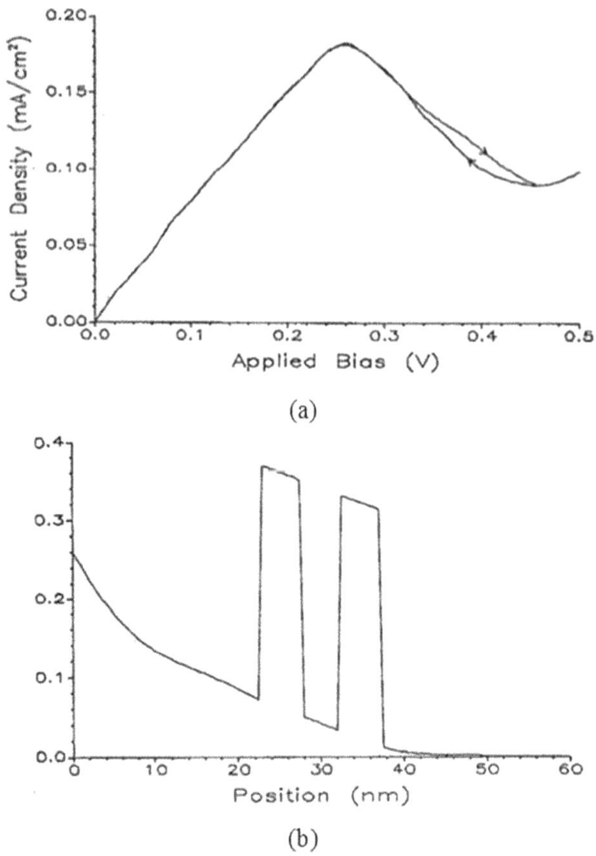

Figure 5.7. (a) The computed current–voltage curve for the resonant tunneling diode whose structure is described in section 5.1. (b) The self-consistent potential found at the peak of the current–voltage curve (~0.26 V). Adapted with permission from N C Kluksdahl, A M Kriman, D K Ferry, and C Ringhofer, *Phys. Rev.* B **39** 7720 (1989), copyright (1989) by the American Physical Society.

(3.5). A quick look at the formation of the nonlocal potential in (3.6) shows that this potential has values in regions where the Wigner function does not really exist (such as outside the device). This creates a numerical problem, as we must ask just how significant are the values which lie outside the device. One method of treating this potential, mentioned previously, is to introduce a coherence length which effectively limits the importance of the potential at some distance from its actual spatial position. Another problem lies in the evaluation of the 'classical' potential from the first derivatives of the total potential. This problem can be illustrated by referring to figure 5.7(b). In this latter figure, there are sharp discontinuities in the potential at the edges of the tunneling barriers. These discontinuities create confusion in the evaluation of the first derivatives, which obviously become infinite. Hence, a question arises as to how much these discontinuities affect the classical potential, and they then create a numerical nightmare. It is hard enough to handle them in Poisson's equation, but they create special problems in the Wigner equation of motion. Many approaches have arisen to ease this problem, including the use of spectral methods and treating the non-classical parts by a collision process, which we discuss in chapter 7. Here, we will discuss the spectral method, following a very recent approach due to Van de Put *et al* [19].

We recognize from (3.5) that the potential energy $V(\mathbf{x})$ creates the nonlocal potential and this affects the momentum change of the Wigner distribution. It is important to note from (3.6) that it is not the potential itself but the difference between two versions of the shifted potential at the points $\mathbf{x} \pm \mathbf{x}'/2$. From this difference in the two potentials, one can then *define* a version of the classical potential as

$$V\left(\mathbf{x} + \frac{\mathbf{x}'}{2}\right) - V\left(\mathbf{x} - \frac{\mathbf{x}'}{2}\right) \equiv -\int_C d\mathbf{s} \cdot \mathbf{F}(\mathbf{s}), \tag{5.24}$$

where the contour C defines a path between the two positions $\mathbf{x} \pm \mathbf{x}'/2$. As we know from electromagnetics, the potential is a conservative potential, so that the choice of the path is not critical and can be any continuous path between the two end points. Now, the basis of the spectral method is to introduce the Fourier transform of this force, as

$$V\left(\mathbf{x} + \frac{\mathbf{x}'}{2}\right) - V\left(\mathbf{x} - \frac{\mathbf{x}'}{2}\right) = -\int_C d\mathbf{s} \cdot \int d^3k \tilde{\mathbf{F}}(\mathbf{k}) e^{i\mathbf{k}\cdot\mathbf{s}}. \tag{5.25}$$

As the path can be any continuous path between the end points, it is convenient to take the linear path in which $\mathbf{s} = \mathbf{x} + \alpha\,\mathbf{x}'$, with α runs over the interval $-1/2$ to $1/2$, consistent with the two end points of the integral. If we insert this value of $\mathbf{s}$ into (5.25), we find

$$\begin{aligned} V\left(\mathbf{x} + \frac{\mathbf{x}'}{2}\right) - V\left(\mathbf{x} - \frac{\mathbf{x}'}{2}\right) &= -\int d^3k \tilde{\mathbf{F}}(\mathbf{k}) \cdot \mathbf{x}' e^{i\mathbf{k}\cdot\mathbf{x}} \int_{-1/2}^{1/2} d\alpha e^{i\alpha\mathbf{k}\cdot\mathbf{x}'} \\ &= i\hbar \int d^3k e^{i\mathbf{k}\cdot\mathbf{x}} \tilde{\mathbf{F}}(\mathbf{k}) \cdot \nabla_{\mathbf{p}} \int_{-1/2}^{1/2} d\alpha e^{i\alpha\mathbf{p}\cdot\mathbf{x}'/\hbar}, \end{aligned} \tag{5.26}$$

where, in the last line, we have introduced the momentum derivative corresponding to the position operator in momentum space. With this, we can now write the Wigner potential as

$$\begin{aligned} W(\mathbf{x}, \mathbf{p}') &= \int d^3\mathbf{k} e^{i\mathbf{k}\cdot\mathbf{x}} \tilde{\mathbf{F}}(\mathbf{k}) \cdot \nabla_{\mathbf{p}'} \int_{-1/2}^{1/2} d\alpha \delta(\alpha\hbar\mathbf{k} - \mathbf{p}') \\ &= \int d^3\mathbf{k} \mathbf{F}_{\mathbf{k}}(\mathbf{x}) \cdot \nabla_{\mathbf{p}'} \int_{-1/2}^{1/2} d\alpha \delta(\alpha\hbar\mathbf{k} - \mathbf{p}'). \end{aligned} \tag{5.27}$$

In the last line, we have incorporated the exponential factor into the definition of the Fourier component $\mathbf{F}_{\mathbf{k}}$. Now, the Wigner equation of motion (3.5) may be rewritten as

$$\frac{\partial f_W}{\partial t} + \frac{\mathbf{p}}{m} \cdot \frac{\partial f_W}{\partial \mathbf{x}} - \frac{1}{h^3} \int d^3\mathbf{k} \mathbf{F}_{\mathbf{k}}(\mathbf{x}) \cdot \frac{\partial}{\partial \mathbf{p}} \int_{-1/2}^{1/2} d\alpha f_W(\mathbf{x}, \mathbf{p} - \alpha\hbar\mathbf{k}, t) = 0. \tag{5.28}$$

This gives us a result that is fully equivalent to the result (3.5), except here the Wigner equation of motion is built upon the spectral components of the force field defined in (5.24). This force field reduces to the classical force field if the spectral components are all smaller than the characteristic size in momentum space over which the Wigner function varies in momentum. That is, if the effect of α is sufficiently small that the two momenta terms can be decoupled, the two integrals can be separated and one reduces mainly to the Boltzmann-like form. It has been argued that for a range of realistic smooth force fields, there exists a natural cutoff which suppresses the high-momentum frequencies that can give rise to some oscillatory behavior [19]. In a sense, this cutoff gives rise to an effective coherence length cutoff related to the decay of the momentum frequencies.

It is also easy to connect this to a particle generation-recombination process which has been adapted in the Monte Carlo approach to be discussed in chapter 6 [68]. We will illustrate this in one dimension for clarity. We again change the variables of integration in (5.28) and rewrite this in the one-dimensional form:

$$\frac{\partial f_W(x, p, t)}{\partial t} + \frac{p}{m}\frac{\partial f_W(x, p, t)}{\partial x} + \int dk \frac{F_k(x)}{\hbar k} \int_{-\hbar k/2}^{\hbar k/2} dp' \frac{\partial f_W(x, p - p', t)}{\partial p} = 0. \quad (5.29)$$

The derivative in momentum can be transferred to the primed momentum coordinate so that the integral is trivially performed, and this leads to

$$\begin{aligned} &\frac{\partial f_W(x, p, t)}{\partial t} + \frac{p}{m}\frac{\partial f_W(x, p, t)}{\partial x} \\ &+ \int dk \frac{F_k(x)}{\hbar k}\left[f_W\left(x, p + \frac{\hbar k}{2}, t\right) - f_W\left(x, p - \frac{\hbar k}{2}, t\right)\right] = 0. \end{aligned} \quad (5.30)$$

Obviously, in the limit of vanishing k, the term in the integral is just the average value of the force field and relates to the classical version of this force field. For other values of k, we can create an equivalent generation (and annihilation) process that transfers weight from one momentum state to another, where the two are separated by $\hbar k$. For this purpose, we define the generation process as [19]

$$G(x, k, t) = \frac{F_k(x)}{\hbar k}. \quad (5.31)$$

Hence, a positive value for the Fourier component of the force field creates Wigner strength at the higher momentum value and annihilates strength at the lower momentum value. The problem arises that a single such process does not conserve probability density. In fact, Van de Put *et al* [19] suggest that two processes A and B will achieve this conservation of probability density, if the two processes are set up as shown in figure 5.8. With this set-up, a symmetric process launches transitions to higher and lower momentum from a central momentum and conserves probability density by construction. The result is that the central two processes in the figure represent the original transition, while the outer two assure that conservation is

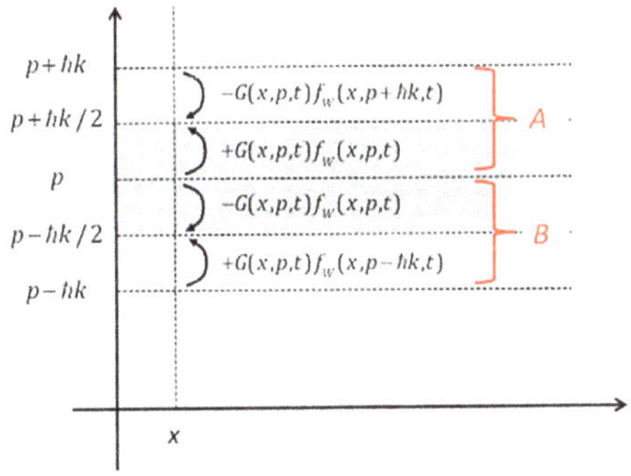

Figure 5.8. The various creation and annihilation processes inherent in (5.31). A and B represent two symmetric processes that ensure conservation of probability density. The grey area corresponds to the original single process in (5.30).

maintained. As noted above, this generation process is quite similar to the concept of signed particles used in the Monte Carlo process.

The above process is clear so long as $p > \hbar k$. If the reverse is true, then the creation and annihilation processes move weight from one momentum direction to the opposite momentum direction. It is true that this is already the case in (3.5), but it is necessary to assure that this does not upset the balance and conservation of the probability density as well as conservation of the current density in the device.

5.3 The resonant tunneling diode: Wigner function simulations

The self-consistent solution of the Wigner equation of motion has been applied to a resonant tunneling diode, whose structure was described in section 5.1 above. A relaxation time approximation was used to simulate the scattering in the device via

$$\left.\frac{\partial f_W(x, p, t)}{\partial t}\right|_{collisions} = -\frac{f_W(x, p, t) - f_W(x, p, 0)}{\tau}, \tag{5.32}$$

where the relaxation time τ was chosen as appropriate for a mobility of 3000 cm^2 V^{-1} s^{-1}. In this scenario, the Wigner function that arises under current flow tries to relax to the initial condition, which is the equilibrium solution found for the actual device structure. It is important to note that this is not the homogeneous thermal equilibrium Wigner distribution as the actual structure will have inhomogeneous density and local charge non-neutrality, although the device as a whole remains charge neutral, even under bias. The steady-state current–voltage of the device is determined by applying an incremental bias potential to the cathode contact, and then letting the Wigner function and self-consistent solution to Poisson's equation result from temporal evolution. It is important to note that the potential is stepped both upward and downward to ensure that the current retraces the same path. This is not true within the negative conductance region, where the quantum well is uncharged during the upsweep but has an accumulation of electrons in the down-sweep. This charge difference causes hysteresis in a portion of the negative

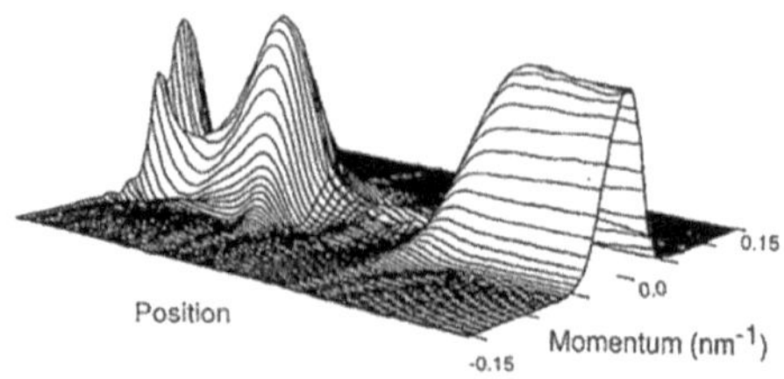

Figure 5.9. The Wigner distribution function for a bias at the peak of the current–voltage curve (~0.26 V). Reprinted with permission from N C Kluksdahl, A M Kriman, D K Ferry, and C Ringhofer, *Phys. Rev.* **B 39** 7720 (1989), copyright (1989) by the American Physical Society.

conductance region. Study of the temporal evolution shows that the self-consistent current reaches the steady state in about 700 fs [69, 70].

Early studies of the RTD assumed that the potential was dropped entirely across the barriers-well structure. In general, this is not the actual situation, as a detailed solution of Poisson's equation will yield a gradual potential drop in order to support the current. This can lead to interesting effects. In figure 5.7(a), we show the current–voltage curve obtained for the structure used. The peak of the current–voltage curve appears at about 0.26 V applied bias (the negative of the potential is applied to the cathode, as mentioned above). In figure 5.7(b), the self-consistent potential energy is plotted at this bias value. Only about one-third of the bias actually appears across the barriers-well structure. A significant fraction of the potential is dropped near the cathode region. There are two contributions to this result. First, there is a significant contact resistance, as described in section 5.2.2. This potential drop leads to the formation of a weak triangular potential well between the cathode 'contact' and the first barrier. This potential well then leads to quantum bound states and a subsequent depletion of carriers in this region. The upward (positive) curvature of the energy signals this electron depletion, and this is reflected in the actual Wigner function at this bias, as shown in figure 5.9. This cathode effect can be reduced by placing undoped layers, typically 5 nm in length, on either side of the barriers [6]. The higher resistance of these undoped layers concentrate the potential more effectively around the barriers-well region.

Other work on the quasi-ballistic behavior of the RTD has appeared in the literature over the years [71–78]. Others have considered the role of the magnetic field and magnetic layers in the RTD [45, 79–82]. There have also been studies of asymmetrical and multi-layer RTDs [83–85].

5.4 Other devices

Simulation of carrier transport in an electron waveguide with the Wigner function has been carried out by Tsuchiya and coworkers [86]. In this work, only elastic scattering by impurities and a low lattice temperature were assumed. With weak scattering, they find the normal transverse quantization that leads to conductance steps expected for quasi-one-dimensional transport evaluated with the Landauer formula. As the strength of the impurity scattering increases, the steps in the conductance are found to deteriorate and eventually to disappear, a result in keeping with traditional quantum solutions of the Schrödinger equation [87]. At large bias voltages, the conductance is observed to deviate from the normal Landauer results, presumably due to carrier heating. Tsuchiya *et al* also investigated the application of a gate across the wire to create a potential barrier, an analogy to a normal tunneling barrier. Isawa and Hatano [88] have also used the Wigner function to study the propagation of electrons in a quantum waveguide. These latter authors considered mainly ballistic transport, but studied the effect of a single point scattering represented by a repulsive delta-function potential. Other waveguide studies of nanowires have subsequently appeared in the literature [89–91].

Miyoshi and coworkers [92] have studied the transport of holes in the potential well that forms at the interface between InGaAs and InP. The Wigner function has also been used to study two-dimensional transport in the quantum point contact [93, 94] and in other materials such as graphene [95]. They have also been applied to studies of the metal-oxide–semiconductor field-effect transistor (MOSFET) [96–99] and to double-gate versions of the device [100, 101].

Miyoshi and coworkers have also used the Wigner function approach to model both electrons and holes in a separate-confinement heterostructure quantum well laser [102, 103], where they find that these two particles are not equally injected into the quantum well. Hence, the conventional gain model needs to be revised. But there has been a vast array of devices to which the Wigner function approach has been applied. Hess and Kuhn have also used the Wigner function for transport, coupled to the Bloch equations for excitation, to study the semiconductor laser [104]; others have also studied these lasers [105, 106]. Other quantum devices, such as the quantum rachet [107] and quantum shuttle [108], have also been studied with the Wigner function. We will discuss other areas of science in later chapters.

References

[1] Ravaioli U, Osman M A, Pötz W, Kluksdahl N and Ferry D K 1985 *Physica* B **134** 36

[2] Mandel L and Wolf E 1965 *Rev. Mod. Phys.* **37** 231

[3] Kumar V and Carroll C W 1984 *Opt. Eng.* **23** 732

[4] Iafrate G J, Grubin H L and Ferry D K 1982 *Phys. Lett.* A **87** 145

[5] Barker J R and Murray S 1983 *Phys. Lett.* A **93** 271

[6] Kluksdahl N C, Kriman A M, Ferry D K and Ringhofer C 1989 *Phys. Rev.* B **39** 7720

[7] Moyal J E 1949 *Proc. Cambr. Phil. Soc.* **45** 99

[8] Levinson I B 1969 *Zh. Eksp. Teor. Fiz.* **57** 660; Levinson I B 1970 *Sov. Phys.—JETP* **30** 362

[9] Sollner T C L G, Goodhue W D, Tannenwald P E, Parker C D and Peck D D 1983 *Appl. Phys. Lett.* **43** 588

[10] Sollner T C L G, Tannenwald P E, Peck D D and Goodhue W D 1984 *Appl. Phys. Lett.* **45** 1319

[11] Chang L L, Esaki L and Tsu R 1974 *Appl. Phys. Lett.* **24** 593

[12] Kriman A M, Kluksdahl N C and Ferry D K 1987 *Phys. Rev.* B **36** 5953

[13] Coon D D and Liu H C 1985 *Appl. Phys. Lett.* **47** 172

[14] Bordone P, Pascoli M, Brunetti R, Bertoni A, Jacoboni C and Abramo A 1999 *Phys. Rev.* B **59** 3060

[15] Kim K Y, Kim J and Kim S 2016 *AIP Adv.* **6** 065314

[16] Ringhofer C 1990 *SIAM J. Numer. Anal.* **27** 32

[17] Furtmaier O, Succi S and Mendoze M 2016 *J. Comput. Phys.* **305** 1015

[18] Xiong Y, Chen Z and Shao S 2016 *SIAM J. Sci. Comput.* **38** B491

[19] van de Put M L, Sorée B and Magnus W 2017 *J. Comput. Phys.* **350** 314

[20] Thomann A and Borzì A 2016 *Numer. Meth. Partial Diff. Eqn.* **33** 62

[21] Ringhofer C 1992 *SIAM J. Numer. Anal.* **29** 679

[22] Ravaioli U, Lugli P, Osman M A and Ferry D K 1985 *IEEE Trans. Electron Dev.* **32** 2097

[23] Barker J R and Murray S 1983 *Phys. Lett.* A **93** 271

[24] Mains R K and Haddat G I 1994 *J. Comp. Phys.* **112** 140

[25] Nedjalkov M, Dimov I, Bordone P, Brunetti R and Jacoboni C 1997 *Math. Comput. Model.* **26** 33

[26] Greiner A, Reggiani L, Kuhn T and Varani L 2000 *Semicond. Sci. Technol.* **15** 1071

[27] Hosseini S E and Faez R 2007 *Jpn. J. Appl. Phys.* **41A** 1300

[28] Cai Z, Fan Y, Li R, Lu T and Wang Y 2012 *J. Math. Phys.* **53** 103503

[29] Trovato M and Reggiani L 2000 *Phys. Rev.* B **61** 16667

[30] Trovato M and Reggiani L 2011 *Phys. Rev.* B **84** 061147

[31] Tsuchiya H, Ogawa M and Miyoshi T 1991 *IEEE Trans. Electron. Dev.* **38** 1246

[32] Kim K-Y and Lee B 1999 *Sol.-State Electron.* **43** 81

[33] Schulz L and Schulz D 2016 *IEEE Trans. Nanotechnol.* **15** 801

[34] Lee J-H and Shin M 2017 *IEEE Trans. Nanotechnol.* **16** 1028

[35] Krieger J B, Kiselev A A and Iafrate G J 2005 *Phys. Rev.* B **72** 195201

[36] Shin G R and Rafelski J 1993 *Phys. Rev.* A **48** 1869

[37] Antonsen F 1997 *Phys. Rev.* D **56** 920

[38] Kim K-Y and Lee B 1999 *J. Appl. Phys.* **86** 5085

[39] Demeio L 2003 *J. Comput. Electron.* **2** 313

[40] Kaniadakis G, Lavagno A and Quarati P 1998 *Phys. Rev.* E **57** 1395

[41] Heller L 2003 *Phys. Rev.* B **67** 094417

[42] Saikin S 2004 *J. Phys. Condens. Matter* **16** 5071

[43] Demeio L, Barletti L, Bertoni A, Bordone P and Jacoboni C 2002 *Physica* B **314** 104

[44] Federov A, Pershin Y V and Piermarocchi C 2005 *Phys. Rev.* B **72** 245327

[45] Kluksdahl N C, Kriman A M and Ferry D K 1989 *High Magnetic Fields in Semiconductors* ed G Landwehr (Berlin: Springer) p 335

[46] Recine G, Rosen B and Cui H-L 2005 *J. Comput. Phys.* **209** 421

[47] Bonosera A, Gulminelli F and Schuck P 1992 *Phys. Rev.* C **46** 1431

[48] Biegel B A and Plummer J D 1996 *Phys. Rev.* B **54** 8070

[49] Bordone P, Bertoni A and Jacoboni C 2002 *Physica* B **314** 123

[50] Nedjalkov M, Kosina H, Ungersboeck E and Selberherr S 2004 *Semicond. Sci. Technol.* **19** S226

[51] Courant R, Friedrichs K and Lewy H 1928 *Math. Ann.* **100** 32

[52] Jansen R-J E, Farid B and Kelly M J 1991 *Physica* B **175** 49

[53] Kim K-Y 2008 *Jpn. J. Appl. Phys.* **47** 358

[54] Shokin Y L 1983 *The Method of Differential Approximation* (Berlin: Springer)

[55] Ames W F 1977 *Numerical Methods for Partial Differential Equations* (New York: Academic)

[56] Ferry D K 2001 *Quantum Mechanics* 2nd edn (Bristol: Institute of Physics Publishing) section 2.9

[57] Kroemer H 1968 *IEEE Trans. Electron Dev.* **15** 819
[58] Zaccaria R P, Iotti R C and Rossi F 2003 *J. Comput. Electron.* **2** 141
[59] Ferrari G, Biacobbi N, Bordone P, Bertoni A and Jacoboni C 2004 *Semicond. Sci. Technol.* **19** S254
[60] Rosati R, Dolcini F, Iotti R C and Rossi F 2013 *Phys. Rev.* B **88** 035401
[61] Kluksdahl N C, Pötz W, Ravaioli U and Ferry D K 1987 *Superlatt. Microstruc.* **3** 41
[62] Engquist B and Majda A 1977 *Math. Comput.* **31** 629
[63] Ringhofer C, Ferry D K and Kluksdahl N C 1989 *Trans. Theory Stat. Phys.* **18** 331
[64] Arnold A 1994 *RAIRO—Math. Mod. Num Anal.* **28** 853
[65] Arnold A 1991 *Proc. NUMSIM '91*
[66] Arnold A and Ringhofer C 1995 *Computational Electronics* ed U Ravaioli and K Hess (Champaign, IL: University of Illinois Press)
[67] Arnold A and Ringhofer C 1996 *SIAM J. Numer. Anal.* **33** 1822
[68] Nedjalkov M and Vasileska D 2008 *J. Comput. Electron.* **7** 22
[69] Kluksdahl N C, Kriman A M, Ringhofer C and Ferry D K 1988 *Sol.-State Electron.* **31** 743
[70] Kluksdah. N C, Kriman A M and Ferry D K 1988 *IEEE Electron Dev. Lett.* **29** 955
[71] Muga J G, Saia R and Brouard S 1995 *Sol.-State Commun.* **94** 877
[72] Taskinen K, Majama T and Kuivalainen P 1997 *Phys. Scr.* **T69** 298
[73] Biegel B A and Plummer J D 1997 *IEEE Trans. Electron Dev.* **44** 733
[74] Garcia-Garcia J J, Oriols X, Martin F and Suñé J 1998 *J. Appl. Phys.* **83** 8057
[75] Dai Z and Ni J 2005 *Phys. Lett.* A **342** 272
[76] Yoder P D and Grupen M 2010 *IEEE Trans. Electron Dev.* **57** 3265
[77] Savio A and Poncet A 2011 *J. Appl. Phys.* **109** 033713
[78] Szydlowski D, Woloszyn M and Spisak B J 2013 *Send. Sci. Technol.* **28** 105022
[79] Wu G Y and Wu K-P 1992 *J. Appl. Phys.* **71** 1259
[80] Wu K-Y and Lee B 1999 *J. Appl. Phys.* **85** 7252
[81] Havu P, Tuomisto N, Väänänen R, Puska M J and Nieminen R M 2005 *Phys. Rev.* B **71** 235301
[82] Wójcik P, Spisak B J, Woloszyn M and Adamowski J 2011 *Acta. Phys. Pol.* A **119** 648
[83] Garcia-Garcia J and Martin F 2000 *Appl. Phys. Lett.* **77** 3412
[84] Kim K-Y and Lee B 2001 *Phys. Rev.* B **64** 115304
[85] Wójcik P, Spisak B J, Woloszyn M and Adamowski J 2009 *Semicond. Sci. Technol.* **24** 095012
[86] Tsuchiya H, Ogawa M and Miyoshi T 1992 *IEEE Trans. Electron Dev.* **39** 2465; Tsuchiya H, Ogawa M and Miyoshi T 1991 *Jpn. J. Appl. Phys.* **30** 3853
[87] Kirczenow G 1989 *Phys. Rev.* B **39** 10452
[88] Isawa Y and Hatano T 1991 *J. Phys. Soc. Jpn.* **60** 3108
[89] Nedjalkov M, Vasileska D, Ferry D K, Jacoboni C, Ringhofer C, Dimov I and Palankovski V 2006 *Phys. Rev.* B **74** 035311
[90] Barraud S 2009 *J. Appl. Phys.* **106** 063714
[91] Barraud S 2011 *J. Appl. Phys.* **110** 093710
[92] Miyoshi T, Tsuchiya H and Ogawa M 1992 *IEEE J. Quantum Electron.* **28** 25
[93] Aronov I E, Berman G P, Campbell D K and Dudiy S V 1997 *J. Phys. Condens. Matter* **9** 5089
[94] Aronov I E, Beletskii N N, Berman G P, Campbell D K, Doolen G D and Dudiy S V 1998 *Phys. Rev.* B **58** 9894

[95] Zamponi N and Barletti L 2011 *Math. Methods Appl. Sci.* **34** 807
[96] Balaban S N, Pokatilov E P, Fomin V M, Gladilin V N, Devreese J T, Magnus W, Schoenmaker W, Van Rossum M and Sorée B 2002 *Sol.-State Electron.* **46** 435
[97] Croitoru M D, Gladilin V N, Formain V M, Devreese J T, Magnus W, Schoenmaker W and Sorée B 2003 *J. Appl. Phys.* **93** 1230
[98] Van Rossum M, Schoenmaker W, Magnus W, De Meyer K, Croitoru M D, Gladilin V N, Fomin V M and Devreese J T 2003 *Phys. Status Solidi* B **237** 426
[99] Croitoru M D, Gladilin V N, Fomin V M, Devreese J T, Magnus W, Schoenmaker W and Sorée B 2008 *Solid State Commun.* **147** 31
[100] Croitoru M D, Gladilin V N, Fomin V M, Devreese J T, Magnus W, Schoenmaker W and Sorée B 2004 *J. Appl. Phys.* **96** 2305
[101] Jiang H and Cai W 2010 *J. Comput. Phys.* **229** 4461
[102] Tsuchiya H and Miyoshi T 1996 *IEEE J. Quantum Electron.* **32** 865
[103] Tsuchiya H, Hayashi Y and Miyoshi T 1996 *Physica* B **227** 411
[104] Hess O and Kuhn T 1996 *Phys. Rev.* A **54** 3347
[105] Weetman P and Wartak M S 2003 *J. Appl. Phys.* **93** 9562
[106] Wartak M S and Weetman P 2003 *Microwave Opt. Technol. Lett.* **38** 369
[107] Zueco D and Garcia-Palacios J L 2005 *Physica* E **29** 435
[108] Jauho A-P, Flindt C, Novotny T and Donarini A 2005 *Phys. Fluids* **17** 100613

IOP Publishing

David K Ferry and Mihail Nedjalkov

Chapter 6

Particle methods

In the previous chapters, we have emphasized how the many-particle problem can be solved by a one-particle distribution function. Beginning in the mid-twentieth century, when computers began to appear, it was realized that one could turn the problem around and use particles to solve the differential equation for the distribution function. While the full history of what is now called the Monte Carlo method is not at all clear, it is known that an early estimate of the value of π could be obtained by dropping needles on the floor, an approach termed Buffon's needle, after the eighteenth-century French mathematician Georges-Louis Leclerc, Comte du Buffon [1]. Enrico Fermi was apparently the first to experiment with using a Monte Carlo method for neutron diffusion, but did not publish this work. In the 1940s, there was further considerable interest in studying the transport of neutrons through solid matter, and Stanislaw Ulam is credited with developing the modern Markov chain approach to Monte Carlo simulations, which was named by his colleague Nicolas Metropolis [2]. Their colleague at Los Alamos, John von Neumann, began to use this approach with the new computer at the University of Pennsylvania for his work on neutron transport [3]. As Metropolis remarked [3], statistical approaches had fallen in disrepute because of their tediousness and slowness, but it was Ulam who recognized that the new computers could change this completely.

In modern parlance, there have emerged two distinct types of Monte Carlo. In one form, the Monte Carlo sampling technique is used to minimize a statistical function, such as the energy in a large ensemble. This has given rise to approaches such as simulated annealing [4] and quantum Monte Carlo [5]. The second type is the so-called kinetic Monte Carlo, in which the statistical physics of moving particles is studied, as was done with the earlier neutron studies. This latter approach was applied to electron transport in semiconductors by Kurosawa [6]. It was then developed further by Jacoboni [7], and others [8, 9]. This approach has also been extended so that the simulation includes the detailed band structure [10]. It is this

doi:10.1088/978-0-7503-1671-2ch6

kinetic Monte Carlo that is generalized to simulate carrier transport for quantum systems [11, 12].

6.1 The classical Monte Carlo technique

Most of the analytical methods used in semiconductor physics are quite difficult to evaluate carefully in the real situation of a semiconductor with non-parabolic energy bands and several complicated scattering processes. An alternative approach is to use the computer to completely solve the transport problem with a stochastic methodology fully in keeping with the retarded Langevin equation for carrier dynamics within an ensemble of carriers. The ensemble Monte Carlo (EMC) technique has been used now for more than five decades as a numerical method to simulate far-from-equilibrium transport in semiconductor materials and devices. It has been the subject of many reviews [13–15]. There is an important point about this approach, and that is the method of Monte Carlo evaluation of the physical averages can be misleading. The first simulations [6] used a single particle trajectory to estimate the physical averages in given sub-domains of the phase space by the relative time spent by the particle in the given sub-domain during the course of the simulation. This approach using temporal recording was initially based upon the ergodicity of the system. Later, it was shown that this time recording method, and thus the single particle Monte Carlo, rely upon the assumption of a stationary system [16]. However, there is no ergodic theory in the non-equilibrium world [17]. In the transient domain, a time average will not give the correct answer, if for no other reason than that the distribution is not stationary. Indeed, in order to study the evolution of the carrier system, one needs to do ensemble averages, and this provides the need to use the EMC approach. There is a second point. Many people believe that the EMC approach actually solves the Boltzmann equation, but this is true only in a long-time limit. For short times, the EMC is actually a more exact approach to the problem [18]. However, we will not dwell on this point here. The EMC is thus widely accepted as a simulated experiment based upon a numerical method for solving the Boltzmann transport equation.

The EMC is built around the general Monte Carlo technique, in which a random walk is generated to simulate the stochastic motion of particles subject to collision processes. These collisions provide both the momentum relaxation process and the random force that would appear in a retarded Langevin equation. Random walks and stochastic techniques may be used to evaluate complicated multiple-dimensional integrals [19]. In the Monte Carlo transport approach, we simulate the basic free flight of a carrier, and randomly interrupt this flight with instantaneous scattering events which shift the momentum (and energy) of the carrier. Here, the length of each free flight and the selection of the appropriate scattering process are both selected by weighted probabilities, with the weights adjusted according to the physics of the transport process, as we will describe below. The scattering process is determined by the Fermi golden rule terms $S_i(\mathbf{k}, \mathbf{k}')$ giving the rate for a transition from a wave vector $\mathbf{k}$ to the wave vector $\mathbf{k}'$. Here, i denotes the particular mechanism responsible for the scattering: phonons, impurities, and so on. These are characterized by the particular scattering out of state $\mathbf{k}$: $\Gamma_i(\mathbf{k}) = \int S_i(\mathbf{k}, \mathbf{k}')\, d^3\mathbf{k}'$. Here we assume that the different scattering rates are uncorrelated with one another, so that the total rates for scattering and out-scattering are $S = \sum_i S_i$ and $\Gamma = \sum_i \Gamma_i$, respectively. The following rate then gives the connection between the transition rates and the probabilities that occur in the Monte Carlo as

$$1 = \sum_i \frac{\Gamma_i(\mathbf{k})}{\Gamma(\mathbf{k})} \frac{\int S_i(\mathbf{k}, \mathbf{k}')\, d^3\mathbf{k}'}{\Gamma_i(\mathbf{k})}. \tag{6.1}$$

These probabilities for selection of the post-scattering final state are selected by the weighted probabilities for selection of a given scattering mechanism. In this way, very complicated physics can be introduced without any additional complexity of the formulation (albeit at much more extensive computing time in most cases). At appropriate times through the simulation, averages are computed to determine quantities of interest, such as the drift velocity, average energy, and so forth. By simulating an ensemble of carriers, rather than the single carrier normally used in a Monte Carlo procedure, the non-stationary time-dependent evolution of the carrier distribution and the appropriate ensemble averages can be determined quite easily. In the following paragraphs, we will outline the general approach for the EMC procedure.

As mentioned above, the dynamics of the particle motion is assumed to consist of free flights interrupted by instantaneous scattering events. The latter change the momentum and energy of the particle according to the physics of the particular scattering process. Of course, we cannot know precisely how long a carrier will drift before scattering, as it continuously interacts with the lattice and we only approximate this process with a scattering rate determined by first-order time-dependent perturbation theory [18]. Within our approximations, we may simulate the actual transport by introducing a probability density $P(t)$, where $P(t)\, dt$ is the joint probability that a carrier will both arrive at time t without scattering (after its last scattering event at $t = 0$), and then will actually suffer a scattering event at this time (i.e. within a time interval dt centered at t). The probability of actually scattering within this small time interval at time t may be written as $\Gamma[\mathbf{k}(t)]\, dt$, where $\Gamma[\mathbf{k}(t)]$ is the total scattering rate of a carrier of wave vector $\mathbf{k}(t)$. This scattering rate represents the sum of the contributions of each scattering process that can occur for a carrier of this wave vector (and energy). The explicit time dependence indicated is a result of the evolution of the wave vector under any accelerating electric (and magnetic) fields. In terms of this total scattering rate, the probability that a carrier has not suffered a collision after time t is given by

$$\exp\left[-\int_0^t \Gamma(\mathbf{k})\, dt'\right], \tag{6.2}$$

where, as mentioned above, the momentum $\mathbf{k}$ is a function of the time within the integral. Thus the probability of scattering within the time interval dt after a free-flight time t, measured since the last scattering event, may be written as the probability

$$P(t)\,dt = \Gamma[\mathbf{k}(t)]\exp\left[-\int_0^t \Gamma(\mathbf{k})\,dt'\right]dt. \tag{6.3}$$

Now, to understand how (6.1) is arrived at, we can begin with the Hamilton's equations of motion for a particle:

$$\frac{\partial \mathbf{p}_i}{\partial t} = -\frac{\partial H}{\partial \mathbf{x}_i}, \quad \frac{\partial \mathbf{x}_i}{\partial t} = \frac{\partial H}{\partial \mathbf{p}_i} = \mathbf{p}_i. \tag{6.4}$$

In these equations, $\mathbf{x}_i$ and $\mathbf{p}_i$ are the position and momentum ($= \hbar\mathbf{k}_i$) of the ith particle. Our interest is in the first of these equations. Here, the Hamiltonian H must include the scattering processes that exist in the system as well as the accelerating electric field. We will deal with these scattering processes in the next chapter, but here they are described by the factor $\Gamma[\mathbf{k}(t)]$ so that we may write the first equation of (6.3), with the field included, as

$$\frac{\partial \mathbf{p}_i}{\partial t} = -e\mathbf{E} - \Gamma(\mathbf{k})\mathbf{p}_i. \tag{6.5}$$

Thus, Γ enters the effective Hamiltonian as a result of the Hamiltonian terms describing the scattering processes. There is an underlying assumption to this equation in which it is tacitly assumed that each and every scattering event is completely localize in space and occurs at the corresponding position $\mathbf{x}_i$. Once again, we note that $\Gamma[\mathbf{k}(t)]$ is not an inverse 'momentum relaxation time', but is the sum of all possible scattering events. Equation (6.4) can now be rewritten, after an integration, as

$$\mathbf{p}_i(t) - \mathbf{p}_i(0) = -\frac{e\mathbf{E}}{\Gamma(\mathbf{k})}\left\{1 - \exp\left[-\int_0^t \Gamma(\mathbf{k})\,dt'\right]\right\}. \tag{6.6}$$

Thus, we find that (6.1) is recovered as an integral part of the accelerative process in the semiconductor. We will return to this important point shortly.

6.1.1 The path integral

To understand better how the above probability keys the EMC process, we want to write the Boltzmann equation in terms of a path integral as a method to illustrate the steps in the EMC process. In this, the streaming terms on the left-hand side will be written as partial derivatives of a general derivative of the time motion along a 'path' in a 6-dimensional phase space; this is then used to develop a closed-form integral equation for the distribution function. This integral has itself been used to develop an iterative technique, but provides one basis of the connection between the Monte Carlo procedure and the Boltzmann equation. In the following, we will switch from using the wave vectors $\mathbf{k}$ to the actual momentum $\mathbf{p}$, in order to better correlate with the formulation of the Wigner equation of motion. To begin, the Boltzmann equation is written as

$$\left(\frac{\partial}{\partial t} + \mathbf{v} \cdot \nabla + e\mathbf{E} \cdot \frac{\partial}{\partial \mathbf{p}}\right) f(\mathbf{r}, \mathbf{p}, t) = -\Gamma_0(\mathbf{k}) f(\mathbf{r}, \mathbf{p}, t) + \int d^3\mathbf{p}' S(\mathbf{p}, \mathbf{p}') f(\mathbf{r}, \mathbf{p}', t), \tag{6.7}$$

where

$$\Gamma_0(\mathbf{k}) = \int d^3\mathbf{p}' S(\mathbf{p}', \mathbf{p}) \tag{6.8}$$

is the total scattering rate *out of* the state $\mathbf{p}$. In (6.7), the right-hand side is made up of the scattering out of this latter state as well as the scatting *into* this state from other states, with this last term weighted by the distribution function at the other state. We refer to these two terms as *out*-scattering and *in*-scattering.

At this point, it is convenient to transform to a variable that describes the motion of the distribution function along a trajectory in phase space. It usually is difficult to think of the motion of the distribution function, but perhaps easier to think of the motion of a typical particle that characterizes the distribution function. For this, the motion is described in a six-dimensional phase space, which is sufficient for the one-particle distribution function being considered here [20]. The coordinate along this trajectory is taken to be s, and the trajectory is rigorously defined by the semi-classical trajectory, which can be found by any of the techniques of classical mechanics (i.e. it corresponds to that path which is an extremum of the action). It is as easy to remember, however, that it follows Newton's laws, and Hamilton's equation (6.4), where the forces arise from all possible potentials—including any self-consistent ones that arise from device simulations. Each normal coordinate can be parameterized as a function of this variable as

$$\mathbf{r} \rightarrow \mathbf{x}^*(s), \quad \mathbf{p} = \hbar\mathbf{k} \rightarrow \mathbf{p}^*(s), \quad t \rightarrow s. \tag{6.9}$$

Then, the partial derivatives appearing in (6.7) may be redefined as

$$\mathbf{v} = \frac{d\mathbf{x}^*}{ds}, \quad e\mathbf{E} = \frac{d\mathbf{p}^*}{dt}, \quad \mathbf{p} = \mathbf{p}^*(t), \quad \mathbf{p}^*(t) = \mathbf{p} + \int_s^t e\mathbf{E}(t')\, dt', \tag{6.10}$$

where the last term is the equivalent integral form of the differential equation initialized by the momentum $\mathbf{p}$ at the start of the drift evolution. Furthermore, we retain only the relevant variables that are necessary for our consideration and assume also that the out-scattering rate is a constant $\Gamma = \Gamma_0$, which may be justified with the introduction of the self-scattering process discussed more fully in the next section. In this way, (6.7) can be rewritten as

$$\frac{df}{ds} + \Gamma_0 f = \int d^3\mathbf{p}'^* S(\mathbf{p}, \mathbf{p}'^*) f(\mathbf{p}'^*, \mathbf{x}^*, t). \tag{6.11}$$

The second term on the left creates the integration factor by which (6.11) can be readily solved. After integration over the variable s in the limits $(0,t)$, we obtain the following result:

$$f(\mathbf{p}^*, t) = f(\mathbf{p}^*, 0)e^{-\Gamma_0 t} + \int_0^t ds \int d^3\mathbf{p}'^* S(\mathbf{p}^*, \mathbf{p}'^*) f(\mathbf{p}'^*, t)e^{-\Gamma_0(t-s)}, \tag{6.12}$$

where $f(\mathbf{p}^*, 0) = f_0(\mathbf{p}, 0)$ is the initial condition and is determined by the equilibrium distribution function prior to application of the electric field and, if we restore the time variables appropriate to the laboratory coordinates, we arrive at

$$f(\mathbf{p}, t) = f_0(\mathbf{p}, 0)e^{-\Gamma_0 t} + \int_0^t ds \int d^3\mathbf{p}' S(\mathbf{p}, \mathbf{p}') f(\mathbf{p}', t)e^{-\Gamma_0(t-t')}. \tag{6.13}$$

If we introduce the normalization of the scattering processes, given in (6.1), we may rewrite this as

$$f(\mathbf{p}, t) = f_0(\mathbf{p}, 0)e^{-\Gamma_0 t} + \int_0^t ds \int d^3\mathbf{p}' [\Gamma_0 e^{-\Gamma_0(t-t')}] \left\{ \frac{S(\mathbf{p}, \mathbf{p}')}{\Gamma_0} \right\} f(\mathbf{p}', t). \tag{6.14}$$

Note that the appearance of the probabilities in the integral form, together with the fact that the distribution within the integral on the right-hand side involves the number of particles in the simulation phase space, allows one to give a phenomenological interpretation to the transport process described by (6.14). The first term gives the contribution of the particles evolving from their initial conditions. This term describes the correlation of the evolving distribution with the initial distribution. Thus, it requires a period of time before the memory of the initial state disappears from the evolving distribution. The second term accounts for the build-up of the new, final state as the particles evolve. These particles are scattered at times t' to the proper trajectory, and then are scattered out at later times.

The use of the constant scattering rate Γ_0 is not required, and (6.13) can be derived for the general case. In particular, the exponentials become

$$\begin{aligned} e^{-\Gamma_0 t} &\rightarrow e^{-\int_0^t \Gamma_0[\mathbf{k}(t')]dt'} \\ e^{-\Gamma_0(t-t')} &\rightarrow e^{-\int_{t'}^t \Gamma_0[\mathbf{k}(t'')]dt''}. \end{aligned} \tag{6.15}$$

As the momentum and position evolve with time, these can be difficult integrals to evaluate, especially with multiple energy-dependent scattering rates.

The integral equation (6.13) involves two different processes described by the terms in the brackets of (6.14). The first is the acceleration and drift of the particles, which is governed by the exponential term. This acceleration and drift is interrupted at a time determined by the probability (6.2). The second process is the scattering process by which the carriers described by $f(\boldsymbol{p}')$ are scattered (by one of the processes within S). These are the two parts of the Monte Carlo algorithm, and it is from such an integral that we recognize that the Monte Carlo method is merely evaluating the two integrals stochastically. The problem with it is that there is no retardation in the scattering process, so that the scattering rate and energy are supposed to respond instantaneously to changes in the momentum along the path $\mathbf{p}' - e\mathbf{F}t'$. That is, the number of particles represented by the distribution function within the integral instantaneously responds during the previous drift. In essence, this is the Markovian assumption, and is true only in the long-time limit.

6.1.2 Free-flight generation

As mentioned above, the dynamics of the particle motion is assumed to consist of free flights interrupted by instantaneous scattering events. The latter change the momentum and energy of the particle according to the physics of the particular scattering process. Of course, we cannot know precisely how long a carrier will drift before scattering as it continuously interacts with the lattice and we only approximate this process with a scattering rate determined by first-order time-dependent perturbation theory. Within our approximations, we may simulate the actual transport by introducing the probability density $P(t)$, given by (6.2). Random flight times may now be generated according to the probability density $P(t)$ by using, for example the pseudo-random number generator available on nearly all modern computers and which yields random numbers in the range [0,1]. Using a simple, direct methodology, the random flight time is sampled from $P(t)$ according to the random number r as

$$r = \int_0^t P(t')\ dt'. \tag{6.16}$$

Since the random number r is uniformly distributed between 0 and 1, we can of course use $1 - r$ instead, and this leads to

$$-\ln(r) = \int_0^t \Gamma[\mathbf{k}(t')]\ dt'. \tag{6.17}$$

A different random number is generated for each particle in the ensemble, and this leads to a different scattering time for each particle.

The set of equation (6.15) that result for the ensemble of particles are the fundamental equations used to generate the random free flight for each carrier in the ensemble. If there is no accelerating field, the time dependence of the wave vector vanishes, and the integral is trivially evaluated. In the general case, however, this simplification is not possible, and it is expedient to resort to another *trick*. Here, we will introduce a fictitious scattering process that has no effect on the carrier. This process is called *self-scattering*, and the energy and momentum of the carrier are unchanged under this process [21]. However, we will assign an energy dependence to this process in just such a manner that the total scattering rate is a constant, as

$$\Gamma_{self}[E(\mathbf{k})] = \Gamma_0 - \Gamma[E(\mathbf{k})], \tag{6.18}$$

where Γ is the is the total scattering rate *out of* the state $\mathbf{p}$ due to all of the real scattering processes and introduced in (6.9). Since the self-scattering process has no effect upon the carrier, it will not change the observable transport properties at all, but its introduction eases the evaluation of the free-flight times, as now the integral in (6.17) becomes trivial to evaluate and leads to

$$t = -\frac{1}{\Gamma_0} \ln(r). \tag{6.19}$$

The constant total scattering rate Γ_{max} is chosen *a priori* so that it is larger than the maximum scattering encountered during the simulation interval. In the simplest case, a single constant is used globally through the simulation (constant gamma method), although other schemes have been suggested that modify the value of Γ_0 at fixed time increments in order to become more computationally efficient.

6.1.3 Final state after scattering

We now consider the next step of the simulation. This is the scattering process appearing in (6.14). We consider that a typical electron arrives at time t (arbitrarily selected arbitrarily the methods of the previous paragraph) in a state characterized by momentum $\mathbf{p}_a$, position $\mathbf{x}_a$, and energy E. At this time, the duration of the accelerated flight has been determined from the probability of not being scattered, given above with a random number r_1, which lies in the interval [0, 1]. At this time, the energy, momentum, and position are updated according to the energy gained from the field during the accelerative period to the values mentioned above—that is, they gain a momentum and energy according to their acceleration in the applied field during the time t. Once these new dynamical variables are known, the various scattering rates can now be evaluated for this particle (in practice, these rates are usually stored as a table to enhance computational speed). A particular rate is selected as the germane scattering process according to a second random number r_2, which is used in the following approach. All scattering processes are ordered in a sequence with process 1, process 2, ..., process $n - 1$, and finally the self-scattering process. The ordering of these processes does not change during the entire simulation. Hence, at time t we can use this new random number r_2 to select the process according to

$$\sum_{i=1}^{s-1} \Gamma_i[E(\mathbf{k})] < r_2\Gamma_0 < \sum_{i=1}^{s} \Gamma_i[E(\mathbf{k})]. \tag{6.20}$$

In this way, process s is selected. Then, the energy and momentum conservation relations for scattering process s are used to determine the post-scattering momentum and energy $\mathbf{p}_2$ and E_2 (that is, $E_2 = E \pm \hbar\omega_0$, depending upon whether the process is absorption or emission, respectively, and the momentum is suitably adjusted to account for the phonon momentum). We may note from (6.20), that the use of the normalized scattering rates that appear in (6.14) makes the selection process of (6.20) more efficient by eliminating the necessary multiplication between the two inequality symbols.

Additional random numbers are used to evaluate any individual parts of the momentum that are not well defined by the scattering process, such as the angles ϑ,

φ associated with the process. For example, there are many processes which are isotropic, e.g. any momentum state lying on the constant energy surface E_2 is equally likely. In this case, the two angles defining the actual momentum state are selected by two additional random numbers r_3 and r_4, which are used to choose the angles as

$$\vartheta = r_3\pi, \qquad \varphi = 2r_4\pi \tag{6.21}$$

in three dimensions (spherical coordinates for the momentum). However, if the scattering is anisotropic, as is the case for processes which are Coulombic in nature, such as impurity and polar optical phonon scattering. In these anisotropic processes, the polar angle is well defined by the $1/q$ variation of the matrix element. On the other hand, the azimuthal angle φ remains random and is not specified by the matrix element, so that φ is randomly selected by (6.21) as $2\pi r_4$. The third random number is now used to select the polar angle, according to the distribution of these angles. Let us consider once more the polar optical scattering as an illustration. The probability of scattering through a polar angle ϑ is provided by the square of the matrix element weighted delta function, which gives the angular probability to be proportional to $1/q^2 = 1/|\mathbf{k}_2 - \mathbf{k}_a|^2$. This is just the un-normalized function

$$P(\vartheta) = \frac{\sin\vartheta}{2E_a \pm \hbar\omega_0 - 2\sqrt{E_a(E_a \pm \hbar\omega_0)}\cos\vartheta}, \tag{6.22}$$

for parabolic energy bands with ω_0 the radian frequency of the optical phonon involved. This distribution function is then used to select the scattering angle ϑ with the random number r_3 through the equation

$$r_3 = \frac{\int_0^{\vartheta} P(\vartheta')d\vartheta'}{\int_0^{\pi} P(\vartheta')d\vartheta'} = \frac{\ln\left(\dfrac{1 - \xi\cos\vartheta}{1 - \xi}\right)}{\ln\left(\dfrac{1 + \xi}{1 - \xi}\right)}, \tag{6.23}$$

where

$$\xi = \frac{2\sqrt{E_a(E_a \pm \hbar\omega_0)}}{\left(\sqrt{E_a} - \sqrt{E_a \pm \hbar\omega_0}\right)^2}. \tag{6.24}$$

Finally, this last expression can be inverted to yield the actual scattering angle selected by this random number as

$$\vartheta = \cos^{-1}\left[\frac{(1 + \xi) - (1 - \xi)^{r_3}}{\xi}\right]. \tag{6.25}$$

This approach is easily extended to non-parabolic bands, but of course becomes a little more complicated. There is a word of caution here for the choice of this polar angle in the anisotropic situation. In computing ϑ, it has been assumed that $\mathbf{k}_a$ defines the polar axis (z direction in momentum). But, this is not the case for the simulation. As a result, the new wave vector $\mathbf{k}_2$ has to be rotated into the laboratory coordinates of the simulation. The angles required for this coordinate rotation are given by the corresponding angles of $\mathbf{k}_a$ defined at the end of the accelerative motion.

The final set of dynamical variables obtained after completing the scattering process are now used as the initial set for the next iteration, and the process is continued for several hundred thousand cycles. This particular algorithm is one that is amenable to full vectorization and is relatively computationally efficient, in that it can utilize the hardware 'scatter-gather' routines available on large vector computers. On high-speed work stations, though, such subtleties are not necessary and the program is quite efficient on the pipelined architecture of most PCs; indeed, modern compilers provide both excellent parallelization on even modest desktop computers these days. In one general variant, the program begins by creating the large scattering matrix in which all of the various scattering processes are stored as a function of the energy; that is, this scattering table may be set up with 1 meV increments in the energy. This includes the self-scattering process. The energy is discretized, and the size of each elemental step in energy is set by the dictates of the physical situation that is being investigated. The initial distribution function is then established—the N electrons actually being simulated are given initial values of energy and momentum corresponding to the equilibrium ensemble, and they are given initial values of position and other possible variables corresponding to the physical structure being simulated. At this point, $t = 0$. Part of the initialization process is also to assign to each of the N electrons a t_i according to (6.19), which is its individual time at which it ends its free flight and undergoes scattering. Then each electron undergoes its free flight and a scattering process, which may be self-scattering. New times are selected for each particle and the process is repeated as long as desired.

One of the advantages of the EMC approach is that we have the multiple particle situation. As a result, it is also possible to incorporate the inter-carrier forces by a molecular dynamics interaction (we come to this later), the initial values of these forces, corresponding to the initial distributions in space, are also computed. Thus it is possible to do a full interacting many-body simulation for the distribution.

6.1.4 Time synchronization

The key problem in treating an ensemble of particles is that each particle has its unique timescale. However, we want to compute ensemble averages for such quantities as the drift velocity and average energy, with the former defined as

$$\mathbf{v}_d(t) = \frac{1}{N}\sum_{i=1}^{N}\mathbf{v}_i(t). \tag{6.26}$$

To achieve accurate results, all the particles need to be aligned at the same time t, which here runs from the beginning of the simulation. Thus, we need to overlay the system with a global timescale, with which each local particle timescale can be synchronized. In practice, this is achieved by introducing a global time variable T, which is discretized into steps as $n\Delta T$. Then at integer multiples of this time step, all particles are stopped in their free flights, and ensemble averages are computed [16].

Each particle has its own time path. It is composed of accelerations and scattering processes. There are N of these time lines all running at the same time. But, we have to introduce 'pauses' $n\Delta T$, at which the entire ensemble is halted and averages computed. In general, these pauses will always occur during the free flights, but it means that we have to keep our accounting carefully in order to properly align these two timescales. This is achieved once the new free-flight time is determined by checking whether or not this free flight will extend beyond the time step boundary. If so, it is broken into two parts so as to give the pause at the time step boundary. This increases the book-keeping but not the difficulty.

If one is incorporating nonlinear effects, such as molecular dynamics for the inter-carrier interactions, or non-equilibrium phonons or degeneracy induced filling of the final states after scattering (next section), then these processes are updated on the pauses of the T timescale as well. In this sense, the imposition of the second timescale synchronizes the distribution and gives the global, or laboratory, timescale of interest in experiments.

6.1.5 Rejection techniques for nonlinear processes

In the case of polar optical phonon scattering, it was possible to actually integrate the angular probability function (6.22). This is not always the case, and one has therefore to resort to other statistical methods. One of these is the so-called rejection technique. Suppose the probability density function for the process, such as (6.22), is quite nonlinear and not easily integrated to get the total probability. Then one can use a pair of random numbers (r_5, r_6) to evaluate the angle. Consider figure 6.1, in which we plot a complicated probability density function. We assume that the maximum coordinate x is unity, so that the range of the function's argument is from zero to one (one can easily use other values, such as π, by proper normalization). The maximum value of the function is also set near unity, and one can always renormalize this to the span of the random numbers. Now, the first random number

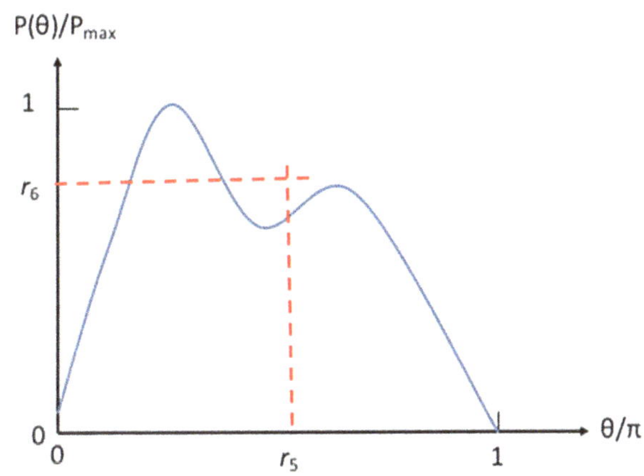

Figure 6.1. The rejection technique for finding the scattering angle for a complicated function of the angle.

r_5 is taken to correspond to the span of the function (the x axis). This determines the argument of the function that is to be evaluated. For example, let us assume that this is $r_5 = 0.65$. We now use the second random number to determine whether $f(r_5) > r_6$. If this relationship holds, then the value r_5 is accepted for the argument of the function, and the scattering process proceeds with this value. Certainly, values of r_5 for which the function is large are more heavily weighted in this rejection process. We consider this in more detail for two important processes: (1) state filling due to the degeneracy of the electron gas, and (2) non-equilibrium phonons.

Degeneracy and Fermi–Dirac statistics have been introduced through the concept of a secondary self-scattering process [22, 23] based upon a rejection technique. We call it secondary self-scattering, because if the condition $f(r_5) > r_6$ is not satisfied, we treat the rejection exactly as a self-scattering process, which was introduced earlier. Each of the scattering processes must include a factor of $[1 - f(E)]$, where $f(E)$ is the dynamic distribution function and represents the probability that the final state after scattering is empty. Rather than recompute the scattering rates as the distribution function evolves in order to incorporate the degeneracy, all scattering rates are computed as if the final states were always empty. A grid in momentum space is maintained and the number of particles in each state is tracked (each cell of this grid has its population divided by the total number of states in the cell, which depends on the cell size, to provide the value of the distribution function in that cell) [23]. The scattering processes themselves are evaluated, but the acceptance of the process depends on the rejection technique. That is, an additional random number is used to accept the process if

$$r_5 < 1 - f(\mathbf{k}_2, t). \tag{6.27}$$

Note that this comparison is made after the final state of scattering is selected. Only then is the check made to determine if that state is available or is already filled. Thus, as the state fills, most scattering events into that state are rejected and treated as a self-scattering process.

The most delicate point of the degeneracy method involves the normalization of the distribution function $f(\mathbf{p})$. The extension of the secondary self-scattering method to the ensemble Monte Carlo algorithm involves the fact that there are N electrons in the simulation ensemble, which represent an electron density of n [23]. The effective volume V of 'real space' being simulated is then N/n. The density of allowed wave vectors of a single spin in $\mathbf{k}$-space is just $V/(2\pi)^3$. In setting up the grid in the three-dimensional wave vector space, the elementary cell volume is given by $\Omega_k = \Delta k_x \Delta k_y \Delta k_z$. Every cell can accommodate at most N_c electrons, with $N_c = 2\Omega_k V/(2\pi)^3$, where the factor of 2 accounts for the electron spin. For example, if the density is taken to be 10^{17} cm^{-3}, $N = 10^4$, and $\Delta k_x \Delta k_y \Delta k_z = (2 \times 10^5 \text{ cm}^{-1})^3$ ($k_F = 2.4 \times 10^6$ cm^{-1} at 77 K), then $V = 10^{-13}$ cm^3 and $N_c = 6.45$. N_c constitutes the maximum occupancy of a cell in the momentum space grid. (Obviously, a more careful choice of parameters would have N_c come out to be an integer, for convenience.) A distribution function is defined over the grid in momentum space by counting the number of electrons in each cell. The distribution function is normalized to unity by dividing the number in each cell by N_c for use in the rejection technique. It should be noted that N_c should be sufficiently large that round-off to an integer (if the numbers do not work out properly, as in the case above) does not create a significant statistical error.

A second usage is for the consideration of non-equilibrium phonon distributions [24]. In the derivations presented above, the assumption was made that the phonons are in equilibrium and characterized by N_q. However, under a number of circumstances, such as the excitation of the semiconductor by an intense laser pulse, the carriers are created high in the energy band, and then decay by a cascade of phonon emission processes. As a result of this cascade the phonon distribution is driven out of equilibrium, and this affects both the emission and absorption processes by which the carriers interact with the phonons. Here, we use **q**, rather than **k**, for the momentum of the phonons. Once again, the momentum space is discretized for the phonon distribution, so that an individual cell in this discretized space has volume $\Delta q_x \Delta q_y \Delta q_z$. This small volume has available a number of states given by $V/(2\pi)^3$, where V is determined by the effective simulation volume N/n, as previously. The difference between state filling for carrier degeneracy and phonon state filling is that there is no limit to the number of phonons that can exist within the state. The basic approach assumes that the phonons are out of equilibrium, and the carrier scattering processes are evaluated with an assumed $N_{max}(\mathbf{q})$. Then, within the simulation, the number of phonons emitted, or absorbed, with wave vector **q** is carefully monitored. At the synchronization times of the global timescale, the phonon population in each cell of momentum space is updated from the emission/absorption statistics that have been gathered during that time step. One must also include phonon decay, which is through a three-phonon process to other modes of the lattice vibrations, so that the update algorithm is simply

$$N(\mathbf{q}, t + \Delta t) = N(\mathbf{q}, t) + G_{net,\, \Delta t} - \left[\frac{N(\mathbf{q}, t) - N_{q0}}{\tau_{phonon}}\right]\Delta t, \tag{6.28}$$

where N_{q0} is the equilibrium distribution, $G_{net,\, \Delta t}(\mathbf{q})$ is the net (emission minus absorption) generation of phonons in the particular cell *during the time step*, and τ_{phonon} is the phonon lifetime. During the simulation, each phonon scattering process is evaluated as if the maximum assumed phonon population were present. Then, a rejection technique is used, by which the phonon scattering process is rejected (and assumed to be a secondary self-scattering process) if

$$r_5 > \frac{N(\mathbf{q}, t)}{N_{max}}, \tag{6.29}$$

where N_{max} is the peak value that was assumed in setting up the scattering matrices. While this is assumed here to be a constant for all phonon wave vectors, this is not required. A more sophisticated approach would use a momentum-dependent peak occupation.

6.2 Paths in quantum mechanics

The immediate problem with moving to quantum transport techniques lies with the use of the particle paths in phase space in the above ensemble Monte Carlo approach. With the orthodox interpretation of quantum mechanics, the uncertainty principle forbids us being able to simultaneously define both a position and momentum for a particle path. It is important to note, however, that this strict interpretation has been finessed by a variety of approaches, many of which clearly endorse some form of a real path for a particle that exists with the wave, as was discussed in chapter 2. In this section, we will discuss some of these paths, which will lead us to our goal of paths for the Wigner function and its interpretation via the ensemble Monte Carlo approach. Let us begin by discussing how such a path may arise merely from the modes in the structure.

As we remarked in chapter 2, Rolf Landauer presented an approach to transport and the calculation of conductance that was dramatically different from the microscopic kinetic theory based on the Boltzmann equation, which had been utilized previously (and is still heavily utilized in macroscopic conductors) [25, 26]. He suggested that one could compute the conductance of low-dimensional systems simply by computing the transmission of the transverse quantized modes from an input reservoir to a similar mode in an output reservoir. The transmission of this probability from one mode to the other was then very similar to the computation of the tunneling probability, except that there was no requirement that the process be one of tunneling. The only real constraint was that of lateral confinement so that the two reservoirs could be discussed in terms of their transverse modes given in (2.13). The key property of the two reservoirs is that they are in equilibrium with any applied potentials. That is, the electrons in the reservoirs are to be described by their intrinsic Fermi–Dirac distributions with any externally applied potentials appearing only as a shift of the relative energies (which would shift one Fermi level relative to the other). While he originally considered that the transport was ballistic, this is not required. Rather, the requirement is that we can assign a definitive mode to the electron when it is in either of the two reservoirs, which means that if scattering is present, it must be described specifically as a transfer of the electron from one internal mode to another within the active region and not in the reservoir. If we consider a potential applied to the right (output) reservoir relative to the left (input) reservoir, then the right reservoir emits carriers into the active region with energies up to the local Fermi level plus the applied bias, E_F+eV_a (note that the energy eV_a will be negative for a positive voltage). The left reservoir emits electrons into the active region with energies only up to E_F. Generally, we will assume that the applied voltage is quite small, although this also is not a stringent requirement.

Building upon the Landauer approach, Markus Büttiker began to address ac conductance in these quantum circuits with a discussion of noise in them [27]. He introduced the idea that particles in the incoming states could be created and annihilated by a set of operators $a^{\dagger}$ and a, respectively. Similarly, one can define a set of operators which create and annihilate a particle in the outgoing (or reflected) states $b^\dagger$ and b, respectively. Then, the scattering matrix provides a unique unitary transformation between the a operators and the b operators. We used this property in the development of section 2.2. As we did in this latter section, we define the scattering matrix $\mathbf{s}_{\alpha\beta}$ as the element of the scattering matrix which describes a wave entering the system from mode β of the input lead and is scattered back into mode α of this lead. Büttiker then determined that the equilibrium noise, equivalently the noise conductance, is determined by the quantity

$$\frac{e^2}{h}\mathrm{Tr}\left\{\mathbf{s}_{\alpha\beta}^\dagger\mathbf{s}_{\alpha\beta}\right\}. \tag{6.30}$$

These bilinear terms are identical to the coefficients determined in section 2.2 to describe the transport coefficients of the dc current. In a sense, this is a form of the Nyquist relation, or equivalently the Kubo formula, in which the fluctuations (noise, arising from a current–current correlation function) are related to the conductance (or current) through the structure.

The connection between the ac response (of the noise) and the dc conductance is interesting. Once we determine the conductance and the scattering matrix, the ac-transport properties are sensitive to the *phases* of the scattering matrix elements [28]. In the latter paper, it is shown that the conductance formula (2.24) can be extended to the frequency dependent conductance via the formula [29]

$$\begin{aligned} g_{\alpha\beta}^{(m)}(\omega) = \frac{e^2}{h}\int dE \cdot \mathrm{Tr}\left\{\mathbf{I}_\alpha^{(m)}(E)\delta_{\alpha\beta} - \mathbf{s}_{\alpha\beta}^{(m)\dagger}(E)\mathbf{s}_{\alpha\beta}^{(m)}(E+\hbar\omega)\right\} \\ \times\frac{f_\beta^{(m)}(E) - f_\beta^{(m)}(E+\hbar\omega)}{\hbar\omega}, \end{aligned} \tag{6.31}$$

where $\mathbf{I}$ is an identity matrix for lead m and mode α, and $f_\beta^{(m)}(E)$ is the Fermi–Dirac distribution appropriate for lead m and mode β, which would reflect the sub-band energy for this mode and the Fermi energy for this lead. In general, this conductance leads to a complex admittance, in which the imaginary part can be related to the effective capacitance (charge storage) of the circuit [30].

The fact that the ac transport is dispersive suggests that the corresponding time dependence is related to a propagation time for the mode through the nanostructure. Indeed, this time delay is referred to as the Wigner–Smith [31, 32] delay time, which may be expressed as [33]

$$\mathbf{Q} = -i\hbar\mathbf{s}^{-1/2}\frac{\partial\mathbf{s}}{\partial E}\mathbf{s}^{-1/2}, \tag{6.32}$$

which is a symmetrized form. A more common form is given as [34]

$$\mathbf{Q} = i\hbar\frac{\partial\mathbf{s}^\dagger}{\partial E}\mathbf{s}. \tag{6.33}$$

An important point is that these matrices are reduced from the full scattering matrix as they only include those modes which are propagating through the system. That is, they ignore the evanescent modes whose sub-band energies lie above the Fermi energy in that lead. Another form in which this delay time arises in a single channel (or mode) conductor is just the derivative of the phase of the scattering matrix element for that mode [35], as

$$\tau_d = \hbar\frac{\partial\varphi(E)}{\partial E}. \tag{6.34}$$

The delay time can be an important approach to gain more information about the properties of a nanostructure. In the conductance through a structure such as an open quantum dot, all transmitted modes contribute equally to the conductance. But if one is interested, for example, in isolating individual modes to determine their respective trajectory through the structure, then the delay time can be quite useful [36]. In particular, a subspace of modes, each of which enters through a single lead and also exits through a single lead, creates a noiseless subspace. These modes can then be separated by a delay time analysis. Rotter *et al* [36, 37] carry out this procedure for a square cavity, in which the leads are offset from one another. The delay time is used to separate these mode structures as each trajectory obviously spends a different amount of time within the cavity. The different states each correspond to a different initial condition in the input lead from which the wave is injected. With a high initial momentum, the wave is nearly classical, so that the results are quite similar to a classical orbit simulation. More information on the calculations can be found in Rotter *et al* [36]. Nevertheless, we reach an important consequence that, if the results are so similar to classical orbits through the structure, then it becomes possible to consider simulation of the mode transmission via semi-classical particle approaches. The hope would be to invert the delay time of (6.34) into a representation of a phase for the particle at a given time during the traversal. Hence, it would seem that there should be a connection between pure Schrödinger equation modes and representations of these by particles. We wish to pursue this idea in this section.

6.2.1 Bohm trajectories

We already noted in chapter 1 that there is a view of quantum mechanics that specifically incorporates particle trajectories. Madelung took careful note of Schrödinger's work, and immediately noted that the probability density had all the appearances of a fluid flow [38]. Kennard quickly learned of the new developments in quantum mechanics as well. At that time, he was interested in hydrodynamics, but quickly became intrigued with the new developments. In 1928, he published his work on the quantum mechanics of a system of particles, showing that the dynamics came directly from the Schrödinger equation [39]. Moreover, he found that the particles would follow normal Hamiltonian dynamics, although the potential would have to be modified through the addition of a quantum potential, discussed in chapter 4. As we noted, this potential is often called the Bohm potential following his resurrection of the Madelung–Kennard hydrodynamic ideas [38, 40]. Here, the Bohm potential provides an addition to the total energy, and provides a non-classical force which guides the wave functions in a self-consistent manner. This can provide a basis to use a particle representation, where the particles move through the presence of both the classical and the quantum forces [39].

Indeed, the use of the so-called Bohm trajectories has developed a rich history since that time [41]. As we noted in chapter 4, this approach has been applied to a study of a resonant tunneling diode (RTD) by Oriols *et al* [42]. The use of Bohm trajectories also has been used in chemistry [43, 44]. It can be shown that the Bohm trajectory approach agrees very well with use of the Schrödinger equation directly [42, 45] and with use of the Wigner phase space distribution function [46]. But what about the uncertainty relationship? Bohm addressed this, by describing the problem of ascertaining exactly the correct initial conditions for the particle trajectories. To Bohm, the uncertainty relation was a description only of the problem of establishing proper initial conditions. Particularly in nonlinear transport, it is known that minor differences in the initial conditions can lead to exponentially divergent results [47]. Wyatt has addressed this by the use of Monte Carlo sampling to establish the Bohmian trajectories [48]. From these approaches, it is hard not to draw the conclusion that Bohmian dynamics, and the corresponding interpretation of the physics, are a valid description of quantum mechanics. Hence, it is absolutely clear that the quantum potential is an essential part of the solutions to quantum mechanics.

6.2.2 Feynman paths

An alternate approach to paths in quantum mechanics seems to have been first suggested by Dirac [49]. Dirac noted that quantum mechanics had emerged from considerations of the Hamiltonian for classical mechanics. He wanted to pursue the alternate formulation of classical mechanics, the Lagrangian approach, and show that this also led to quantum mechanics. In this latter approach, one works with position and velocity rather than position and momentum, although the two approaches are obviously closely related to one another. His rationale, however, grew from the recognition that the Lagrangian method would allow one to collect all of the multi-particle equations of motion and 'express them as a stationary property of a certain action function'. So, Dirac would then introduce the set of classical paths and produce a sum over them, but he did not provide just how this would be accomplished. Nor did he show that this would lead to the Schrödinger equation or the commutation relationship. It was Feynman who made the leap forward in this regard [50]. Feynman showed that the quantum action was almost always just equivalent to the classical action, which becomes just the phase acquired by the quantum evolution of a 'particle' in moving from the initial point to the endpoint. Then, quantum mechanics could be recovered if a set of postulates were put forward. The first, like the Schrödinger equation, was that the probability would arise from the squared magnitude of a complex number. In the Schrödinger case, this was just the wave function itself. For Feynman, however, this complex number was given by summing all the paths through configuration space, and weighting each path by its action $\exp(iS/\hbar)$, where the action S is given as the time integral over the Langrangian function. The proper sum is found by integrating over these weights for each path that is possible. The exponential weighting by the action introduces the relative probability for a given path. The resulting integral is the Feynman path integral [51]. There is a cautionary part of this approach, and that is the need for the Hamiltonian to be independent of time so that equilibrium can be assumed to be established. Going away from this point creates difficulties. In addition, it is important to note that the Feynman path integral is a very different beast than the path integral discussed in section 6.1.1.

Nevertheless, it is clear that the use of classical paths can be brought forward to the study of quantum systems. Much like the previous section, there may be additional quantum potentials involved, such as the effective potential discussed in chapter 4. Nevertheless, an approach based upon 'particles' having clear and well-defined paths in space is a viable approach to the study of quantum mechanics. Mason and Hess have used the Feynman path integral to study hot carrier transport in semiconductors, a typical open quantum system [52]. In this study, they used non-perturbative approaches to incorporate both the electric field and the electron–phonon interaction. By using the Feynman–Vernon influence functional [53], they are able to integrate out the phonon coordinates and reach a tractable form for the path integral, and they are able to evaluate the resulting path integral by Monte Carlo techniques. Nevertheless, they are limited to times on the order of a few scattering processes due to the high dimensionality of the integrand, and the continuous growth in the number of paths required for the open system. The use of their approach has more in common with the kinetic Monte Carlo procedures discussed in section 6.1, rather than the so-called quantum Monte Carlo used in statistical minimization approaches incorporating a variational treatment.

Needless to say, many workers have tried to extend the Feynman path integral approach to the situation with Wigner functions. These have the same troubles as the Feynman path integrals when applied to open quantum systems and transport. We will not discuss these approaches here, as we want to focus on the Chambers path integrals such as (6.12) and the modifications that are necessary for the Wigner function. We turn to this in the next section.

6.2.3 Wigner paths

The development of the idea of paths with the Wigner function lies in the closeness of the Wigner equation of motion to the Boltzmann equation. If we compare the Wigner equation of motion (3.5) with the inclusion of scattering terms, which will be developed in the next chapter, with the Boltzmann equation (6.7), it is apparent that the major difference lies in the Wigner potential term (3.6). If we are to develop a Chambers-type path integral corresponding to (6.14), we have either to assume the potential is of less than quadratic form, or else we must separate out a classical-like term to force evolution along the phase space path. Let us rewrite this Wigner potential term in the equation of motion as [54]

$$W(\mathbf{x}, \mathbf{p}) = \frac{1}{h^3}\int d^3\mathbf{r}\, e^{-i\mathbf{p}\cdot\mathbf{r}}\frac{1}{i\hbar}\left[V\left(\mathbf{x} + \frac{\mathbf{r}}{2}\right) - V\left(\mathbf{x} - \frac{\mathbf{r}}{2}\right)\right]. \tag{6.35}$$

Now, it is clear that, if the potential is of quadratic or less order, then the appropriate term in the Wigner equation of motion can be written in the classical form

$$-e\mathbf{E}\cdot\frac{\partial}{\partial\mathbf{p}}f_W = -\nabla V_0\cdot\frac{\partial}{\partial\mathbf{p}}f_W. \tag{6.36}$$

In this last form, we define the classical-like part V_0 of the Wigner potential. Hence, this suggests to split the potential that appears in (6.35) as

$$V = V_0 + V', \tag{6.37}$$

where V' describes only the quadratic or higher-order terms of the potential. Then, we can rewrite the Wigner equation of motion in the form

$$\begin{aligned}\frac{\partial f_W}{\partial t} + \frac{\mathbf{p}}{m}\cdot\nabla f_W + e\mathbf{E}\cdot\frac{\partial f_W}{\partial\mathbf{p}} &= -\tilde{\Gamma}_0 + \frac{1}{h^3}\int d^3\mathbf{p}'\tilde{W}(\mathbf{x},\mathbf{p}-\mathbf{p}')f_W(\mathbf{x},\mathbf{p}')\\ &+\int d^3\mathbf{p}'\tilde{S}(\mathbf{p},\mathbf{p}')f_W(\mathbf{x},\mathbf{p}',t),\end{aligned} \tag{6.38}$$

where $\tilde{\Gamma}_0$ and $\tilde{S}$ are the Wigner–Weyl transformed versions of the scattering processes (which are discussed in the next chapter). The terms Γ_0 and S include the self-scattering terms and correspond to their earlier usage in this chapter. Here, the extra Wigner potential term now only includes the terms arising from V'. This extra term has the form of an extra scattering term which arises explicitly from the Wigner potential. In this sense, it has the form of an additional '*in*-scattering' term [54, 55]. This approach has been discussed more recently by Brosens and Magnus [56].

A somewhat different result can be obtained if we use the spectral methods discussed in section 5.2.4 [57]. If we pull the classical term from the Fourier transform (5.26), we can then write the new Wigner equation of motion as

$$\begin{aligned}\frac{\partial f_W}{\partial t} + \frac{\mathbf{p}}{m}\cdot\nabla f_W + e\mathbf{E}\cdot\frac{\partial f_W}{\partial\mathbf{p}} &= -\tilde{\Gamma}_0\\ &+\frac{1}{h^3}\int d^3\mathbf{k}\tilde{\mathbf{F}}_k\cdot\frac{\partial}{\partial\mathbf{p}}\int_{-1/2}^{1/2}d\alpha f_W(\mathbf{x},\mathbf{p}-\alpha\hbar\mathbf{k},t)\\ &+\int d^3\mathbf{p}'\tilde{S}(\mathbf{p},\mathbf{p}')f_W(\mathbf{x},\mathbf{p}',t).\end{aligned} \tag{6.39}$$

The difference here is that the classical force has been removed, prior to the Fourier transformation. This is signified by the tilde over the $\mathbf{F}$. One might at first think that this means only the $\mathbf{k} = 0$ term, but this would be a mistake. The average potential, which gives this $\mathbf{k} = 0$ Fourier component, produces no average force; rather, it is just an energy shift. Similarly, it is not the term linear in $\mathbf{k}$ that produces the classical force. Rather, the classical force is inhomogeneous, whereas the term linear in $\mathbf{k}$ is a constant electric field throughout the device. So, one has to proceed with care in using this spectral approach. Nevertheless, the spectral approach has already introduced an effective scattering process into the system, so that treating this extra term as a scatterer is already included in the concept.

So far, we have essentially not differentiated between Wigner trajectories and Wigner paths. Here, we note that the trajectory corresponds to evolutionary movement of the Wigner function, while the paths describe the motion of individual particles that make up the distribution function. The subtle difference has been discussed by Pascoli *et al* [58]. Let us assume, for the moment, that the initial Wigner function is composed of a summation over a set of particles, each of which is a δ-function in phase space. Then, these particles will follow a set of classical trajectories if the quantum potential contribution to (6.38) or (6.39) is not present. As a result, the Wigner function itself will follow the same path as the representative classical system, since the Wigner equation of motion now reduces to the ballistic Boltzmann equation. In essence, we are saying that the composite (initial) Wigner function is composed of a set of δ-function individual particle Wigner functions whose evolutionary Wigner path is the classical trajectory. This, of course, changes when collisions are present. Between the collisions, the classical paths still survive, but the transitions that occur during the collisions can be distinct from those of the classical system. Hence, a major deviation from classical behavior occurs during the collision. This is an important point, and arises whether the 'collision' is a true collision with a phonon or is an effective collision arising from the quantum potential terms. As mentioned above, we return to a detailed treatment of the collisions in the next chapter.

The idea that the Wigner function can be a summation of classical functions was actually proposed earlier [59]. This approach follows in a similar manner as the later spectral approach discussed above and in the previous chapter. In this approach, we begin once more with (3.5) and (3.6). The potential term is defined in terms of a continuous set of quasi-classical type forces **F(x,s)** through the relation

$$\mathbf{F}(\mathbf{x},\mathbf{s})\cdot\mathbf{s} = V\left(\mathbf{x}-\frac{\mathbf{s}}{2}\right) - V\left(\mathbf{x}+\frac{\mathbf{s}}{2}\right). \tag{6.40}$$

In the limit $\mathbf{s}\to 0$, we recover the classical force. Now, we may write (3.5) as

$$\begin{aligned}&\left(\frac{\partial}{\partial t}+\frac{\mathbf{p}}{m}\cdot\nabla\right)f_W(\mathbf{x},\mathbf{p},t)\\&+\frac{1}{h^3}\int\int d^3\mathbf{p}'d^3\mathbf{s}\,\cos\left(\frac{\mathbf{s}\cdot\mathbf{p}'}{\hbar}\right)F(\mathbf{x},\mathbf{s})\cdot\frac{\partial f_W(\mathbf{x},\mathbf{p}+\mathbf{p}',t)}{\partial\mathbf{p}'}=0\end{aligned} \tag{6.41}$$

in the absence of collisions. To reach this result, we have used the generalized divergence theorem:

$$\int d^3\mathbf{p}'A(\mathbf{p}')\cdot\frac{\partial\varphi(\mathbf{p}')}{\partial\mathbf{p}'} = -\int d^3\mathbf{p}'\varphi(\mathbf{p}')\frac{\partial}{\partial\mathbf{p}'}\cdot A(\mathbf{p}') + \oint d^2\mathbf{p}'A\varphi, \tag{6.42}$$

which allows us to reach the gradient of the Wigner function on the primed momentum (the surface integral vanishes). This unified formulation now allows one to define the generalized quantum Liouville operator $\mathrm{L_Q}$ as

$$\mathcal{L}_Q = \frac{\partial}{\partial t}+\frac{\mathbf{p}}{m}\cdot\nabla+\mathbf{F}(\mathbf{x},\mathbf{s})\cdot\frac{\partial}{\partial\mathbf{p}'} = \frac{\partial}{\partial t}+S_Q. \tag{6.43}$$

Using this, (6.38) can be rewritten as

$$\iint \frac{d^3\mathbf{p}' d^3\mathbf{s}}{h^3} e^{i\mathbf{p}'\cdot\mathbf{s}/\hbar} \mathcal{L}_Q f_W(\mathbf{x}, \mathbf{p}+\mathbf{p}', t) = 0. \tag{6.44}$$

If we regard $\mathbf{p}'$ and $\mathbf{s}$ as parameters, then we have just the classical Liouville equation for an evolving distribution function in the force field, that then satisfies the evolution equation

$$g(\mathbf{x}, \mathbf{p}, t) = \hat{T} \exp\left(-\int_0^t dt' \, S_Q(t')\right) g(\mathbf{x}, \mathbf{p}, 0) \tag{6.45}$$

for $t > 0$. This result is basically that of the interaction representation, which is discussed further in chapter 7; moreover, there is a factor of $i/\hbar$ that appears in the interaction representation, but which is considered to be part of S_Q in (6.45). Here, $\hat{T}$ is the Dyson time-ordering operator. The time dependence of the linear operator in the exponential function derives solely from the time dependence of the potential itself as it evolves in the self-consistent process. This suggests that the formal solution of the Wigner function evolution is given by

$$f_W(\mathbf{x}, \mathbf{p}, t) = \iint \frac{d^3\mathbf{p}' d^3\mathbf{s}}{h^3} \cos\left(\frac{\mathbf{p}'\cdot\mathbf{s}}{\hbar}\right) g(\mathbf{x}, \mathbf{p}+\mathbf{p}', t), \tag{6.46}$$

where g is the classical distribution which evolves from the initial Wigner distribution via the classical local Liouville equation. Thus, (6.46) tells us that the quantum Wigner function evolves from a set of classical distributions. Each classical distribution evolves under one of the forces $\mathbf{F}(\mathbf{x},\mathbf{s})$ which are an unfolding of the total force generated from the potential according to (6.40).

While the above treatment, which is due to Barker and Murray [59], seems to be very attractive, it has been claimed that the approach really only works for the case of linear and quadratic potentials [60]. In this latter case, the evolution is said to be truly classical, so that the approach is incomplete if applied to more extensive potentials. The latter authors have then generated a new effective Lagrangian operator in which derives their results from a function integral of the Feynman type. Nevertheless, this allows one to continue in the spirit of the above equations. Barker [61] has addressed some of the concerns with a slightly different connection between the classical distribution and the quantum Wigner function. Whether or not these approaches can truly provide the basis for simulation of quantum phase space functions remains an open question that needs much further investigation. In a sense, part of the issue is that of the non-positive parts of the Wigner function, as addressed in the next section.

6.3 Using particles with the Wigner function

As we have seen, the phase space representation of the Wigner function facilitates the comparison between complex quantum dynamics and the corresponding classical motion [62, 63]. Because the density matrix from which the Wigner function is described can have a non-unitary evolution and off-diagonal terms in the density matrix, it can readily lead to non-classical propagation of the Wigner function. Indeed, quantum coherence is reflected in oscillatory behavior of the Wigner function and non-positive-definite regions in phase space [64]. It has been shown that the classical δ-function representation of the particle evolves in an extended region of phase space, like a 'spot', that has its own evolutionary behavior and exhibits oscillatory Airy function interference fringes [63, 65]. If pairs of trajectories are considered [66, 67], then it was shown that the oscillatory behavior can be considered as arising from interference between these pairs of trajectories. The inclusion of these oscillatory, non-positive parts of the Wigner function are a challenge to the use of normal particles and must be faced. In this section, we explore some methods of ensuring that these non-positive-definite portions of the evolving Wigner function are properly represented.

6.3.1 Weighted Monte Carlo

The weighted Monte Carlo approach was formally introduced in 1992 [68]. The paths considered for this approach are based upon a series expansion of the Chambers integral (6.12), or the equivalent path integral for the Wigner function. The problem that they addressed was the fact that there are regions in phase space in which the distribution function has little weight. This means that the particles in the Monte Carlo simulation have low probabilities of reaching these areas and the consequent solutions are very noisy in these phase space areas. In some sense, this approach arises from older ideas of variations in importance sampling to reduce variances in Monte Carlo [69]. The approach is also similar to that of particle-splitting [70–73], in which a particle arriving at a low probability region is split into a number of particles, each having a reduced weight governed by the number of particles into which the original is split. The weighted Monte Carlo procedure has been shown to be especially useful in the backward Monte Carlo, where one uses a series expansion to guide a propagation path from the final state back to the initial state [68].

The weighted Monte Carlo can also be used for the Wigner equation of motion. The path integral is again expanded in its Neumann series as above [74–76]. The use of this series then corresponds to a set of paths that contain a different number of scattering interaction events. The summation over these paths is then similar to a Feynman path integral, with the notoriously slow convergence. This is overcome by the weighted Monte Carlo method: a large number of paths is chosen, each weighted by a suitable factor. Since the paths involve a number of scattering events, each scattering event introduces a phase proportional to

$$2\cos\left[q(x_1 - x_2) - \omega_q(t_1 - t_2)\right], \tag{6.47}$$

where q and ω_q are the wave number and radian frequency of the phonon. Presumably, after many scattering events, multiple paths can be summed to provide the negative regions of the Wigner function.

The problem with this approach lies in the fact that a Wigner function may have negative excursions in its value even without the presence of scattering. This is known to occur merely in the passage of a localized Wigner packet through a tunneling barrier, but also when the Wigner function is used to describe two entangled particles [64] in free propagation. While the passage of a particle through a tunneling barrier involves a small delay time (which is somewhat controversial in its own right [77]), it is a local effect although there may well be a spatial variation in the spectral approach. But it is hard to fathom how the weighted Monte Carlo will provide the negative values to these problems by using purely phenomenological considerations. The next section addresses this problem, and in section 6.3.3 below, negative weights are introduced by the concept of signed particles, derived from the Neumann expansion of the integral form of the Wigner equation.

6.3.2 Introducing an *affinity* parameter

The *affinity* method is an extension of the weighted Monte Carlo in which each particle carriers an affinity, which may be negative. This immediately solves the problem discussed in the last paragraph, since a negative part of the Wigner function is clearly represented by particles whose affinity is also negative. The Monte Carlo approach is set up so that two systems are solved simultaneously. The first system is the particle system, which resembles a standard classical EMC. The second system is the wave properties of the particles—the affinity. That is, all particles in the system are treated classically as whole particles. They are scattered using normal EMC scattering techniques, and are drifted and accelerated using the standard field term derived from the solution of the Poisson equation in the presence of the real potential, such as the tunneling barriers. However, the discontinuities in the potential are handled as boundary conditions on the solutions of the Poisson equation. That is, the potential jump is introduced in matching solutions from different regions of the solution space for the Poisson equation. Once the above operations have been completed, the Wigner distribution function is calculated from the particle's position and affinity according to [78–80]

$$f_W(\mathbf{x}, \mathbf{p}) \sim \sum_i \delta(\mathbf{x} - \mathbf{x}_i)\delta(\mathbf{p} - \mathbf{p}_i)A_i, \tag{6.48}$$

where $\mathbf{p}$, $\mathbf{x}$, and A_i are the momentum, position and affinity, respectively, and the sum i runs over the set of particles used in the simulation. There are two points here. By using this in the Wigner equation of motion, we see that one needs to update both the classical properties (position, momentum, etc) and the quantum properties (affinity). The first of these updates is done by the normal Monte Carlo technique discussed in section 6.1. The second is the affinity update. By constructing the distribution function, we are then able to utilize the Wigner potential (6.35) to calculate the updated Wigner potential term for the next time step. In a sense, this is a time-splitting approach, in which the particle properties are updated, then the quantum properties are updated by the change in the total potential, in which the solution to Poisson's equation is utilized as discussed above. *This change in potential now determines the change of affinity that each particle experiences due to the quantum structure in the system.* That is, it is the Wigner potential term of the Wigner equation of motion, (6.38) for example in the collision case, that updates the wave-like properties of the particles through the update of their affinity. This can be summarized as follows. All particles in the system are drifted, accelerated and scattered, regardless of their affinities. The particle affinities are changed by the change in the Wigner potential. It is clear that all the quantum mechanics is incorporated into the method via the nonlocal potential and the variation of the particle affinity. An alternate interpretation of the method is to recognize that the particles themselves do not see the quantum barriers, only their affinities 'see' the barriers.

It is important to note that the affinity method here differs from the weighted Monte Carlo of the previous section. The weighted Monte Carlo procedure has been shown to be especially useful in the backward Monte Carlo, where one uses the series expansion to guide a propagation path from the final state back to the initial state [68]. Weighted forward trajectories differ from those of the normal EMC counterpart by the different evolution probabilities used to guide particles towards rarely visited regions. Such trajectories carry weight determined by the ratio of the biased and natural probabilities, which is then taken into account in the evaluation of the physical averages. Thus, the EMC is a particular case of the weighted concept where the weights are equal to 1 [81]. The affinity approach differs from this, in that it is a method to incorporate the negative values of the Wigner function.

As alluded to previously, this remains a quantum ensemble Monte Carlo method, in that we retain the full-particle nature of the EMC technique. We are able to utilize full ensemble statistics by noting that any ensemble average takes the form

$$\langle Q \rangle = \frac{\sum_{i=1}^{N} A_i Q_i}{\sum_{i=1}^{N} A_i}. \tag{6.49}$$

In the classical case, each affinity would be unity, and the sum in the denominator would just be the number of particles used in the simulation. What changes here is the weight assigned to each particle to account for the quantum effects.

Because this is a weighted Monte Carlo, the total number of electrons, for example, that are included in the system is just given by the denominator of (6.49). However, because some $A_i < 0$, the actual number of simulation particles must be larger than the number of electrons. To achieve this in one approach, we introduced a maximal envelope (ME), which is a larger particle distribution to define N_T, the total number of particles [82]. In this approach, the ME at any phase point is larger than the magnitude of the Wigner function itself, as would be the case if all $A_i = 1$. The particle density initially is then spread physically over the ME, according to its variation in position and momentum. The initial affinity is assigned by the following procedure: if a particle has position $\mathbf{x}$ and momentum $\mathbf{p}$, then its affinity is given value 0 if that position and momentum is not occupied in the initial Wigner distribution at $t = 0$. That is, a number of particles in a small region $\Delta\mathbf{x}\Delta\mathbf{p}$ are assigned affinity 1 according to the number determined from integrating the initial Wigner distribution over this small region of phase space. If more particles are within this region, the excess particles are given affinity 0. In the equilibrium situation, the Wigner function is positive-definite, so we do not have negative affinities initially (this may be different in the presence of entanglement, but it is easy to extend the discussion to this case in a straightforward manner). The affinities are then updated by the procedure described above. A further importance of using the ME is to ensure that enough particles are present to not only gain and lose affinity due to the Wigner potential time evolution, but also to sample the entire phase space domain of the Wigner function itself. As the Wigner potential term acts nonlocally, correlation, reflection and transmission of density can occur where the Wigner function is zero or negative. Since the particles act as charge carriers, particles need to correctly sample the phase space domain to correctly incorporate the nonlocal updates.

In this approach, care must be taken in establishing an extended set of boundary conditions, so that not only are the Monte Carlo and Wigner function boundary conditions satisfied, but also the ME must have its own set of boundary conditions which are consistent with the former set [82]. This is achieved by randomly distributing particles in the contact region during injection to satisfy the Monte Carlo condition, then randomly distributing these particles in momentum according to the ME and, finally, assigning the particle affinities based on a thermal distribution function in the contact such that charge neutrality at the contacts is met. This last step incorporates not only the Wigner function boundary condition but also the need for charge neutrality, such as it may be. Previous work has suggested the need for something like a drifted Maxwellian boundary conditions (see section 5.2.2). Absorbing boundary conditions are required when solving for the Wigner distribution to prevent spurious reflections. However, the Monte Carlo and ME conditions on the boundary naturally include this. Due to the absorbing nature of the boundary conditions and the need to update the ME to assure proper sampling of the entire phase space, current is calculated from the probability current, which is known quantum mechanically to be

$$\mathbf{J}(\mathbf{x}, t) = q \mid \psi(\mathbf{x}, t)|^2 \mathbf{v}(\mathbf{x}, t). \tag{6.50}$$

By using the definition of the affinity and the properties of the Wigner distribution function, it is easily shown that the current in the device may be obtained from summation over the particles as (for a linear sample of length L and cross-section S)

$$I(t) = \frac{q}{mLS}\sum_{i=1}^{N_T} A_i p_i(t). \tag{6.51}$$

In this last equation, the momentum is the longitudinal momentum along the current direction. One might simply assume that the sum over the transverse momentum might vanish, but this can be misleading or even wrong in situations where there are local inhomogeneities within the device structure. An example is the structure in Shifren *et al* [83], where the transverse momentum might vanish globally, but not locally (see figure 4.3).

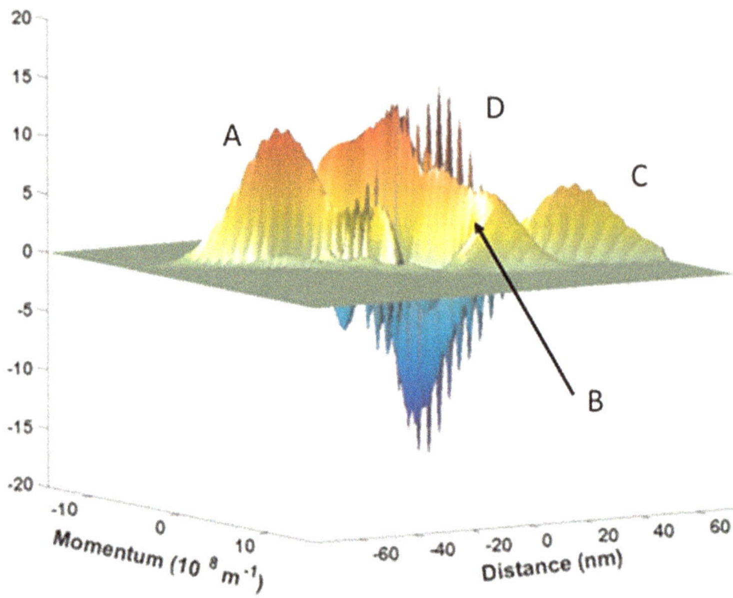

Figure 6.2. Tunneling of a single Wigner wave packet through a single barrier in the GaAs system. One can see the reflected (A), transmitted (C), transition (B), and entanglement (D) parts of the interacting packet. Adapted from [84].

In figure 6.2, we illustrate the situation of a Gaussian Wigner function tunneling through a single-barrier structure [84]. The structure is composed of GaAs active regions and an AlGaAs barrier of 0.3 eV height and 5 nm thickness. There are several parts of this figure which are interesting. The incident Gaussian arrives and the barrier and various parts of it (B) 'slide' along the barrier in momentum, from the positive momentum of the incident packet to the negative momentum of the reflected packet (A). The transmitted packet is (C). A rapid oscillatory part (D) forms as soon as the packet nears the barrier and remains after the transmitted and reflected packets emerge. We have to remember that, in the absence of scattering, both the reflected and transmitted packets are formed by the decay of the incident packet. In this sense, the rapid oscillations reflect the entanglement of the transmitted and reflected packets. Note also that the transition part (B) also contains significant negative values especially near the barrier during the 'slide'. It is this entanglement that allows the two packets to reform into a single packet if the time is reversed [80]. In fact, if we erase this oscillatory part at a later time, and then reverse the time, the two packets each form their own transmitted and reflected packets, because we have erased the memory that the two came from an original single packet.

The affinity approach has been extended in more recent studies. One perceived problem is the computation of average quantities over the various cells of the real and momentum space grids. If there are no particles in the cells, then convergence and noise can become significant problems. Hence, it has been suggested that one inject particles with zero affinity into cells that are normally empty to improve the simulation approach [85–87]. Whereas earlier work primarily injected particles from the contacts, this new approach smooths the simulation with injection of these zero-weight particles throughout the computational domain. These authors have then used this approach to simulate more meaningful devices, such as the double-gate MOSFET [85] and nano-wire MOSFETs [87], as well as RTDs. Several reviews of this work have appeared [88, 89], and they have also studied carefully the phonon-induced decoherence of the memory effects in the Wigner function [90]. The affinity method has also been adopted in Japan [91].

6.3.3 Signed particles

In the above, both the weight and the affinity are really artificial numerical quantities whose purpose is to simulate the quantum phase interference that occurs in the Wigner function during real propagation in a quantum system. A melding of these two concepts was pursued in an alternative approach [92, 93]. To illustrate the idea we redefine Γ_0 in (6.38) and rewrite the equation as

$$\frac{\partial f_W}{\partial t} + \frac{\mathbf{p}}{m} \cdot \nabla f_W = -\gamma_0 f_W + \frac{1}{h^3}\int d^3\mathbf{p}'[W^+(x, \mathbf{p} - \mathbf{p}') - W^-(x, \mathbf{p} - \mathbf{p}') + \gamma_0 \delta(\mathbf{p} - \mathbf{p}')], \tag{6.52}$$

where

$$\begin{aligned} W^+(\mathbf{x}, \mathbf{p}) &= \max(0, W(\mathbf{x}, \mathbf{p})) \\ W^-(\mathbf{x}, \mathbf{p}) &= W^+(\mathbf{x}, -\mathbf{p}) \\ \gamma_0(\mathbf{x}) &= \int d^3\mathbf{p} W^+(\mathbf{x}, -\mathbf{p}). \end{aligned} \tag{6.53}$$

Here, W is given by (6.35) and we have used the anti-symmetric properties of the Wigner potential to replace W by $W^+ - W^-$. Equation (6.52) formally resembles the zero-field version of (6.7) and thus can be processed in the same way to obtain a counterpart to (6.14). In the square bracket of the latter, we now have the term

$$\gamma_0 e^{-\gamma_0 (t-t')}$$

that now has the full meaning, as before, of a probability for the free-flight duration, in complete analogy with (6.3). This probability is now determined by the spatial distribution of the 'power of the Wigner potential' given by γ_0, given by (6.53). The square brackets of (6.52) contains the three terms that constitute the scattering interpretation. If integrated over the momentum, these terms yield a value of unity exactly as in (6.1). However, the minus sign of the middle term precludes interpreting it as a probability for selection of the after-scattering state. We can try to interpret the sign as a weight and associate it to the trajectory. Thus, we consider

$$\left(\frac{W^+}{\delta_0} + \frac{W^-}{\delta_0} + \delta\right) = 3\left(\frac{W^+}{3\delta_0} + \frac{W^-}{3\delta_0} + \frac{\delta}{3}\right). \tag{6.54}$$

Now the terms on the right can be given a probabilistic interpretation: with a probability of 1/3 we choose the type of interaction with the Wigner potential according to one of the three terms and then select the final state after scattering according to the term chosen. Since the replacement of the original term in (6.52) by that of (6.54) modifies the kernel, this needs to be compensated by a corresponding weight. If the term W^- is chosen, the weight is multiplied by −1. An evolving trajectory then accumulates weight, which after each scattering event is multiplied by 3 to account for the pre-factor on the right of (6.54).

In this approach, the weights are accompanied with signs associated with them. This leads to high variance in the results, so that a weight decomposition approach, which limits the value of a weight by storing part of it on a phase space grid, is necessary. This significantly improved the variance during simulations with this method. The variance problem mentioned above finally leads to the adoption of a pure sign convention [94, 95]. Since each of the three terms on the left of (6.54) corresponds to a probability distribution, we use them to generate three different after-scattering states. This corresponds to a branching of the trajectory that can be conveniently interpreted as generation of two novel new particles. For the δ function term, the final state coincides with the initial state, which means that the interacting particle survives unchanged. As the weight of each particle is now given by its sign, this sign is taken into account in the physical averaging process. The classical sum over the particles now involves a prefactor of the sign, just as the affinity that appears in (6.49). This gives rise to a very important property of the signed particles. Two particles of opposite sign will annihilate each other if they arrive at the same phase space point, and this arises primarily due to their common probabilistic future. The annihilation greatly reduces the computational burden. This creates a model in which the Wigner function is considered to include a Boltzmann-like scattering term, but which now includes a generation term. The quantum information is carried by the sign of the particles. This approach is most useful when treating the interaction with the nonlocal potential as a scattering event, as in the Wigner path method. When a scattering even from the potential occurs, two new particles are created, one with the momentum increased by q and one where it is reduced by q, with q determined randomly from the probability distribution of the potential's spatial Fourier transform, much in line with the spectral decomposition method discussed above. The sign on one of these new particles is taken to be the same as the incident particle, while the sign on the other is the opposite. These signs are taken into account in each averaging process that is used to find average values. Equivalent particles with opposite signs annihilate one another when they meet in phase space. In accordance with the properties of the Wigner equation, the signed-particle approach maintains the anti-symmetry of the solution during the evolution.

The signed-particle Monte Carlo approach has been used to study a number of small semiconductor devices, including the role of interface roughness [96]. In later work, they also adopted a potential decomposition in order to improve the method [97]. Here, the actual nonlocal potential was decomposed into its classical part and a remaining quantum-mechanical nonlocal part. The classical part gives just the classical electric field that enters the Boltzmann equation, and this can lead to a normal ensemble Monte Carlo approach. The quantum part of the potential is then the difference between the actual nonlocal potential and the classical potential and is treated by the scattering method discussed above. Scattering by the phonons and the quantum potential both involve the generation of particles and the use of the particle sign to impart quantum interference. This approach has been extended to the many-body problem in which one deals with interacting electrons [98]. For example, consider two identical electrons trapped in a one-dimensional box. These two electrons interact with one another through the Coulomb potential. The system starts at $t = 0$ with the Wigner function being generated by a proper anti-symmetric wave function. If we take the positions of the two particles as x_1 and x_2, then we can write the Wigner function as

$$
\begin{aligned}
f_W(x_1, x_2; p_1, p_2) &= \frac{1}{(\pi\hbar)^2}\int\int dx'_1 dx'_2\, e^{i(x'_1 p_1 + x'_2 p_2)} \\
&\quad \times \psi_0^*\left(x_1 + \frac{x'_1}{2}, x_2 + \frac{x'_2}{2}\right)\psi_0\left(x_1 - \frac{x'_1}{2}, x_2 - \frac{x'_2}{2}\right)
\end{aligned}
\tag{6.55}
$$

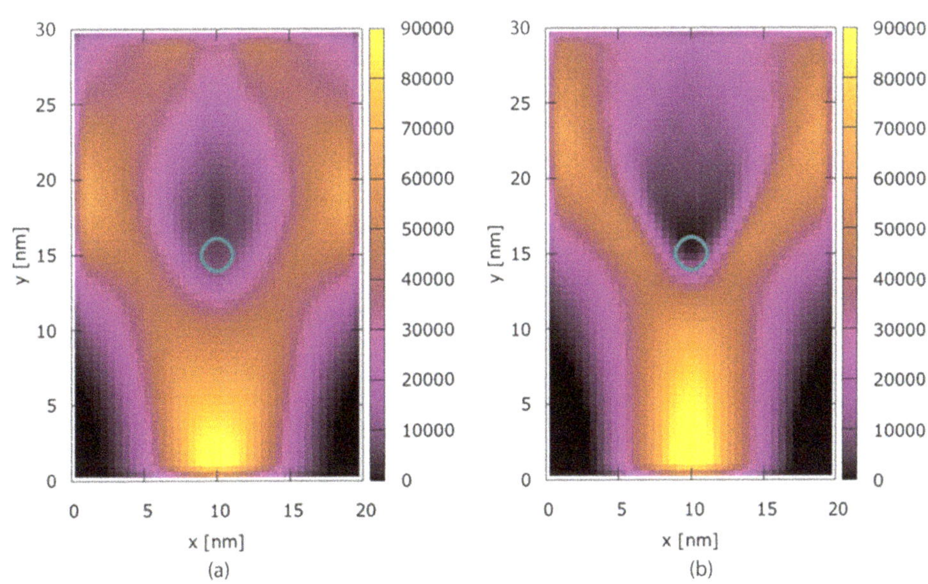

Figure 6.3. (a) Particle simulation for the quantum Wigner function for transport around an impurity (circle) which is embedded in a waveguide. (b) Classical particle simulation for the same system.

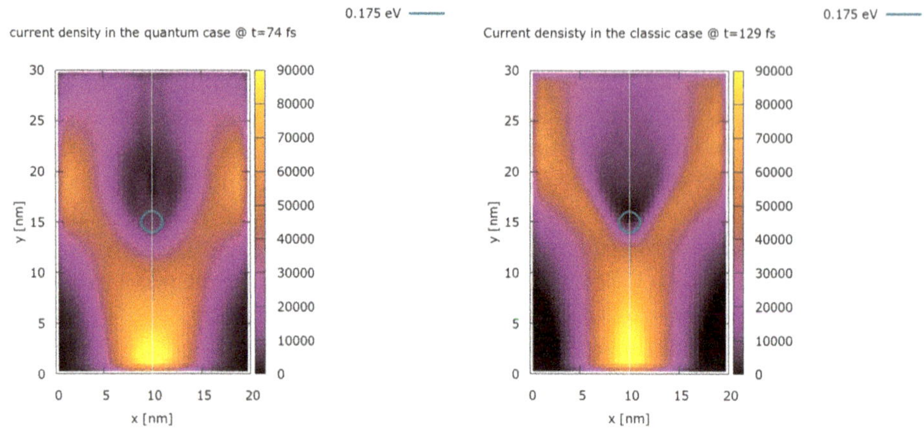

Figure 6.4. Videos showing the quantum (left) and classical (right) evolution of the wave packets shown in figure 6.3. A video is available online at http://iopscience.iop.org/book/978-0-7503-1671-2.

where

$$\psi_0(x_1, x_2) = \begin{vmatrix} \varphi_1(x_1) & \varphi_2(x_1) \\ \varphi_1(x_2) & \varphi_2(x_2) \end{vmatrix} \tag{6.56}$$

is a proper Slater determinant for the two electron system. The individual wave functions are given by

$$\varphi_1(x) = N_1 e^{-(x-x_{10})^2/2\sigma^2} e^{ip_{10}x/\hbar}, \quad \varphi_2(x) = N_2 e^{-(x-x_{20})^2/2\sigma^2} e^{ip_{20}x/\hbar}. \tag{6.57}$$

Here N_1 and N_2 are normalization factors, x_{10} and x_{20} are the initial positions of the two particles, and p_{10} and p_{20} are their initial momenta.

As an example of the signed-particle approach, we show in figure 6.3 the results for a single impurity in a narrow electron waveguide. In panel (a), we show the results for the quantum simulation, while in panel (b), we show the results for the classical simulation. Clearly, the quantum simulation shows that the splitting of the Wigner function has some memory, which is not visible in the figures but which leads to the reforming of the single Wigner function once past the impurity (see also the video in figure 6.4).

References

[1] Leclerc G-L 1735 *Histoire de l'Académie royale des sciences* Ann. 1733 (Paris: Royal Impr.) p 43

[2] Metropolis N and Ulam S 1949 *J. Am. Stat. Assoc.* **247** 335

[3] Metropolis N 1987 *Los Alamos Sci. special issue* p 125

[4] Khachaturyan A, Semenovskaya S and Vainshtein B 1979 *Sov.-Phys. Crystall.* **24** 519

[5] Caffarel M and Claverie P 1988 *J. Chem. Phys.* **88** 1088

[6] Kurosawa T 1966 *Proc. Intern. Conf. on Phys. Semicond., Kyoto J. Phys. Soc. Jpn.* **21** 424

[7] Jacoboni C and Reggiani L 1970 *Phys. Lett.* A **33** 333

[8] Curby R C and Ferry D K 1971 *Phys. Lett.* A **35** 64
[9] Littlejohn M A, Hauser J R and Glisson T H 1975 *Appl. Phys. Lett.* **26** 625
[10] Shichijo H and Hess K 1981 *Phys. Rev.* B **23** 4197
[11] Jensen K L and Buot F A 1989 *Inst. Phys. Conf. Ser.* **99** 137
[12] Rossi F and Jacoboni C 1992 *Europhys. Lett.* **18** 169
[13] Jacoboni C and Reggiani L 1983 *Rev. Mod. Phys.* **65** 645
[14] Jacoboni C and Lugli P 1989 *The Monte Carlo Method for Semiconductor Device Simulation* (Vienna: Springer)
[15] Hess K 1991 *Monte Carlo Device Simulation: Full Band and Beyond* (Boston, MA: Kluwer)
[16] Kosina H, Nedjalkov M and Selberherr S 2003 *J. Appl. Phys.* **93** 3553
[17] Price P 1979 *Semiconductors and Semimetals* vol 10 ed R K Willardson and A C Beer (New York: Academic) pp 249–308
[18] Ferry D K 1991 *Semiconductors* (New York: Macmillan)
[19] Kalos M H and Whitlock P A 1986 *Monte Carlo Methods* (New York: Wiley)
[20] Budd H 1966 *J. Phys. Soc. Jpn.* **21** 424
[21] Rees H D 1969 *J. Phys. Chem. Sol.* **30** 643
[22] Bose S and Jacoboni C 1976 *J. Phys.* C **9** 315
[23] Lugli P and Ferry D K 1985 *IEEE Trans. Electron Dev.* **32** 2431
[24] Lugli P, Jacoboni C, Reggiani L and Kocevar P 1987 *Appl. Phys. Lett.* **50** 1251
[25] Landauer R 1957 *IBM J. Res. Develop.* **1** 223
[26] Landauer R 1970 *Philos. Mag.* **21** 863
[27] Büttiker M 1992 *Phys. Rev.* B **46** 12485
[28] Prêtre A, Thomas H and Büttiker M 1996 *Phys. Rev.* B **54** 8130
[29] Büttiker M, Prêtre A and Thomas H 1993 *Phys. Rev. Lett.* **70** 4114
[30] Büttiker M, Thomas H and Prêtre A 1993 *Phys. Lett.* A **180** 364
[31] Wigner E P 1955 *Phys. Rev.* **98** 145
[32] Smith F T 1960 *Phys. Rev.* **118** 349
[33] Brouwer P W, Frahm K M and Beenakker C W J 1997 *Phys. Rev. Lett.* **78** 4737
[34] Reichl L 2004 *The Transition to Chaos* 2nd edn (New York: Springer)
[35] Ringel Z, Imry Y and Entin-Wohlman O 2008 *Phys. Rev.* B **78** 165304
[36] Rotter S, Ambichl P and Libisch F 2011 *Phys. Rev. Lett.* **106** 120602
[37] Rotter S, Ambichl P and Libisch F 2011 Generating Particlelike Scattering States in Wave Transport
[38] Madelung E 1926 *Z. Phys.* **40** 322
[39] Kennard E H 1928 *Phys. Rev.* **31** 876
[40] Bohm D 1952 *Phys. Rev.* **85** 166
[41] See e.g. ed Oriols X and Mompart J 2012 *Applied Bohmian Transport: From Nanoscale Systems to Cosmology* (Singapore: Pan Stanford)
[42] Oriols X, Garcia-Garcia J J, Martin F, Suñé J, Gonález T, Mateos J and Pardo D 1998 *Appl. Phys. Lett.* **72** 806
[43] Leopreore C L and Wyatt R E 1999 *Phys. Rev. Lett.* **82** 5190
[44] Wyatt R E 1999 *J. Chem. Phys.* **111** 4406
[45] Shifren L, Akis R and Ferry D K 2000 *Phys. Lett.* A **274** 75
[46] Colomés E, Zhan Z and Oriols X 2015 *J. Comput. Electron.* **14** 894
[47] Devaney R L 1986 *An Introduction to Chaotic Dynamical Systems* (Menlo Park, CA: Benjamin)
[48] Coffey T M, Wyatt R E and Schieve W C 2008 *J. Phys.* A **41** 335304

[49] Dirac P A M 1933 *Phys. Z. Sowjetun.* **3** 64
[50] Feynman R P 1948 *Rev. Mod. Phys.* **20** 367
[51] Feynman R P and Hibbs A R 1965 *Quantum Mechanics and Path Integrals* (New York: McGraw-Hill)
[52] Mason B A and Hess K 1989 *Phys. Rev.* B **39** 5051
[53] Feynman R P and Vernon Jr F L 1963 *Ann. Phys.* **24** 118
[54] Bordone P, Bertoni A, Brunetti R and Jacoboni C 2003 *Math. Comput. Simul.* **62** 307
[55] Lozovik Y E and Filinov A V 1999 *Zh. Eksp. Teor. Fiz.* **115** 1872
Lozovik Y E and Filinov A V 1999 Tr. in *J. Exp. Theor. Phys.* **88** 1026
[56] Brosens F and Magnus W 2010 *Sol. State Commun.* **150** 2102
[57] van de Put M L, Sorée B and Magnus W 2017 *J. Comput. Phys.* **350** 314
[58] Pascoli M, Bordone P, Brunetti R and Jacoboni C 1998 *Phys. Rev.* B **58** 3503
[59] Barker J R and Murray S 1983 *Phys. Lett.* A **95** 271
[60] Sels D, Brosens F and Magnus W 2012 *Phys. Lett.* A **376** 809
[61] Barker J R 2010 *J. Comput. Electron.* **9** 243
[62] Heller E J 1976 *J. Chem. Phys.* **65** 1544
[63] Dittrich T, Viviescas C and Sandoval L 2006 *Phys. Rev. Lett.* **96** 070403
[64] Zurek W H 2001 *Nature* **412** 712
[65] Berry M V and Balazs N L 1979 *J. Phys.* A **12** 625
[66] Rios P P de M and Ozorio de Almeida A M 2002 *J. Phys.* A **35** 2609
[67] Ozorio de Almeida A M and Brodier O 2006 *Ann. Phys.* **321** 1790
[68] Rossi F, Poli P and Jacoboni C 1992 *Semicond. Sci. Technol.* **7** 1017
[69] Kalos M H and Whitlock P A 1986 *Monte Carlo Methods Vol. 1: Basics* (New York: Wiley)
[70] MacDonald J L and Cashwell E D 1976 *Trans. Am. Nucl. Soc.* **24** 198
[71] Prettyman T H, Gardner R P and Verghese K 1990 *Nucl. Instr. Meth. Phys. Res.* A **299** 516
[72] Painter S 1993 *Comput. Phys. Commun.* **77** 142
[73] Hirayama H 1995 *J. Nucl. Sci. Technol.* **32** 1201
[74] Vitanov P, Nedjalkov M, Jacoboni C, Rossi F and Abramo A 1994 *Advances in Parallel Algorithms* ed B I Sendov and I Dimov (Amsterdam: IOS Press) pp 117–28
[75] Dimov I T 1996 *Monte Carlo Methods for Applied Scientists* (Singapore: World Scientific)
[76] Jacoboni C and Bordone P 2004 *Reports Prog. Phys.* **67** 1033
[77] Jonson M 1992 *Quantum Transport in Semiconductors* ed D K Ferry and C Jacoboni (New York: Plenum) pp 193–237
[78] Shifren L and Ferry D K 2001 *Phys. Lett.* A **285** 217
[79] Shifren L and Ferry D K 2002 *Physica* B **314** 72
[80] Shifren L and Ferry D K 2002 *J. Comput. Electron.* **1** 55
[81] Nedjalkov M and Vitanov P 1990 *Sol.-State Electron.* **33** 407
[82] Shifren L, Ringhofer C and Ferry D K 2003 *IEEE Trans. Electron Dev.* **50** 769
[83] Shifren L, Akis R and Ferry D K 2000 *Phys. Lett.* A **274** 75
[84] Shifren L 2002 PhD Thesis *unpublished*
[85] Querlioz D, Saint-Martin J, Do V-N, Bournel A and Dollfus P 2006 *IEEE Trans. Nanotechnol.* **5** 737
[86] Querlioz D, Dollfus P, Do V-N, Bournel A and Nguyen V L 2006 *J. Comput. Electron.* **5** 443
[87] Querlioz D, Saint-Martin J, Huet K, Bournel A, Aubry-Fortuna V, Chassat C, Galdin-Retailleau S and Dollfuss P 2007 *IEEE Trans. Electron Dev.* **54** 2232

[88] Querlioz D, Nguyen H-N, Saint-Martin J, Bournel A, Galdin-Retailleau S and Dollfus P 2009 *J. Comput. Electron.* **8** 324
[89] Querlioz D, Saint-Martin J and Dollfus P 2010 *J. Comput. Electron.* **9** 224
[90] Querlioz D, Saint-Martin J, Bournel A and Dollfus P 2008 *Phys. Rev.* B **78** 165306
[91] Koba S, Aoyagi R and Tsuchiya H 2010 *J. Appl. Phys.* **108** 064504
[92] Nedjalkov M, Kosik R, Kosina H and Selberherr S 2002 *Proc. 7th Intern. Conf. on Simulation of Semiconductor Processes* (New York: SISPAD) pp 187–90
[93] Nedjalkov M, Kosina H and Selberherr S 2004 *Large Scale Scientific Computing* vol 2907 ed I Lirkov *et al* (Heidelberg: Springer) pp 178–84
[94] Nedjalkov M, Kosina H, Ungersboeck E and Selberherr S 2004 *Semicond. Sci. Technol.* **19** S226
[95] Nedjalkov M, Kosina H, Selberherr S, Ringhofer C and Ferry D K 2004 *Phys. Rev.* B **70** 115319
[96] Sellier J M, Nedjalkov M, Dimov I and Selberherr S 2013 *J. Appl. Phys.* **114** 174902
[97] Sellier J M, Nedjalkov M, Dimov I and Selberherr S 2015 *Math. Comput. Simul.* **107** 108
[98] Sellier J M, Nedjalkov M and Dimov I 2015 *Phys. Reports* **577** 1

IOP Publishing

The Wigner Function in Science and Technology

David K Ferry and Mihail Nedjalkov

Chapter 7

Collisions and the Wigner function

The collision of the particles with other entities, such as impurities, lattice vibrations (phonons), or other particles, may easily be added to the Hamiltonian with the appropriate energy terms. Evaluating these terms for use in transport or other studies is not usually so straightforward, or even easy. Thus, we need an approximation scheme to introduce the scattering. The standard approach to this is via the interaction representation, which gives us a prescription for developing the perturbation series. As one might infer from our previous chapters, this will lead to multiple, messy integrals. Fortunately, the Feynman diagrams give us an elegant and pictorial method of looking at the various terms in the perturbation series. As a result, they also give us a much more intuitive method of examining the various terms. It is important to note, however, that it is basically impossible to include all possible perturbative terms: the number of these terms is just too massive. In addition, one must be sure that the perturbation series actually converges, although it is not always possible to do so. This presents a problem in some cases, where it is necessary to push ahead while not being confident that the approach actually converges to the correct, or even to any, answer. In practice, one normally only keeps certain sets of terms. The decision as to which sets of terms one retains is based upon intuition, and upon the fact that, if the series is to converge, the low-order terms are larger than higher-order terms. Hence, one normally proceeds upon a basis of faith that the most important terms are the ones that are kept. This can be a self-fulfilling prophecy, but it can also result in misleading results. So, care must be taken as one moves ahead with this perturbative approach.

In spite of the skepticism above, the treatment of scattering in most semiconductors can be handled with the lowest-order interactions, if for no other reason than that it is known that scattering is weak in most semiconductors. But, Wigner functions are applied far beyond mere semiconductor transport. The degree to which the scattering dominates the property of interest has to be examined in each individual case. So, the perturbation series that derives from the interaction

doi:10.1088/978-0-7503-1671-2ch7

representation is a guide to this evaluation; and it is known that higher-order terms are important in many situations, even in semiconductors. For example, the simplest, lowest-order term for impurity scattering leads only to the normal Drude scattering approximation, and does not include the inhomogeneous nature of impurity scattering that is vital to understanding momentum relaxation in semiconductors. As we remarked already in chapter 2, including the correction for momentum relaxation requires including a set of higher-order terms, known as the ladder diagrams, via the Bethe–Salpeter equation [1]. So, a complicated integral equation is required in the quantum case merely to recover the classical behavior in semiconductors. Another example is the case of weak localization, which is observed in many semiconductors and even metals at low temperatures. This reduction in conductivity arises from interference between time-reversed paths for back-scattering by a set of impurities, and is represented by a set of maximally-crossed ladder diagrams [2].

In this chapter, we will develop the form of the electron–phonon interaction to low order for the special situation of the Wigner equation of motion. In the next section, we outline the interaction representation. Following that, we outline the formation of the basic matrix element for various electron–phonon processes. Then we turn to the collision integral for the Wigner functions. Finally, we will discuss how these scattering events have been introduced into simulations using the ensemble Monte Carlo processes.

7.1 The interaction representation

We generally begin directly with the Schrödinger equation itself. Here, we keep the Hamiltonian as an operator, and write the general solution to the time-dependent Schrödinger equation as

$$\psi(\mathbf{x}, t) = e^{-iHt/\hbar}\psi(\mathbf{x}, 0). \tag{7.1}$$

It is clear from a general consideration that the Hamiltonian itself contains a great many different operators. We assume that these operators are not explicit functions of time (although this will change when we get to the electron–phonon interaction). The result (7.1) is often referred to as the Schrödinger representation, or picture, in which the time variation is placed with the wave function. An alternative view is the Heisenberg representation, or picture, in which the operators are written as

$$i\hbar\frac{\partial A(t)}{\partial t} = [A(t), H] \Rightarrow A(t) = e^{iHt/\hbar}A(0)e^{-iHt/\hbar}. \tag{7.2}$$

In the interaction representation, or picture, we combine these by assuming that the Hamiltonian has a base form H_0 and a perturbing potential V. Then, (7.2) is modified to depend only upon the base form H_0. Then, the wave function can be written as

$$\psi(\mathbf{x}, t) = e^{iH_0t/\hbar}e^{-iHt/\hbar}\psi(\mathbf{x}, 0). \tag{7.3}$$

This implies that the wave function remains essentially stationary, but changes only with the perturbing potential V. In general, one must be very careful as this point, because in general these two exponentials cannot easily be combined. In general, this is handled through the Baker–Hausdorf formula [3]:

$$e^A e^B = e^{A+B} e^{-[A,\,B]/2}, \tag{7.4}$$

in the case in which $[A,B]$ is a c-number. With the interaction representation, we can write matrix elements for two arbitrary basis states as

$$\begin{aligned} \left\langle \hat{\psi}_1^\dagger(t) A(t) \hat{\psi}_2(t) \right\rangle &= \left\langle \hat{\psi}_1^\dagger(0) e^{iHt/\hbar} e^{-iH_0t/\hbar} \left(e^{iH_0t/\hbar} A(0) e^{-iH_0t/\hbar}\right) e^{iH_0t/\hbar} e^{-iHt/\hbar} \hat{\psi}_2(0) \right\rangle \\ &= \left\langle \hat{\psi}_1^\dagger(0) e^{iHt/\hbar} A(0) e^{-iHt/\hbar} \hat{\psi}_2(0) \right\rangle, \end{aligned} \tag{7.5}$$

which gives us the correct time dependence. That is, we can now assign the exponentials either to the wave functions for the Schrödinger picture or to the operator for the Heisenberg picture.

We recognize that the perturbing potential is the term that is going to upset this nice balance, and we need to generate an approach for handling this term. To begin, we can write the time derivative of the field operator as

$$\frac{\partial \hat{\psi}(t)}{\partial t} = \frac{i}{\hbar} e^{iH_0t/\hbar} (H_0 - H) e^{-iHt/\hbar} \hat{\psi}(0) = -\frac{i}{\hbar} V(t) \hat{\psi}(0). \tag{7.6}$$

Let us consider the unitary operator obtained by the two exponentials in (7.3), which we may define as

$$U(t) = e^{iH_0t/\hbar} e^{-iHt/\hbar}. \tag{7.7}$$

We may use the very same differential expansion that appears in (7.6) to show that this has the time variation

$$\frac{\partial U(t)}{\partial t} = -\frac{i}{\hbar} V(t) U(t). \tag{7.8}$$

This last equation can be formally integrated to give

$$U(t) = 1 - \frac{i}{\hbar} \int_0^t V(t') U(t')\, dt', \tag{7.9}$$

where we have used the fact, from (7.7), that $U(0) = 1$. This last equation formally defines a basis for iterating through the solutions for $U(t)$ as a series in the powers of the interaction potential $V(t)$. Thus, for example, the first few terms are given as

$$\begin{aligned} U_0(t) &= 1 \\ U_1(t) &= 1 - \frac{i}{\hbar} \int_0^t V(t')\, dt' \\ U_2(t) &= 1 - \frac{i}{\hbar} \int_0^t V(t') dt' + \left(\frac{i}{\hbar}\right)^2 \int_0^t dt' \int_0^{t'} dt'' V(t') V(t''). \end{aligned} \tag{7.10}$$

In the last equation, it is always required that $t'' < t'$. This generates the entire perturbation series. This series is often rewritten as an exponential function as

$$U(t) = T\left[\exp\left(-\frac{i}{\hbar}\int_0^t V(t')\,dt'\right)\right], \tag{7.11}$$

where T is the time-ordering parameter.

7.2 The electron–phonon interaction

Probably the most important dissipative process in conducting systems is that due to the electron–phonon interaction. The approach is quite similar to that for impurity scattering except for a few important differences [1, 4]. First, the Coulomb interaction is instantaneous, whereas the electron–phonon is not. This means that there will be an additional time integration and a propagator for the phonon itself. This propagator corresponds to the propagation of the phonon along the time path of this additional time integration. How this additional time enters into considerations of the Wigner function is an important point, since the latter is a single time function. This has been handled by a variety of approaches, which we will deal with below. Here, we will look particularly at the matrix element that enters time-dependent perturbation theory and will be central to the development of the Wigner form of the collision integral. In general, one can expand the electron–phonon interaction Hamiltonian as a power series in the phonon wave vector q. There are terms which vary as $1/q$ such as the polar optical phonon and piezoelectric interactions. There are also terms which are independent of q such as the non-polar optical phonon interaction. Finally, there are terms linear in q, such as the acoustic phonon interaction. There are, of course, higher-order terms as well, but these low-order terms are the most important and give rise to the majority of scattering in semiconductors. We will begin with the last of these, treating these different terms in reverse order.

7.2.1 Acoustic phonons

Scattering by the acoustic modes of the lattice vibrations is one of the most common interactions. The acoustic modes have frequencies which vanish as the wave vector **q** goes to zero and, as may be expected, the wave velocity corresponds to one of the sound velocities in the crystal. The sound velocity depends upon the crystal direction and the polarization of the wave as it moves through the crystal. The wave creates a local strain in the crystal and this strain perturbs the energy bands. This provides a scattering potential which is referred to as the *deformation potential.* This is typically expressed as [5]

$$V_1 = \delta E \equiv \Xi_1 \nabla \cdot \mathbf{u}_\mathbf{q}, \tag{7.12}$$

where Ξ_1 is the deformation potential for a particular band in which the carrier is located, Δ is the dilation of the crystal lattice created by the acoustic wave of Fourier amplitude $\mathbf{u}_\mathbf{q}$. Any static displacement of the lattice would be a movement of the crystal as a whole and does not contribute, so only this first-order term contributes. We recall that the phonons arise from a Fourier transform of the atomic oscillations in the lattice, and that each Fourier mode is represented by a harmonic oscillator with the phonon frequency. Thus, the acoustic wave Fourier amplitude is a relatively uniform value over the entire crystal and the wave itself can be written from the harmonic oscillator basis as

$$\mathbf{u}_{\mathbf{q}} = \left(\frac{\hbar}{2\rho_m \Omega \omega_{\mathbf{q}}}\right)^{1/2} \left[a_{\mathbf{q}} e^{i\mathbf{q}\cdot\mathbf{x}} + a_{\mathbf{q}}^{\dagger} e^{-i\mathbf{q}\cdot\mathbf{x}}\right] \mathbf{e}_{\mathbf{q}} e^{-i\omega_{\mathbf{q}} t}, \tag{7.13}$$

where the plane wave factors have been incorporated along with the normalization factor for completeness. The factors $a_{\mathbf{q}}$ and $a_{\mathbf{q}}^{\dagger}$ are the annihilation and creation operators for the quantized harmonic oscillator representation of the phonon mode. The quantity $\mathbf{e}_{\mathbf{q}}$ is the polatization vector for the wave. The quantity ρ_m is the mass density of the crystal and Ω is the volume of the crystal. Because the divergence operation in (7.12) produces a result only for propagation along the polarization direction, this gives rise normally to the longitudinal acoustic mode interaction. A different result can be obtained in ellipsoidal bands, but we ignore this complication. The fact that the interaction is now first-order in $\mathbf{q}$ leads to this term being called a first-order mode.

Quite generally, scattering is handled with the Fermi golden rule, in which the scattering rate is proportional to the square of the matrix element and a delta function which conserves energy. It is the latter delta function which is broadened in quantum transport. Our interest here is to look at the appropriate matrix elements. We can calculate the matrix element by considering the proper sum over both the electron and the phonon wave functions. The second term of (7.13), the term for the emission of a phonon by the carrier, leads to the matrix element

$$\begin{aligned} M(\mathbf{k}, \mathbf{q}) = i\mathbf{q} \cdot \mathbf{e}_{\mathbf{q}} \left(\frac{\hbar \Xi_1^2}{2\rho_m \Omega \omega_{\mathbf{q}}}\right)^{1/2} \int d^3\mathbf{x} \int d^3\mathbf{x}' \\ \times \psi_{\mathbf{k}}^{\dagger}(\mathbf{x}) \varphi^{\dagger}(\mathbf{x}') a_{\mathbf{q}}^{\dagger} e^{-i\mathbf{q}\cdot\mathbf{x} - i\omega_{\mathbf{q}} t} \varphi(\mathbf{x}') \psi_{\mathbf{k}}(\mathbf{x}), \end{aligned} \tag{7.14}$$

where the ϕ are the harmonic oscillator functions for the lattice parts of the total wave function and the ψ are electronic Bloch functions. In the case of confined particles, for example in a quantum well, the Bloch wave function is also accompanied by the envelope function describing confinement [6]. Normally one might expect a single integration, since the space of the electrons and phonons is the same. However, the waves and the electrons are considered to be independent quantitites. The creation operator excites the harmonic oscillator to a higher energy state by one quantum of vibrational energy. Nevertheless, it is assumed here that the phonon distribution remains in equilibrium, although it may be handled in the non-equilibrium state, as discussed in the Monte Carlo treatment below in section 7.4. This now leads to the Bose–Einstein distribution entering the expression as $N_{\mathrm{q}} + 1$, where N_{q} is this distribution. The next step is to split the integration over the lattice variables into a summation over the set of lattice cells and an integration over a unit cell itself. This summation produces the conservation of momentum which causes the electron wave vector to be changed by the phonon wave vector. When all of these factors are taken together, (7.14) reduces to

$$M(\mathbf{k}, \mathbf{q}) = i\mathbf{q} \cdot \mathbf{e}_{\mathbf{q}}\left(\frac{\hbar\Xi_1^2}{2\rho_m\Omega\omega_{\mathbf{q}}}\right)^{1/2}\sqrt{N_{\mathbf{q}} + 1}\int d^3\mathbf{x}u_{\mathbf{k}-\mathbf{q}}^{\dagger}(\mathbf{x})u_{\mathbf{k}}(\mathbf{x}), \tag{7.15}$$

with the last integration being carried out only over the unit cell, and the two functions are the cell periodic parts of the Bloch functions. This last integral is called the overlap integral and is usually unity in parabolic bands. In the presence of the **k·p** interaction, the cell periodic wave function has sp^3 hybrid admixtures which are energy dependent, so that the overlap integral has an energy-dependent part. This latter interaction also leads to non-parabolic bands. Generally, we ignore the non-parabolic corrections, but they are straightforward, if messy. With the non-parabolic assumption, the overlap integral (squared) becomes

$$|M(\mathbf{k}, \mathbf{q})|^2 = \frac{\hbar\Xi_1^2 q^2}{2\rho_m\Omega\omega_{\mathbf{q}}}(N_{\mathbf{q}} + 1). \tag{7.16}$$

The result for the absorption term is the same except the factor of 1 in the parantheses does not appear. Normally, in the acoustic mode interaction, the temperature is sufficiently high that $N_{\mathrm{q}} \gg 1$, and the difference between the two situations can be ignored. The Bose–Einstein distribution itself can be approximated as

$$N_{\mathbf{q}} = \frac{1}{\exp\left(\hbar\omega_{\mathbf{q}}/k_BT\right) - 1} \sim \frac{k_BT}{\hbar\omega_{\mathbf{q}}} = \frac{k_BT}{\hbar q v_s} \gg 1, \tag{7.17}$$

where we have introduced the linearity of the acoustic dispersion through the sound velocity v_{S}. Using these relations, we can now write (7.16) as

$$|M(\mathbf{k}, \mathbf{q})|^2 = \frac{\hbar\Xi_1^2 k_BT}{\rho_m\Omega v_s^2}. \tag{7.18}$$

The approximation in (7.17) is known as the equi-partition approximation, and must be questioned at low temperatures. Typically, the acoustic phonon energy is at most a few millivolts, but when the temperature is low, the expansion of the exponential is not valid.

7.2.2 Piezoelectric scattering

In materials such as GaAs, which lack a center of inversion symmetry, it is possible for the piezoelectric interaction to lead to scattering of carriers, particularly at low temperatures. This interaction arises from the distortion of the lattice by acoustic modes which leads to the generation of an electric field via the piezoelectric tensor. This new electric field can act upon the carriers as a scattering process. Normal semiconductors such as Si have this inversion symmetry and thus show a piezoelectric interaction only under rare conditions. In the III–V semiconductors, it is primarily the d_{14} element of the piezoelectric tensor which leads to the interaction [7, 8]. The interaction energy shift can be found from

$$\delta E = -\varepsilon_s \mathbf{E} \cdot \mathbf{P} \approx i\frac{ed_{14}}{\varepsilon_s}\frac{q^2}{q^2 + q_{sc}^2}, \tag{7.19}$$

where **F** is the electric field, $/\varepsilon_s$ is the low-frequency dielectric constant, and **P** is the polarization of the lattice. The latter is determined by the stress–strain parameters and arises from the presence of the acoustic wave as $\mathbf{P} \sim (d_{14}q/\varepsilon_s)\mathbf{u_q}$. The electric field is induced by this polarization via the piezoelectric interaction, and is of Coulomb form and screened by the Debye screening wave vector q_{sc}. With this form of the perturbing potential, the integration over wave functions follows exactly (7.14), and the matrix element becomes

$$|M(\mathbf{k}, \mathbf{q})|^2 = \frac{4e^2 d_{14}^2 k_B T}{\varepsilon_s^2 \rho_m \Omega v_s^2}\frac{q^2}{\left(q^2 + q_{sc}^2\right)^2}. \tag{7.20}$$

This result has also used the equi-partition approximation to expand the Bose–Einstein distribution, and so is subject to the same limitations as above.

7.2.3 Non-polar optical and intervalley phonons

When there is more than a single atom per unit cell, the optical modes of the lattice vibration are allowed. In these vibrations, the two (or more) atoms per unit cell vibrate with a relative motion between them. As a result, these phonons are rather energetic, with energies on the order of 30–100 meV (and even larger in many materials). The interactions of these phonons with the carriers are *inelastic* due to the large energy exchange from one particle to the other. Although one normally thinks of scattering occurring solely within the principle valley of the conduction (or valence) band, the zone edge modes can also contribute to scattering between inequivalent sets of valleys in the band. For example, these phonons also cause scattering among the six equivalent ellipsoids of the conduction band of Si, but also between the Γ and the L or X valleys of the conduction band of many III–V compounds.

The matrix element generally is found with the use of a deformable ion model, in which the two sub-lattices move relative to one another. This causes the potential field of each set of atoms to be displaced slightly, which causes a shift in the bond charges. This leaves a small excess of charge where the ions have moved apart and a small deficit of charge where they have moved together. These effects lead to a deformation field D, which is usually given in units of eV/cm. The interaction itself is a zero-order interaction in that there is no explicit term in the phonon wave vector in the interaction term itself, although there is of course still a momentum conservation condition, which can be seen in the energy-conserving delta function of the Fermi golden rule. Because the interaction is zero order, we can write the perturbing potential as

$$\delta E = Du_{\mathbf{q}}, \tag{7.21}$$

where, as before, $u_{\mathbf{q}}$ is the phonon amplitude. Then, following the same procedure as above, the square of the matrix element can be written as

$$\begin{aligned}|M(\mathbf{k}, \mathbf{q})|^2 = \frac{\hbar D^2}{2\rho_m \Omega \omega_{\mathbf{q}}} &\left[N_{\mathbf{q}} \delta\left(E(\mathbf{k}) - E(\mathbf{k} + \mathbf{q}) + \hbar\omega_{\mathbf{q}}\right)\right. \\ &\left. + (N_{\mathbf{q}} + 1)\delta(E(\mathbf{k}) - E(\mathbf{k} - \mathbf{q}) - \hbar\omega_{\mathbf{q}})\right]\end{aligned}, \tag{7.22}$$

where the first term in the square brackets is for the absorption of a phonon by the electron (or hole) and the second term is for the emission of a phonon by the electron (or hole). Obviously, with the high energy of the optical phonon, the equi-partition approximation cannot be used and the distinction between emission and absorption of the phonon becomes critical.

7.2.4 Polar optical phonons

When the two (or more) atoms per unit cell are different, particularly when they come from different columns of the periodic table, then the zone-center optical modes can take on a Coulombic nature as the effective charge on the two atoms per unit cell is different. This polarization is important in the dielectric function as well. Because the interaction is Coulombic, it is a strong interaction and is the dominant scattering processes in most III–V materials. The vibration of the two (or more) dissimilar atoms leads to a polarization, described by the difference in the movement of the two atoms per unit cell in, for example, the III–V materials. This polarization can be expressed as

$$\mathbf{P}_{\mathbf{q}} = \left(\frac{\hbar}{2\gamma\Omega\omega_{\mathbf{q}}}\right)^{1/2} \mathbf{e}_{\mathbf{q}}\left(a_{\mathbf{q}}^{\dagger} e^{-i\mathbf{q}\cdot\mathbf{x}} + a_{\mathbf{q}} e^{i\mathbf{q}\cdot\mathbf{x}}\right) e^{-i\omega_{\mathbf{q}} t}, \tag{7.23}$$

where

$$\frac{1}{\gamma} = \omega_{\mathbf{q}}^2 \left(\frac{1}{\varepsilon(\infty)} - \frac{1}{\varepsilon(0)}\right) \tag{7.24}$$

is the effective coupling constant, which is written in terms of the high-frequency dielectric constant $\varepsilon(\infty)$ and the low-frequency dielectric constant $\varepsilon(0)$. These two values differ by the polarization contribution that arises from this lattice vibration. Since these are known from experiment, they serve as a good value to determine the strength of the polar interaction with the carriers. The interaction energy is given by the first expression in (7.19), but with this much stronger polarization term. In the presence of screening of the Coulombic potential, the square of the matrix element can be written as

$$|M(\mathbf{k}, \mathbf{q})|^2 = \frac{\hbar e^2}{2\gamma\omega_{\mathbf{q}}}\left(\frac{q}{q^2 + q_{sc}^2}\right)^2\Big[N_{\mathbf{q}}\delta\big(E(\mathbf{k}) - E(\mathbf{k} + \mathbf{q}) + \hbar\omega_{\mathbf{q}}\big) + \big(N_{\mathbf{q}} + 1\big)\delta\big(E(\mathbf{k}) - E(\mathbf{k} - \mathbf{q}) - \hbar\omega_{\mathbf{q}}\big)\Big], \tag{7.25}$$

where again the first term is for the absorption of the phonon and the second term is for the emission of the phonon by the carrier.

7.2.5 A precautionary comment

As we commented above, while the energy-conserving delta functions have been included within the matrix elements in the above equations, this is the semi-classical approach. Instead, in keeping with comments made earlier, one has to worry about the possible two-time behavior of the scattering and just how this is incorporated into the Wigner approach. These are naturally just the imaginary part that arises in the limit of the resonance when we are on the energy shell. In actual usage in quantum transport, the replacement with the proper off-shell contributions will be important points in the developments of the rest of this chapter.

7.3 The Wigner scattering integrals

The generally accepted formulation of the scattering terms in the Wigner equation of motion is that originally given by Levinson [9]. He worked out the situation for the presence of both an electric and a magnetic field, although he ignored the diffusion term as a simple addition. The magnetic field is incorporated naturally in terms of the vector potential **A**. In the absence of the diffusion term, one can view the density matrix and the Wigner function as being translationally invariant and satisfying the equation

$$\langle \mathbf{x} + \mathbf{a} \mid \hat{\rho}(t) | \mathbf{x}' + \mathbf{a}\rangle = \exp\left[\frac{ie}{\hbar}\mathbf{A}(\mathbf{a}) \cdot (\mathbf{x} - \mathbf{x}')\right]\langle \mathbf{x} \mid \hat{\rho}(t) | \mathbf{x}'\rangle. \tag{7.26}$$

Levinson observes that any matrix elements obtained by parallel transport through the variables $\mathbf{x}$ and $\mathbf{x}'$ contain no new information about the system. This form becomes more transparent when it is transformed into the Wigner function as

$$\begin{aligned} f_W(\mathbf{x}, \mathbf{p}, t) &= \frac{1}{h^3}\int d^3\mathbf{s}e^{i\mathbf{p}\cdot\mathbf{s}/\hbar}\left\langle \mathbf{x} - \frac{\mathbf{s}}{2} \mid \hat{\rho}(t) | \mathbf{x} + \frac{\mathbf{s}}{2}\right\rangle \\ &= \frac{1}{h^3}\int d^3\mathbf{s}e^{i(\mathbf{p}-e\mathbf{A})\cdot\mathbf{s}/\hbar}\left\langle -\frac{\mathbf{s}}{2} \mid \hat{\rho}(t) | \frac{\mathbf{s}}{2}\right\rangle, \end{aligned} \tag{7.27}$$

where, in the second line, we have introduced (7.26). This leads to an important observation: in the homogeneous system, the Wigner function does not depend upon $\mathbf{x}$ and $\mathbf{p}$ separately, but only depends upon them through the kinetic momentum in the magnetic field:

$$\hbar \mathbf{k} = \mathbf{p} - e\mathbf{A}. \tag{7.28}$$

This means that we can write the Wigner function as

$$f_W(\mathbf{x}, \mathbf{p}, t) = f_W(\mathbf{k}, t), \tag{7.29}$$

exactly as in the classical case.

The resulting equation of motion may then be developed as in chapter 4, and in the presence of the magnetic field, we find

$$\left\{\frac{\partial}{\partial t} + \frac{e}{\hbar}[\mathbf{E}(t) + \mathbf{v}(t) \times \mathbf{B}(t)] \cdot \frac{\partial}{\partial \mathbf{k}}\right\} f_W(\mathbf{k}, t) = I\left(f_W | \mathbf{k}, t\right), \tag{7.30}$$

where $\mathbf{v} = \hbar\mathbf{k}/m$. This equation is, of course, the quantum analogue of the classical kinetic equation for the distribution function. We notice, however, that only the classical electric field has been extracted from the full Wigner potential. Hence, either the first term in the square brackets, for this classical field, can be replaced by the full nonlocal Wigner potential, or the non-classical part of the Wigner potential can be considered as a scattering term, as suggested in chapter 6. We adopt the former approach, as we want to only consider here the true scattering processing in the function I. This latter collision term is far different from the classical form and can be written in greater detail as

$$I\left(f_W | \mathbf{k}, t\right) = \int_0^t dt' \int d^3\mathbf{k}' f_W(\mathbf{k}', t) I(\mathbf{k}, \mathbf{k}'; t, t'). \tag{7.31}$$

In fact, the kernel term involves four separate terms: there are terms for the emission and absorption of a phonon out of the state **k**, terms for the emission and absorption into the state **k** from another state. These, of course, are the *out*-scattering and *in*-scattering terms we discussed in chapter 6 in developing the integral equation for the Wigner function. In addition, the scattering kernel must be summed over the types of phonons present in the system. The out-scattering terms for emission and absorption may be written as [9]

$$A^{(\pm)}(\mathbf{k}, \mathbf{k}'; t, t') = \int d^3\mathbf{q} W_\mathbf{q}^{(\pm)} \Delta_\mathbf{q}\left(\mathbf{k} \mp \frac{\mathbf{q}}{2}, \mathbf{k}' \mp \frac{\mathbf{q}}{2}; t, t'\right), \tag{7.32}$$

where the spectral functions are given by

$$\begin{aligned}\Delta_\mathbf{q}(\mathbf{k}_1, \mathbf{k}_2; t, t') = \frac{1}{h}\int d^3\mathbf{x} \int d^3\mathbf{x}' e^{-i\left[\omega_\mathbf{q}(t-t') - \mathbf{q}\cdot(\mathbf{x}-\mathbf{x}')/2\hbar\right]} U(-\mathbf{x}|t, t') U(-\mathbf{x}'|t', t) \\ \times \delta(\mathbf{p} - \mathbf{p}' - e\mathbf{E}(t - t')) e^{-i(\mathbf{p}+\mathbf{p}')\cdot(\mathbf{x}-\mathbf{x}')/2\hbar},\end{aligned} \tag{7.33}$$

where

$$\mathbf{p} = \hbar\mathbf{k}_1 + e\mathbf{A}(\mathbf{x}, t), \quad \mathbf{p}' = \hbar\mathbf{k}_2 + e\mathbf{A}(\mathbf{x}', t). \tag{7.34}$$

$W_{\mathbf{q}}^{(\pm)} = 2\pi \, |M(\mathbf{k}, \mathbf{q})|^2/\hbar$ are the matrix elements defined in section 7.2. The unitary operators that appear in (7.33) are given by (7.11) and evaluated at the position indicated. Equivalently, the in-scattering terms may be written as

$$B^{(\pm)}(\mathbf{k}, \mathbf{k}'; t, t') = \int d^3\mathbf{q} W_{\mathbf{q}}^{(\pm)} \Delta_{\mathbf{q}}\left(\mathbf{k} \pm \frac{\mathbf{q}}{2}, \mathbf{k}' \mp \frac{\mathbf{q}}{2}; t, t'\right), \tag{7.35}$$

where here, and in (7.32), the upper sign is for phonon emission and the lower sign is for phonon absorption. The presence of the fields E and H in the spectral functions importantly tells us that there is an influence of these fields on the scattering process. Scattering occurs between states that are evolving in these applied fields and not between plane waves of a free electron. This scattering between evolving states has often been called the intra-collisional field effect [10, 11]. This effect reflects the evolution of the states during the actual collision process over a time termed the collision duration. It is found in approaches using Green's functions [12] and the collision duration is found to be a few femtoseconds in some semiconductors [13]. The presence of this effect and the two time variables that appear in (7.33) tells us that the scattering process is, in reality, a two-time process, and goes beyond the single time variable of the Wigner function itself. That is, the Wigner function is actually evolving itself during the collision [14], a fact that is included in the path integral forms developed in chapter 6.

7.4 Collisions in the Monte Carlo approach

A number of early steps toward a full particle simulation of the Wigner equation in a complex device have appeared. In one approach, the simulation uses particles to handle the transport, but the quantum effects of tunneling through a set of barriers are introduced by solving for the energy-dependent tunneling probabilities and then using these to decide whether or not a particle passes the barrier [15]. In this regard, the tunneling is treated as a scattering process, as has been suggested by Barker [16]. Another approach used the particles to simulate the Wigner equation of motion and study a process by which particles could be created (multi-particle production) [17]. In this approach, they weighted the individual particles by a process that was introduced earlier in classical Monte Carlo to study rare events [18], as discussed in the previous chapter. While these approaches bring particles into the collision process, they do not fully solve the quantum coherence problem in quantum transport. Here, we present the historical approach on Wigner paths that uses weighted particles evolving forward in time. However, there is an alternative approach based upon the backward Monte Carlo method, where particles evolving backward in time are associated with the Levinson and the Barker–Ferry models of electron–phonon interaction. The latter goes one step further than the two-time Levinson model by taking into account the finite lifetime of the particle states due to the scattering process [19]. The homogeneous Levinson and Barker–Ferry equations have been generalized to account for the spatial evolution in a quantum wire [20]. Observed are effects related to the initial build-up of the energy-conserving delta function and the evolution of the Wigner state during the process of collision—the

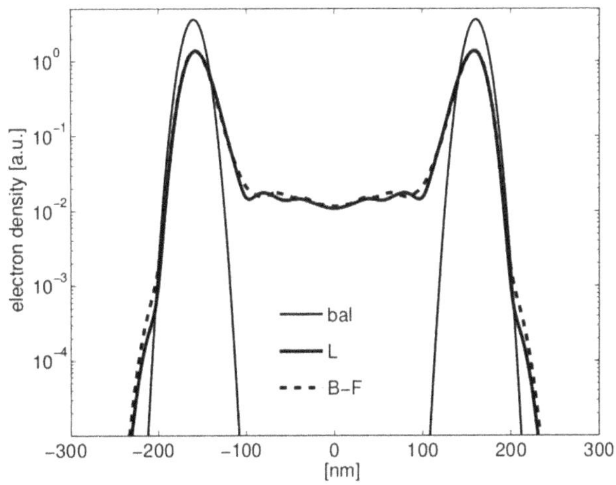

Figure 7.1. Electron density after 175 fs evolution following excitation of carriers via a short-pulse laser. The fastest classical electrons form the fronts of the two ballistic peaks (solid line), center around ± 200 nm. The fastest quantum electrons of the Levinson (bold line) and Barker–Ferry (dashed line) models reach distances further from the centers of the Gaussians describing the laser excitation [20].

intra-collisional field effect. A particular spatial effect of the uncertainty introduced by the collisional broadening is demonstrated in figure 7.1, in which the evolution of an initial spread in the momentum Wigner state, described by two Gaussians in momentum and position, interacts with a single polar optical phonon (with the appropriate phonon energy). At low temperatures, the highly non-equilibrium Wigner state is assumed to be created by a short-pulse laser excitation that introduces carriers approximately 150 meV above the band minimum. Because of the low temperature, these carriers can only emit the phonon. According to the classical model, and the need for energy and momentum conservation, the fastest components of the state (which can be associated with these classical particles) are those that travel ballistically. In the quantum models, uncertainty arising from the collision durations leads to broadening of a manner in which there are electrons which gain energy and can move faster than in the ballistic case.

How do we know that it is valid to impose classical trajectories onto the Wigner equation of motion? It turns out that this has been studied under the guise of Wigner *paths* [21–24], discussed in the previous chapter. The concept of a Wigner path is useful in that it provides a pictorial representation of the quantum evolution, and is quite useful for numerical simulation of the Wigner equation of motion. Such a path is the path followed by a small sample of the Wigner function as it evolves through phase space. Moreover, this path evolves quite like a classical group of particles, except for scattering and for rapidly varying potentials. That is, like classical particles, this small sample evolves with a conservation of its phase space volume, except in these occasions. But the existence of these paths and their conservation properties means that we can really use a kinetic Monte Carlo approach to simulate the evolution of the Wigner function as described by the kinetic equation [24]. As before, the usual integro-differential equation that includes the full scattering integral is converted to a path integral formulation, which gives an integral equation in place of the standard integro-differential equation. In this integral equation, we no longer have in- and out-scattering, but consider only the incoming contributions from a variety of sources. Because this is a quantum system evolution, the integrand assumes complex values as a result of the quantum interference and the fact that the system is subjected to the uncertainty principle, as we discussed in the previous section. As we recall from our earlier discussions, the negative parts of the Wigner function are direct indications of the regions where uncertainty is important. An extended version of the weighted Monte Carlo [18], or the equivalent affinity approach [25], is now used to generate a set of Wigner paths. The sum over these paths is a numerical estimate of the Wigner function itself. In order to generate a Wigner path, there is an entire sequence of random variable selections that must be made. Some of these are [24] the following:

1. The number of interactions to be considered for the particular path. For electron–phonon interactions or potential interactions (such as from impurities), the number of vertices must be even, to account for the magnitude squared of the matrix element. Hence, each vertex involves the matrix element.
2. The times at which each of the interaction vertices occurs. These must lie between the initial time and the observation time at which the Wigner function is being considered.
3. The type of the interaction at each vertex during the simulated path.
4. The momentum change at each vertex. For the electron–phonon interaction, one half of the phonon momentum modifies the wave vector at each vertex. This is described further below.
5. The initial value of the path, along with the 'free flights' between the vertices. This can be done for either forward or backward (in time) evaluations of the path, as discussed in chapter 6. In addition, it is critical to consider carefully the phase of the wave function that is the basis of the path. Again, it is the phase interference that gives rise to the negative parts of the Wigner function.

As an example, let us consider the second-order vertices that may contribute. In figure 7.2, some of the paths that can arise are shown; there are only five of these diagrams because it is the diagonal terms of the density matrix that properly correlate with the density. Accordingly, these five paths all involve two interactions. In (a), we show two interactions with a local potential, such as an impurity or the nonlocal part of the Wigner potential. In (b), we see a path with phonon emission at both vertices (upper portion) or phonon absorption at both vertices (lower part). Finally, in (c) we illustrate a path with phonon emission at the first vertex and phonon absorption at the second vertex (upper panel), and a path with phonon absorption at the first vertex and phonon emission at the second vertex (lower panel). The paths in (c) are virtual processes, in which no phonon is really emitted or absorbed. In the classical world, where the collision occurs instantaneously, the two vertices would coalesce and the process would be non-existent. But, in the quantum world, it is possible for the scattering to begin at one vertex and then decide not to

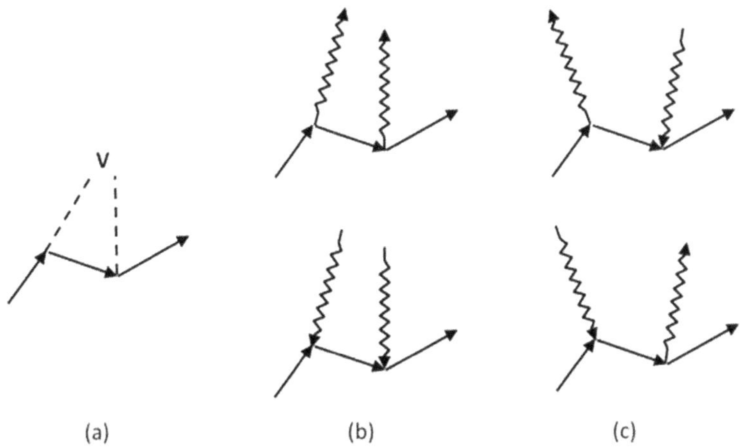

Figure 7.2. Interaction diagrams for types of scattering. (a) Potential scattering by an impurity or Wigner potential, (b) phonon emission (upper) and absorption (lower), and (c) virtual phonon emission (upper) and absorption (lower). Time is considered to be the horizontal axis and space the vertical axis for these diagrams.

occur with the second vertex [26]. These virtual events still contribute important phase shifts and lead to a result that can give a negative value for the Wigner function. Of course, for a given path and a longer time of propagation, a great many scattering events can occur during the path [27]. Importantly, it is possible to also consider multiple scattering processes. That is, there may be a path for the emission of phonon A and then one for the absorption of phonon B followed by these same events in the same sequence, which would be the emission of two phonons of different energy and momentum in a high-order process. One can note that the path diagrams are very similar to the Feynman diagrams that appear in Green's function theory. This is not an accident as the sources are very similar. In this regard, one may treat the horizontal axis as time and the vertical axis as space.

In the previous section dealing with the Levinson forms, it was remarked that one half of the phonon momentum was passed to the path at each vertex. For the virtual process, this one half of the momentum is then withdrawn at the second vertex, so no net momentum is passed to the path. But if the scattering process is completed, the net momentum change is $\hbar q$, just as in the classical case. The passing of half the momentum at each vertex can be seen in the Levinson formulas above. So, this is a correct interpretation of the manner in which the scattering events are inserted into the paths.

Another factor that is missing in this approach was already discussed in chapter 2. In the density matrix, and in the Wigner function, there are two wave functions. Perturbation theory discusses the corrections to a single wave function as it evolves in time. This can be seen in the development of the interaction representation above in section 7.2. But, in chapter 2, we noted that the conductivity bubble is composed of *two* wave function paths. Importantly, scattering events that transition between these two paths are known to be important [21]. That is, the two interaction vertices may be on different paths. This effect is known to be important for impurity

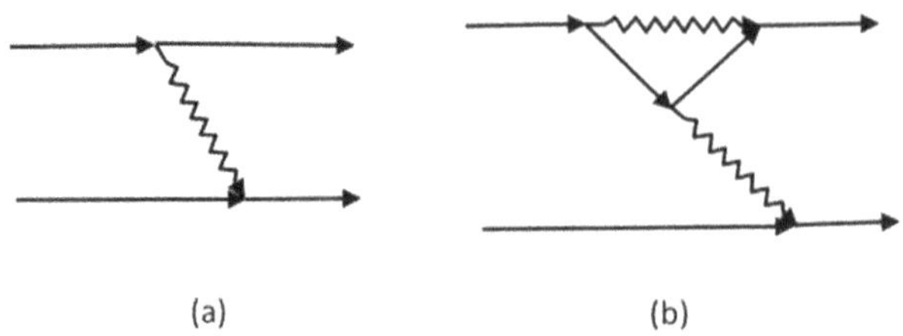

Figure 7.3. Scattering events which span the two Wigner function paths. (a) A simple diagram which would be part of the ladder diagram series composed of multiple such scatterings, which is common for impurity scattering. (b) A complicated multiple scattering in which the times overlap. Such a scattering even is exceedingly difficult to incorporate in Green's function theory, but is readily done with the Monte Carlo approach.

scattering, as it gives rise to the famous $(1 - \cos \vartheta)$ term found in classical transport theory for anisotropic scattering. It has been less studied in the case of phonon scattering. In particular, in the case of semiconductors, these terms have not been shown to be important primarily because transport here involves weak scattering. However, such interactions are depicted in figure 7.3, where both a simple interaction and a more difficult one are shown.

It is important to note that, in figure 7.2, there are two times and two positions [28]; that is, the scattering is nonlocal and involves a time duration [29]. This was evident in the development that lead to the Levinson forms. It is not something that is often discussed in connection with the Wigner function, but we have to recall that the momentum in the Wigner formulation arises from the difference in the two positions of the wave functions. The momentum is connected to the direction normal to the main diagonal in the density matrix. Hence, the time duration is connected to that required to transit between the two positions at the corresponding momentum. These different variables are thus tied together in a type of self-consistent manner. Many methods have been suggested to include the collisional displacement as a modification to the use of semi-classical scattering within the Monte Carlo method [21, 22, 30]. In the case of [30], it was suggested to replace the energy-conserving delta function with the joint spectral density, represented by a Lorentzian line centered around the classical energy conservation point. This was not very successful due to the very long tails of this function, which led to arbitrarily high energies. To properly include the role of the finite collision duration, we need to account for two effects: (1) the collision duration itself and the resulting energy shift, and (2) the position change during the collision. These are the momentum space shift and the real space shift, respectively.

It is the time between the two vertices that is important, as the 'particle' can be accelerated by the field for this time, thus the field effect between the vertices or intra-collisional field effect. This is an important quantum effect. For example, it has clearly been seen that this causes displacements of the particle both in real space and momentum space [31, 32]. We can illustrate this effect by considering a Gaussian Wigner function centered at a relatively high energy [31], as shown in figure 7.4. In

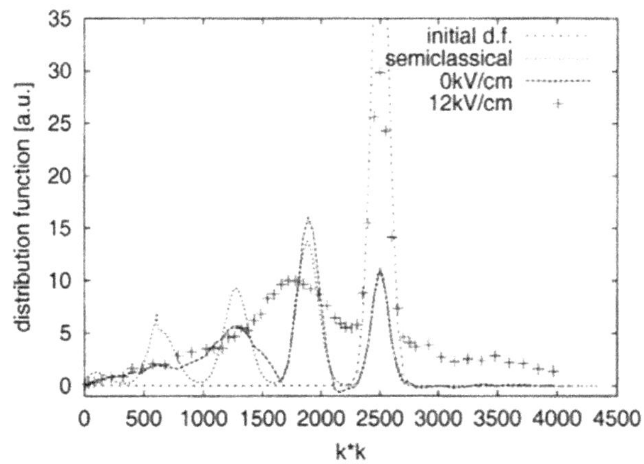

Figure 7.4. Simulation of an initial Gaussian Wigner function (dashed curve) after 300 fs. The solid curve is the simulation with a semi-classical distribution, while the dotted curve is the Wigner function with zero applied electric field. The crosses represent the Wigner distribution when a field of 12 kV cm^{-1} is applied. The broadening and shift is a result of the intra-collisional field effect. Reprinted from Nedjalkov *et al* [31] with permission from Elsevier.

the semi-classical simulation, and that at zero applied bias, there are a series of phonon peaks below the initial Wigner distribution. These peaks represent the emission of one, two, three, and four optical phonons in GaAs. Similar behavior is observed in femtosecond laser spectroscopy of GaAs and semi-classical Monte Carlo simulations of this process [33]. In this latter case, the phonon replicas are eventually washed out due to electron–electron and electron–hole interactions. These processes were not present in the Wigner function study. Instead, as can be seen in the 12 kV cm^{-1} curves, the general well-formed peaks are blurred together by the momentum space shifts caused by the intra-collisional field effect. Of great significance is the fact that the first phonon peak is shifted in position relative to the field-free case, and this is a direct result of the increase in the emission energy as the field is in the direction of the phonon emission process. As pointed out, there is also a spatial displacement of the electrons as well. This occurs since the carrier is accelerated or decelerated by the electric field, and its spatial position will be different from that expected for the classical particle. Other discussion on the difference between the quantum scattering and the classical scattering has been given by Bertoni *et al* [34].

The collision duration itself is typically only a few femtoseconds [13]. In this time, the change in position that occurs as a result of the collision duration may be written approximately as [32]

$$\delta \mathbf{x} = \left(\mathbf{v}(t) + \frac{e\mathbf{E} \cdot \mathbf{v}(t)}{mv(t)} \frac{\tau_c}{2} \right) \tau_c, \tag{7.36}$$

where $\mathbf{E}$ is the electric field that is experienced during the collision, and τ_c is the collision duration. During the interaction, there is an overlap between the wave packet at **x** and that at **x**+δ**x**. We can assume that these wave functions are Wigner packets which are Gaussian in shape, which leads to an overlap that varies as

$$I(\delta \mathbf{x}) = \frac{1}{\sqrt{4\pi\sigma^2}} e^{-(\delta x)^2/4\sigma^2}, \tag{7.37}$$

where

$$\sigma = \sqrt{\frac{\hbar^2}{12mk_BT}} \sim 1.9 \text{ nm}, \tag{7.38}$$

is the de Broglie thermal wave length, although other appropriate lengths have been discussed in chapter 4. That is, the numerical factors come from the same discussion as that of the effective potential in the chapter 4. Then, the probability for a given shift in position is

$$P(\delta x) = |I(\delta x)|^2 = \frac{1}{4\pi\sigma^2} e^{-(\delta x)^2/2\sigma^2}, \tag{7.39}$$

with the direction of the shift given by the relation between the velocity and the electric field. This provides a method of incorporating the displacement in time and space when using the semi-classical (instantaneous) approximation for the scattering process. When the collision duration is known, one can estimate the position shift from (7.36) and then evaluate the probability for such a shift from (7.39). As the approach is for the Monte Carlo, a decision on whether or not to accept the shift can be made naturally. In a sense, this physical shift, and the evolution of the Wigner function during the collision, introduces a non-Markovian behavior in the equation of motion [35].

The non-zero collision duration and the resulting position displacement can lead to some interesting effects beyond those normally expected. Consider, for example, the structure in figure 7.5 [36]. In panel (a), we have a particle approaching a barrier of modest height, such as may be found in the RTD. If there is no collision, then it is likely that the particle will be reflected backwards, although there is a small probability that it will penetrate the barrier. In the absence of any positional displacement, this particle could also absorb or emit a phonon without affecting the fact that it will likely be reflected. In panel (b), however, we consider the effect of a particle undergoing the intra-collisional field effect near the barrier. Now, if the field is sufficiently high, the particle may gain enough energy to surpass the barrier and wind up in a state which goes directly over the barrier. This may be accompanied by a positional displacement, which is indicated in the figure. During this collision, the particle may emit or absorb a phonon. If it emits a phonon, then it is obvious that it will wind up with a lower kinetic energy in its final state. But it may also absorb a phonon and still wind up with a lower kinetic energy in its final state. The difference in these two processes is merely a difference in the energy that must be absorbed from the electric field during the collision. Thus, our energy conservation requirement becomes

$$E(\mathbf{k} \pm \mathbf{q}) - E(\mathbf{k}) \mp \hbar\omega_{\mathbf{q}} = V(\mathbf{x}) - V(\mathbf{x} + \delta\mathbf{x}). \tag{7.40}$$

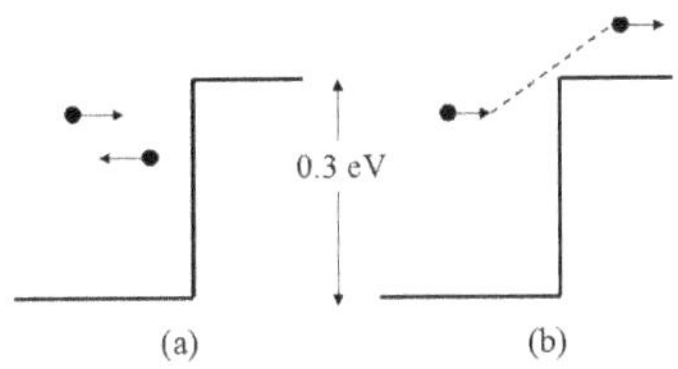

Figure 7.5. (a) Electrons classically impinging upon a barrier and being reflected. (b) Electrons hopping over the barrier due to energy gained during the intra-collisional field effect and the emission or absorption of an optical phonon.

As we note in the figure, one has to carefully handle the kinetic energy in the presence of complicated potentials, particularly when using the Monte Carlo approach. If one treats the nonlocal parts of the Wigner potential as a scattering process, then the events in figure 7.4 must be handled by a complicated higher-order scattering vertex set.

References

[1] Fetter A L and Walecka J D 1971 *Quantum Theory of Many-particle Systems* (New York: McGraw-Hill)
[2] Enz C P 1992 *A Course on Many-body Theory Applied to Solid-State Physics* (Singapore: World Scientific)
[3] Messiah A 1961 *Quantum Mechanics* vol 1 (New York: Interscience) p 442
[4] Mahan G D 1981 *Many-Particle Physics* (New York: Plenum)
[5] Shockley W and Bardeen J 1950 *Phys. Rev.* **77** 407
Shockley W and Bardeen J 1950 *Phys. Rev.* **80** 72
[6] Ferry D K, Goodnick S M and Bird J 2009 *Transport in Nanostructures* 2nd edn (Cambridge: Cambridge Univ. Press) section 2.7.2
[7] Kittel C 1986 *Introduction to Solid State Physics* 6th edn (New York: Wiley)
[8] Seeger K 1989 *Semiconductor Physics* 4th edn (Berlin: Springer)
[9] Levinson I B 1970 *Zh. Exp. Teor. Fys.* **57** 660
Tr. in Levinson I B 1970 *Tr. Sov. Phys. JETP* **30** 362
[10] Thornber K K and Feyman R P 1970 *Phys. Rev.* B **1** 4099
Thornber K K and Feyman R P 1971 *Phys. Rev.* B **4** 674
[11] Barker J R 1973 *J. Phys.* C **6** 2663
[12] Bertoncini R, Kriman A M and Ferry D K 1990 *Phys. Rev.* B **41** 1390
[13] Bordone P, Vasileska D and Ferry D K 1996 *Phys. Rev.* B **53** 3846
[14] Filinov V S, Thomas P, Varga I, Meier T, Bonitz M, Fortov V E and Koch S W 2003 *J. Phys.* A **36** 5905
[15] Jensen K L and Buot F A 1991 *IEEE Trans. Electron Dev.* **38** 2337
[16] Barker J R 1985 *Physica* B **134** 22
[17] Bialas A and Krzywicki A 1995 *Phys. Lett.* B **354** 134
[18] Rossi F, Poli P and Jacoboni C 1992 *Semicond. Sci. Technol.* **7** 1017
[19] Barker J R and Ferry D K 1979 *Phys. Rev. Lett.* **42** 1779
[20] Nedjalkov M, Vasileska D, Ferry D K, Jacoboni C, Ringhofer C, Dimov I and Palankovski V 2006 *Phys. Rev.* B **74** 035311
[21] Pascoli M, Bordone P, Brunetti R and Jacoboni C 1998 *Phys. Rev.* B **58** 3503

[22] Bertoni A, Bordone P, Brunetti R and Jacoboni C 1999 *J. Phys. Condens. Matter* **11** 5999
[23] Bordone P, Bertoni A, Brunetti R and Jacoboni C 2001 *VLSI Des.* **13** 211
[24] Jacoboni C, Bertoni A, Bordone P and Brunetti R 2001 *Math. Comput. Simul.* **55** 67
[25] Shifren L, Ringhofer C and Ferry D K 2003 *IEEE Trans. Electron Dev.* **50** 769
[26] Brunetti R, Jacoboni C and Rossi F 1989 *Phys. Rev.* B **39** 10781
[27] Bordone P, Pascoli M, Brunetti R, Bertoni A, Jacoboni C and Abramo A 1999 *Phys. Rev.* B **59** 3060
[28] Brunetti R, Bertoni A, Bordone P and Jacoboni C 2001 *VLSI Design* **13** 375
[29] Iotti R C, Dolcini F and Rossi F 2017 *Phys. Rev.* B **96** 115420
[30] Reggiani L, Lugli P and Jauho A P 1987 *Phys. Rev.* B **36** 6602
[31] Nedjalkov M, Kosina H, Kosik R and Selberherr S 2002 *Microelectron. Eng.* **63** 199
[32] Shifren L and Ferry D K 2003 *Phys. Lett.* A **306** 332
[33] Osman M A and Ferry D K 1987 *Phys. Rev.* B **36** 6018
[34] Bertoni A, Bordone P, Brunetti R, Jacoboni C and Sano N 1999 *Physica* B **272** 299
[35] Nettel S and Beck H 2014 *Physica* B **436** 91
[36] Shifren L and Ferry D K 2002 *Physica* B **314** 72

IOP Publishing

The Wigner Function in Science and Technology

David K Ferry and Mihail Nedjalkov

Chapter 8

Entanglement

Quantum mechanics began, of course, with the work of Max Planck [1, 2]. It evolved and grew through the work of many others, but its most important attribute, entanglement, was not recognized until 1935. The arguments between Einstein and Bohr perhaps reached their climax in this latter year with the EPR paradox [3]. In this latter work, the authors discussed the idea of two particles being simultaneously created by a single decay process, following which they would move away from each other with equal and opposite momentum. It is not our point to revisit the arguments for and against EPR, but merely to establish that the two particles existed in the thought experiment. It was always more or less assumed that when two bodies interacted quantum mechanically, they would be linked much as the correlation that builds up between two bodies that collided classically. It was Schrödinger who really put the concept together in terms of the wave function, and it was he who coined the term *entanglement* to explain the interaction. As he says [4]:

> 'If two separated bodies, each by itself known maximally, enter a situation in which they influence each other, and separate again, then there occurs regularly that which I have just called entanglement of our knowledge of the two bodies.'

Generally, it is possible to know maximal information about each of the two bodies. If there is no interaction, the two bodies have no way in which to 'learn' information about the other. A measurement of one can furnish no information about the other. The important thing about the entanglement is that when they interact, the two bodies are formed into a single system in which each body leaves traces of itself on the other body. After the interaction [4]

> '… the knowledge remains maximal, but at its end, if the two bodies have again separated, it is not again split into a logical sum of knowledges (*sic*)

doi:10.1088/978-0-7503-1671-2ch8

about the individual bodies. What remains of *that* may have become less than maximal, even very strongly so.'

The important point to be made here is that, even after the two bodies have quit interacting and have moved some distance away from one another, they remain a single system, and the maximal knowledge of this entire single entity is likely more than the available knowledge of the original two bodies. The interaction itself may be classical or quantum mechanical and occurs when the bodies are relatively close to one another. Even when they are far apart, the memory of the interaction is retained until some form of decoherence occurs.

The single system of the two bodies remains so until one or the other, or both, undergo interaction with another system. This may cause further entanglement or it may destroy the entanglement of the first system. The latter can be called a decoherence effect, after which the system again becomes two separate bodies. But, we no longer have full knowledge of these two bodies, as we have not specified the nature of the secondary interaction, and the nature of the decoherence likely causes additional loss of information.

Later, Bohm would again raise the EPR argument. While EPR had used two simple particles and let them interact, Bohm [5, 6] suggested another method of phrasing the paradox. He suggested that we consider a diatomic molecule, which is split apart by a method that does not affect the spin of each of the two atoms. In the molecule, the total spin is zero as the two atoms have equal and opposite spin. In the splitting process, this does not change. Once the two atoms have moved apart, there is no further interaction. However, each molecule has both a position and a momentum. Yet, Bohm suggests that the total wave function for the two atoms can be written simply as

$$\psi(1, 2) = \frac{1}{\sqrt{2}}\left[\psi_{\uparrow}(1)\psi_{\downarrow}(2) - \psi_{\downarrow}(1)\psi_{\uparrow}(2)\right], \tag{8.1}$$

where the arrows indicate the spin direction and the numbers refer to the two particles or atoms. But this is an incomplete wave function, and only describes the spin states of the two atoms. (The failures of such a trivial wave function are discussed in connection with the Schrödinger's cat paradox in ref [7].) One could say that this is a minimal information wave function, but it has in recent years become the de facto wave function for the EPR problem. For example, we know that each individual atom has a position and a momentum, but this wave function *can tell us nothing about these individual quantities* even if a measurement is made on the system seeking the position or the momentum. Indeed, the positions and the momenta of the two atoms must be considered as hidden variables, as no experiment on the system as described by (8.1) can yield any information about the probability distribution of these quantities. Another issue is that a measurement of the spin of either atom can yield only information about a single measurement. It gives either up or down! If we are to believe that the proper wave function gives us the probability distribution of possible values, then (8.1) is a failure. Again, it is not our aim in the present work to dig deeper into arguments over the sufficiency of any wave function. Instead, it is to point out that the Wigner distribution function provides a phase space representation of position and momentum, and has been shown to describe the existence of the entanglement at the same time. In this sense, it provides a form which answers many of the failures of (8.1) in diverse measurement situations.

8.1 An illustration of entanglement

As an illustration of the entanglement of a total wave function, we want to consider the entanglement of two Gaussian wave packets, which begin at $x = 0$ at $t = 0$. It is sufficient to do this in one dimension to illustrate the point, but we want to make sure that we have the proper moving wave packet. Hence, let us begin with a single wave packet in momentum space that is centered around a momentum $p_0 = \hbar k_0$ and which is thus given as

$$\varphi(k) = \left(\frac{2\sigma^2}{\pi}\right)^{1/4} e^{-\sigma^2(k-k_0)^2}, \tag{8.2}$$

where σ will be related to the spread of the Gaussian in real space. We can now Fourier transform this Gaussian into real space as

$$\begin{aligned}\psi(x, t) &= \frac{1}{\sqrt{2\pi}}\int_{-\infty}^{\infty} dk\ \varphi(k)e^{i(kx-\omega t)} = \frac{1}{\sqrt{2\pi}}\left(\frac{2\sigma^2}{\pi}\right)^{1/4}\int_{-\infty}^{\infty} dk\ e^{-\sigma^2(k-k_0)^2+i(kx-\omega t)} \\ &= \left(\frac{1}{2\pi\sigma^2}\right)^{1/4} e^{i(k_0x-\omega t)}e^{-x^2/4\sigma^2},\end{aligned} \tag{8.3}$$

and we see that 2σ is the standard deviation of the Gaussian wave packet. The leading exponential provides the propagation wave nature of the packet. Of course, the variable x is really a shifted variable in time, where it is properly $x - vt$, where $v = \hbar k_0/m$.

Now, if we assume two wave packets, one moving with $k_0 > 0$ and one moving with $k_0 < 0$, each will form its own Wigner function in the Wigner transformation. With the transformation, the two wave functions now become

$$f_W^{\pm}(x, p, t) = \frac{2}{h}e^{-2\sigma^2(p\mp p_0)^2-x^2/2\sigma^2}, \tag{8.4}$$

where the upper sign is for the wave moving to the right and the lower sign is for the wave moving to the left, and $p = \hbar k$, $p_0 = \hbar k_0$. Hence, the two wave packets, each formed from its own wave function packet, have independent identities. If, however, we assume that they have interacted so that they form a single wave function of the form

$$\psi_T(x, t) = \frac{1}{\sqrt{2}}[\psi_1(x - x_0, k_0) + \psi_2(x + x_0, -k_0)], \tag{8.5}$$

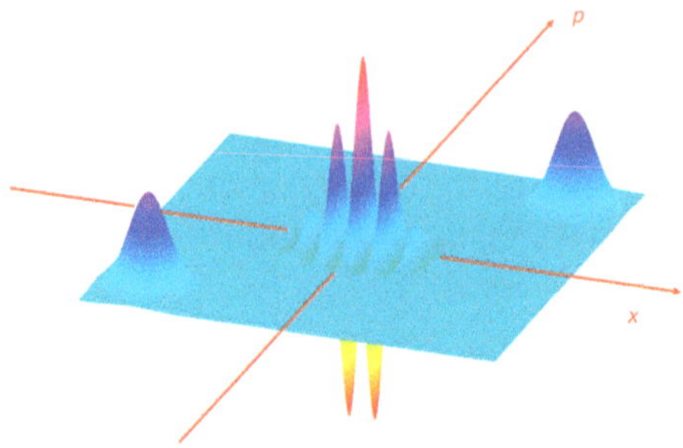

Figure 8.1. Two Gaussian wave packets are shown, under the assumption that they form a single wave function. Depicted is the resulting Wigner function, with the entanglement exhibited around $x = p = 0$.

where $x_0 = vt$, then we will obtain a much different result. Obviously, ψ_1 is a wave packet moving toward the positive x direction while ψ_2 is a wave packet moving toward the negative x direction. Now, when we perform the Wigner transform, we encountered terms that are cross-products of the two wave packets, and this leads to the result that

$$f_W(x, p, t) = \frac{1}{4}\left[f_W^+(x, p, t) + f_W^-(x, p, t) + 2e^{-x^2/2\sigma^2-2\sigma^2p^2}\cos(2k_0x)\right]. \quad (8.6)$$

Obviously, the third term in the brackets is a new term that represents the memory interference that arises from the entanglement between the two wave packets [8]. This term is centered around $x = p = 0$, and oscillates rapidly along the x axis. In figure 8.1, we plot the Wigner function of (8.6) at a time for which the two original packets have moved away from $x = 0$. The central entanglement term is still seen to be present.

In chapter 6, we illustrated a tunneling Gaussian wave packet through a single potential barrier. There, we discussed the connection between the transmitted and reflected wave packets. The latter two packets arise from the splitting of a single incident wave packet. Hence, they should be entangled as this contains the memory that they were formed from a single process. Reversing time allows the reformation of the original packet only if this entanglement behavior is retained in the full system. In figures 8.2 and 8.3, we show the equivalent wave packet from figure 6.1 at a later time [9], so that the transmitted and reflected packets are nearly fully formed. Yet there remains a rapid oscillatory part near the center of the barrier where the decomposition of the incident packet occurs. Again, this is centered around $p = 0$, and is rapidly oscillating in x, reflecting the same behavior as in (8.6) and figure 8.1.

8.2 Entanglement in harmonic oscillators

The fundamental dynamics of two quantized oscillators, which interact with one another, provides another insight into entanglement and its measurement. The quantized oscillators are important in another regard, and Planck originally described his black body radiation in terms of these oscillators [1] and provided

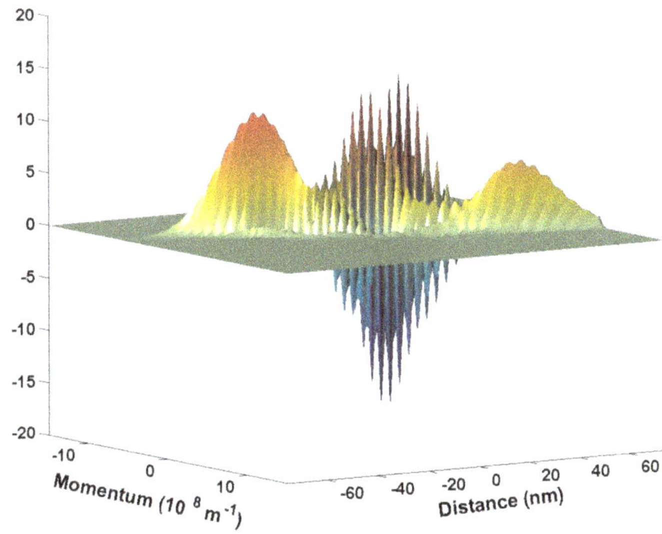

Figure 8.2. (a) The tunneling packet of figure 6.1 is shown at a later time, where the transmtted packet (on the right) and the reflected packet (on the left) are nearly fully formed. The memory that these came from a single incident packet is retained in the entanglement around the barrier and centered at $p = 0$.

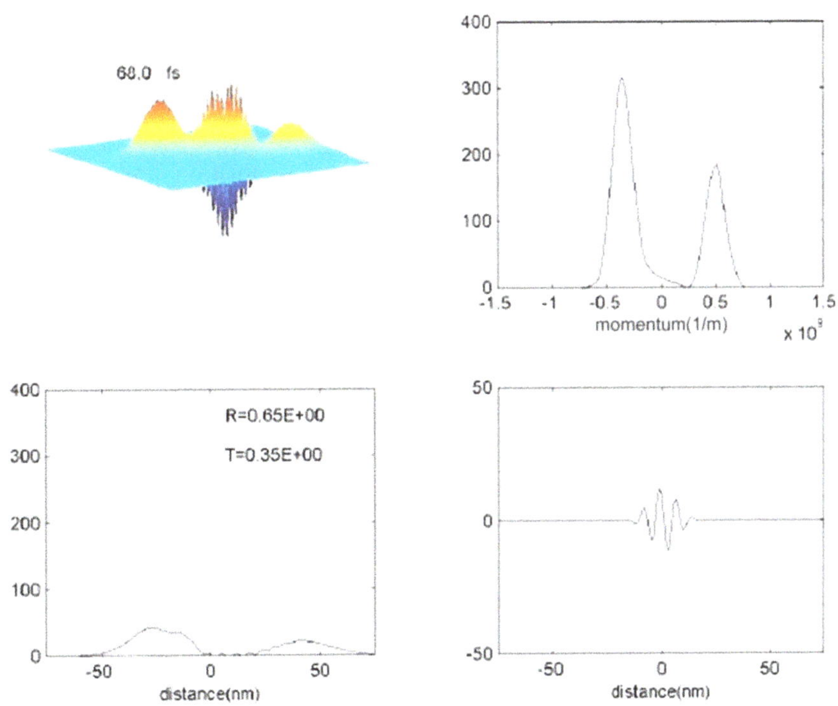

Figure 8.3. (b) A video showing the tunneling packet, with the projection onto the momentum and position axes, as well as illustration of the entanglement in the lower-right corner. A video is available online at http://iopscience.iop.org/book/978-0-7503-1671-2.

the quantization conditions. And, of course, we use these oscillators in a great variety of scientific fields, including in the description of phonons in condensed matter physics and quantized electromagnetic fields. The problem at hand has been discussed previously by many, but we follow the approach of Kim and Iafrate [10], and particularly that of Iafrate [11]. The Hamiltonian for the two coupled harmonic oscillators can be written as [10]

$$H = \hbar\omega_1\left(a_1^\dagger a_1 + \frac{1}{2}\right) + \hbar\omega_2\left(a_2^\dagger a_2 + \frac{1}{2}\right) + \hbar\kappa(a_1^\dagger a_2 + a_1 a_2^\dagger), \tag{8.7}$$

where the creation and annihilation operators for each of the two harmonic oscillators give us the occupation $n_\nu = a_\nu^\dagger a_\nu$ with $\nu = 1,2$. Without loss of generality, we can assume that $\omega_1 \geqslant \omega_2$. While two such harmonic oscillators can be used to define a two-dimensional harmonic oscillator, here we remain in a more general orientation, in which the angular momentum can be defined as

$$\begin{aligned} L_x &= \frac{\hbar}{2}(a_1^\dagger \otimes a_2 + a_1 \otimes a_2^\dagger) \\ L_y &= \frac{\hbar}{2}(a_1^\dagger \otimes a_2 - a_1 \otimes a_2^\dagger) \\ L_z &= \frac{\hbar}{2}(n_1 - n_2) \\ L_T &= \frac{\hbar}{2}(n_1 + n_2). \end{aligned} \tag{8.8}$$

The first two terms are defined in terms of a tensor product. Each creation or annihilation operator describes a combination of a position and a momentum operator, so defines a sense of direction for rotation. The angular momentum can then be understood, for example, by looking at the two-dimensional harmonic oscillator, where one works with complex combinations of L_x and L_y, which are the traditional raising and lowering operators for the pair of oscillators.

Each of the two oscillators can be described by a one-oscillator state defined by its number operator, as $|n_\nu\rangle$ and the coupled two-oscillator state may be defined by

$$|n_1 n_2\rangle = e^{-iL_y\gamma/\hbar}|n_1\rangle|n_2\rangle, \tag{8.9}$$

where

$$\tan(\gamma) = 2\kappa/\Delta\omega, \quad \Delta\omega = \omega_1 - \omega_2 \geqslant 0. \tag{8..10}$$

Now, the energy may be written as [10]

$$\begin{aligned} E_{n_1 n_2} &= \frac{\hbar}{2}(n_1 + n_2 + 1)(\omega_1 + \omega_2) + \frac{\hbar}{2}(n_1 - n_2)\Omega \\ \Omega &= \sqrt{(\omega_1 - \omega_2)^2 + (2\kappa)^2}. \end{aligned} \tag{8.11}$$

Since the exponential operator in (8.9) is unitary, the two particle states are orthonormal,

$$\langle n_1' n_2' \mid n_1 n_2 \rangle = \langle n_1' | \langle n_2' | | n_1 \rangle \mid n_2 \rangle = \delta_{n_1' n_1} \delta_{n_2' n_2}, \tag{8.12}$$

and form a complete set for all values of the number operators.

We want to now consider any arbitrary function described by these two oscillator states. That is, for any quantum function f that is written as an expansion in the two oscillator wave functions, it will satisfy

$$i\hbar \frac{\partial f}{\partial t} = [H, f], \tag{8.13}$$

where H is the Hamiltonian (8.7). We can now develop the Wigner function for this quantum function by the normal Wigner transform as

$$f_W(\mathbf{x}, \mathbf{p}) = \frac{1}{h^3} \int d^3\mathbf{s} \left\langle \mathbf{x} - \frac{\mathbf{s}}{2} \middle| f \middle| \mathbf{x} + \frac{\mathbf{s}}{2} \right\rangle e^{i\mathbf{p}\cdot\mathbf{s}/\hbar}. \tag{8.14}$$

Let us now assume that we go ahead and write f in terms of two oscillator states so that

$$\begin{aligned} f_W(\mathbf{x}, \mathbf{p}) &= \frac{1}{h^3} \int d^3\mathbf{s} \sum_{n_1, n_2} \sum_{n_1', n_2'} \left\langle \mathbf{x} - \frac{\mathbf{s}}{2} \mid n_1 n_2 \right\rangle \\ &\quad \langle n_1, n_2 | f | n_1', n_2' \rangle \left\langle n_1', n_2' | \mathbf{x} + \frac{\mathbf{s}}{2} \right\rangle e^{i\mathbf{p}\cdot\mathbf{s}/\hbar} \\ &= \frac{1}{h^3} \int d^3\mathbf{s} \sum_{n_1, n_2} \sum_{n_1', n_2'} \psi^{\dagger}_{n_1 n_2}\left(\mathbf{x} - \frac{\mathbf{s}}{2}\right) \\ &\quad \langle n_1, n_2 | f | n_1', n_2' \rangle \psi_{n_1', n_2'}\left(\mathbf{x} + \frac{\mathbf{s}}{2}\right) e^{i\mathbf{p}\cdot\mathbf{s}/\hbar}, \end{aligned} \tag{8.15}$$

where, in the second line, we have introduced the spatial wave functions that arise from the first and third terms in the first line, by which the scalar products transition to a different basis set. The central expectation value that remains in the second line is a c-number and can be removed from the integral. Hence, we can rewrite (8.15) as

$$\begin{aligned} f_W(\mathbf{x}, \mathbf{p}) &= \sum_{n_1, n_2} \sum_{n_1', n_2'} \langle n_1, n_2 | f | n_1', n_2' \rangle \\ &\quad \frac{1}{h^3} \int d^3\mathbf{s}\; \psi^{\dagger}_{n_1 n_2}\left(\mathbf{x} - \frac{\mathbf{s}}{2}\right) \psi_{n_1', n_2'}\left(\mathbf{x} + \frac{\mathbf{s}}{2}\right) e^{i\mathbf{p}\cdot\mathbf{s}/\hbar} \\ &\equiv \sum_{n_1, n_2} \sum_{n_1', n_2'} \langle n_1, n_2 | f | n_1', n_2' \rangle T_{n_1', n_2', n_1 n_2}(\mathbf{x}, \mathbf{p}). \end{aligned} \tag{8.16}$$

Here, the T-function introduced in the second line set represents a two-oscillator wave functions in phase space. One can use the results of Moyal [12]: it is easy to show that these T-functions satisfy a form of orthogonality as

$$\int d^3\mathbf{x} \int d^3\mathbf{p} T^*_{n_1''n_2'',n_1'''n_2'''}(\mathbf{x}, \mathbf{p}) T_{n_1'n_2', n_1n_2}(\mathbf{x}, \mathbf{p}) = \frac{1}{h^3} \delta_{n_1'n_1'''} \delta_{n_2'n_2'''} \delta_{n_1''n_1} \delta_{n_2''n_2}. \tag{8.17}$$

As a consequence, it is also easy to determine the value of the matrix element leading the integral in (8.16) from the Wigner function, once it is determined, by integrating the latter with the appropriate T-function.

The importance of the above development lies in the fact that we can now define a measure of the entanglement between the two harmonic oscillators as [10]

$$M_{en} = 1 - \left| f^T_{n_1n'_1} \right|^2, \tag{8.18}$$

where

$$f^T_{n_1n'_1} = \sum_{n_2} \langle n_1n_2 \,|\, f \,|\, n'_1n_2 \rangle \tag{8.19}$$

is the trace over the quasi-diagonal matrix element. The matrix element terms in (8.19) can be written as

$$\begin{aligned} f^T_{n_1n'_1} &= \int d^3\mathbf{x} \int d^3\mathbf{p} \sum_{n_2} \int d^3\mathbf{s} \left[\psi^\dagger_{n'_1n_2}\left(\mathbf{x} + \frac{\mathbf{s}}{2}\right) \psi_{n_1n_2}\left(\mathbf{x} - \frac{\mathbf{s}}{2}\right) e^{i\mathbf{p}\cdot\mathbf{s}/\hbar} \right]^* f_W(\mathbf{x}, \mathbf{p}) \\ &= \int d^3\mathbf{x} \int d^3\mathbf{p} K_{n_1n'_1}(\mathbf{x}, \mathbf{p}) f_W(\mathbf{x}, \mathbf{p}), \end{aligned} \tag{8.20}$$

in which we have introduced in the second line the kernel that projects out of the Wigner function, or another function f, those parts which do not contribute to the entanglement measure. That is, it projects out those parts which could be considered part of the pseudo-diagonal contributions. Let us consider this kernel term a little further. From (8.20), the kernel is defined to be

$$K_{n_1n'_1}(\mathbf{x}, \mathbf{p}) = \sum_{n_2} \left[\int d^3\mathbf{s} \psi^\dagger_{n'_1n_2}\left(\mathbf{x} + \frac{\mathbf{s}}{2}\right) \psi_{n_1n_2}\left(\mathbf{x} - \frac{\mathbf{s}}{2}\right) e^{i\mathbf{p}\cdot\mathbf{s}/\hbar} \right]^*. \tag{8.21}$$

Now, let us expand the two wave functions in a Taylor series about the position $\mathbf{x}$, so that they become

$$\begin{aligned} \psi^\dagger_{n'_1n_2}\left(\mathbf{x} + \frac{\mathbf{s}}{2}\right) &= \sum_{j=0}^{\infty} \left(\frac{\mathbf{s}}{2} \cdot \nabla\right)^j \psi^\dagger_{n'_1n_2}(\mathbf{x}) \equiv \sum_{j=0}^{\infty} f_j(\mathbf{x}, \mathbf{s}) \\ \psi_{n'_1n_2}\left(\mathbf{x} - \frac{\mathbf{s}}{2}\right) &= \sum_{k=0}^{\infty} \left(-\frac{\mathbf{s}}{2} \cdot \nabla\right)^k \psi^\dagger_{n'_1n_2}(\mathbf{x}) \equiv \sum_{k=0}^{\infty} g_k(\mathbf{x}, \mathbf{s}), \end{aligned} \tag{8.22}$$

where we have defined two new functions in the last term of each line. The product of the two wave functions appearing in (8.21) can now be written as [11]

$$\begin{aligned}\psi^{\dagger}_{n'_1n_2}\left(\mathbf{x}+\frac{\mathbf{s}}{2}\right)\psi_{n_1n_2}\left(\mathbf{x}-\frac{\mathbf{s}}{2}\right) &= \sum_{j=0}^{\infty}\sum_{k=0}^{\infty} f_j(\mathbf{x},\mathbf{s})g_k(\mathbf{x},\mathbf{s}) \\ &= \sum_{j=0}^{\infty}\frac{1}{j!}\sum_{k=0}^{\infty}\frac{j!}{k!(j-k)!} \\ &\quad\times\left[\left(\frac{\mathbf{s}}{2}\cdot\nabla\right)^{j-k}\psi^{\dagger}_{n'_1n_2}(\mathbf{x})\right]\left[\left(-\frac{\mathbf{s}}{2}\cdot\nabla\right)^{k}\psi_{n'_1n_2}(\mathbf{x})\right].\end{aligned} \tag{8.23}$$

If we use this in the integral over $\mathbf{s}$ that appears in (8.21), we can then introduce some short-hand properties, which arise from the facts that

$$\begin{aligned}\int d^3\mathbf{s}e^{i\mathbf{p}\cdot\mathbf{s}/\hbar} &= \delta(\mathbf{p}) \\ \mathbf{s}e^{i\mathbf{p}\cdot\mathbf{s}/\hbar} &= -i\hbar\frac{\partial}{\partial\mathbf{p}}e^{i\mathbf{p}\cdot\mathbf{s}/\hbar}.\end{aligned} \tag{8.24}$$

If we insert these properties into the term in square brackets in (8.21), we can rewrite the equation as

$$\begin{aligned}\psi^{\dagger}_{n'_1n_2}\left(\mathbf{x}+\frac{\mathbf{s}}{2}\right)\psi_{n_1n_2}\left(\mathbf{x}-\frac{\mathbf{s}}{2}\right) &= \sum_{j=0}^{\infty}\frac{1}{j!}\left(\frac{i\hbar}{2}\right)^{j}\sum_{k=0}^{\infty}\frac{j!}{k!(j-k)!}(-1)^{j-k} \\ &\quad\times\left[\left(\frac{\partial}{\partial\mathbf{p}}\cdot\nabla\right)^{j-k}\psi^{\dagger}_{n'_1n_2}(\mathbf{x})\right]\left[\left(-\frac{\partial}{\partial\mathbf{p}}\cdot\nabla\right)^{k}\psi_{n'_1n_2}(\mathbf{x})\right]\delta(\mathbf{p}).\end{aligned} \tag{8.25}$$

Using this in (8.21), we arrive at the form for the kernel to be

$$\begin{aligned}K_{n_1n'_1}(\mathbf{x},\mathbf{p}) &= \sum_{j=0}^{\infty}\frac{1}{j!}\left(\frac{i\hbar}{2}\right)^{j}\sum_{k=0}^{\infty}\frac{j!}{k!(j-k)!}(-1)^{k} \\ &\quad\times\sum_{n_2}\left[\left(\frac{\partial}{\partial\mathbf{p}}\cdot\nabla\right)^{j-k}\psi_{n'_1n_2}(\mathbf{x})\right]\left[\left(-\frac{\partial}{\partial\mathbf{p}}\cdot\nabla\right)^{k}\psi^{\dagger}_{n_1n_2}(\mathbf{x})\right]\delta(\mathbf{p}).\end{aligned} \tag{8.26}$$

As an illustration of the effect, let us look at the series expansion in terms of powers of $\hbar$. The leading terms to this order are given as

$$\begin{aligned}K_{n'_1n_1}(\mathbf{x},\mathbf{p}) &\simeq \sum_{n_2}\psi^{\dagger}_{n_1n_2}(\mathbf{x})\psi_{n'_1n_2}(\mathbf{x})\delta(\mathbf{p}) \\ &\quad+\frac{i\hbar}{2}\sum_{n_2}\left[\psi^{\dagger}_{n_1n_2}(\mathbf{x})\nabla\psi_{n'_1n_2}(\mathbf{x})-\psi_{n'_1n_2}(\mathbf{x})\nabla\psi^{\dagger}_{n_1n_2}(\mathbf{x})\right]\cdot\frac{\partial}{\partial\mathbf{p}}\delta(\mathbf{p}) \\ &= \rho_{n'_1n_1}(\mathbf{x})\delta(\mathbf{p})+\mathbf{J}_{n'_1n_1}(\mathbf{x})\cdot\frac{\partial}{\partial\mathbf{p}}\delta(\mathbf{p})+\dots,\end{aligned} \tag{8.27}$$

where we have introduced the density matrix

$$\rho_{n'n}(\mathbf{x}) = \sum_{n_2} \psi^{\dagger}_{nn_2}(\mathbf{x})\psi_{n'n_2}(\mathbf{x}) \tag{8.28}$$

and the current density

$$\mathbf{J}_{n'n}(\mathbf{x}) = \frac{i\hbar}{2}\sum_{n_2}\left[\psi^{\dagger}_{nn_2}(\mathbf{x})\nabla\psi_{n'n_2}(\mathbf{x}) - \psi_{n'n_2}(\mathbf{x})\nabla\psi^{\dagger}_{nn_2}(\mathbf{x})\right]. \tag{8.29}$$

We note that (8.29) differs from the normal probability current found in textbooks, as the mass term that usually appears is not present here. Hence, this current is a momentum probability current. If we now use the last line of (8.27) in the second line of (8.20), we arrive at

$$f^{T}_{n_1n'_1} = \int d^3\mathbf{x}\, \rho_{n_1n'_1}(\mathbf{x}) f_W(\mathbf{x}, 0) - \int d^3\mathbf{x}\, \mathbf{J}_{n_1n'_1}(\mathbf{x}) \cdot \left.\frac{\partial f_W(\mathbf{x}, \mathbf{p})}{\partial \mathbf{p}}\right|_{\mathbf{p}=0} + \dots \tag{8.30}$$

If we were to keep all orders of the expansion in $\hbar$, (8.30) becomes a series in the powers of the momentum derivatives of the Wigner function, or any other function used, all of which are evaluated at $\mathbf{p} = 0$. The coefficients of the series all depend upon the spatial derivatives of the eigenstate basis functions that are used in the system.

8.3 Measures of entanglement

In the previous section, we introduced a measure of entanglement with (8.18). Here, the measure was formed by subtracting the fraction of the total system that occupied the quasi-diagonal states which left those so-called off-diagonal states. In a sense, such a measure goes to the early statistical idea that entanglement was associated with the off-diagonal states of the density matrix. In fact, this might be too simple a definition of the entanglement. As we pointed out in the introduction to this chapter, the ideas of quantum entanglement arose from the EPR paradox [3]. Schrödinger described entanglement as something that identified the intrinsic order of statistical relations between subsystems of a compound quantum system, such as the two particles in EPR [4]. But even this leaves a thread of doubt about the understanding of just what constitutes entanglement. Modern approaches focus upon our equation (8.1), due to Bohm [6], to more clearly explain what is meant by entanglement. While this latter equation is written in terms of the eigenstates of each of the two particles, it cannot be separated into just a product of states of the two particles. In that sense, the wave function is an entangled state. That is, (8.1) describes what may be called mixed entangled states that have a form such that there is no possible decomposition into product states [13]. In this sense, the quasi-diagonal states of (8.18) are the separable states, and what is left is the non-separable part of the system—the so-called entangled part. The value 1 that leads in (8.18) arises from the orthonormality of the combined two harmonic oscillator system.

But (8.18) and the discussion above is just one possible measure of entanglement. The advent of what we now call quantum information theory and quantum computation has lead to a plethora of various measures of entanglement [14]. Many of these are based upon ideas from statistical physics and thermodynamics, in particular the ideas of entropy. In fact, this approach probably relies upon the original work of Shannon in describing measures of the fidelity of information in classical information theory [15]. The normal story goes, that Shannon, in a discussion with von Neumann, commented on his theory [16]:

> 'The theory is in excellent shape, except that I need a good name for "missing information".'

Von Neumann is said to have replied [16]:

> 'Why don't you call it entropy. In the first place, a mathematical development very much like yours already exists in Boltzmann's statistical mechanics, and in the second case, no one understands entropy very well, so in any discussion you will be in a position of advantage.'

So, there we have it. Von Neumann tells us that no one understands entropy, and Feynman tells us that no one understands quantum mechanics [17]. This suggests that quantum entropy is double cursed. But it seems to survive.

In statistical physics and thermodynamics, one normally considers systems which are not isolated but are not strictly in equilibrium as well. One describes the system not only with averages of position and momentum, but also via averages of energy, temperature, and entropy. This, of course, does not tell us what entropy is or is not. Rather, one normally thinks about entropy as a measure of disorder in the system and/or a measure of the *inability* to be able to convert the thermal energy into work done by the system. With such a definition, or description, we can recognize that entropy is additive, and then one can write that the change in entropy is described by

$$dS = dS_e + dS_i, \tag{8.31}$$

where dS_e arises from interactions and exchange of entropy between the system and its environment and dS_i represents the internal production, or destruction, of entropy. The external entropy production can be positive, negative, or zero depending upon how the system interacts with its environment. On the other hand, the internal entropy production is always greater than or equal to zero. Thus, for an isolated system $dS \geqslant 0$, which is just the second law of thermodynamics.

A rapidly expanding field is quantum information theory, which depends upon the underlying concept of entanglement and quantum coherence to achieve secure communications. It is thus surprising that a rigorous theory for the quantification of entanglement/coherence only recently has been suggested [18]. This theory follows the ideas based upon the quantification of entanglement. Moreover, it recently has been demonstrated that the two concepts, coherence and entanglement, are quantitatively equivalent [19, 20]; that is, any non-zero amount of coherence in a system can be converted into an equal amount of entanglement between that system and another initially incoherent one. This means that the two concepts which describe very different physics have a common mathematical foundation. Thus, talking about entanglement implies that we also have in mind the idea of quantum coherence. The mathematical approach has been developed in the framework of operator theory in Hilbert space, complete with proper basis sets and tensor product spaces. The Wigner form has been used to present the basic notions of the quantification theory of coherence in phase space terms [21].

In a non-equilibrium system, one hopes that it evolves to a steady-state system in which a local entropy is the same function of the various thermodynamic variables as the true equilibrium state entropy is of the equilibrium thermodynamic variables. This implies that there exists a fundamental differential form for the entropy that can be described by [22]

$$dS = \frac{1}{T}dU + \frac{p}{T}dV - \sum_{i=1}^{c}\frac{\mu_i}{T}dn_i, \tag{8.32}$$

where U is the internal energy of the system, T is the temperature, p is the pressure, V the volume, c the number of chemical components in the system, and μ_i is the chemical potential of the ith component of mole number n_i. This implies that there exists what we have called local equilibrium, or steady-state behavior, for the system, and that the local equation of state is independent of field gradients. The description of the local steady-state behavior is said to be the venue of non-equilibrium statistical mechanics. We have encountered this several times in previous chapters, where we invoked a drifted Maxwellian or a drifted Fermi–Dirac, described by parameters such as the drift velocity and electron temperature. Indeed, a formal method of describing such a parameterized distribution function is via its entropy and entropy production terms [23].

If we have a distribution in phase space described by an $f(x,p)$, the description of the system can be considered as being composed of a set of microstates which make up a macrostate describable by the set of thermodynamic variables for the local macrostate. Then, one definition of the entropy is

$$S = k_B \ln \Omega = k_B \ln[Z_{gr} - \beta(E - \mu N)], \tag{8.33}$$

where the first form is due to Boltzmann and the second form is the adaption to a grand canonical distribution found in condensed matter physics. Here, Ω is the number of microstates that compose the macrostate, k_B is Boltzmann's constant, Z_{gr} is the grand canonical partition function, $\beta = 1/k_BT$, E is the mean energy and N is the density. An important connection is that, if we define a probability of a microstate to be p_i, the first form can be rewritten as [23]

$$S = -\sum_{i=1}^{c} p_i \ln p_i, \tag{8.34}$$

where the negative sign arises from the fact that the probability is < 1. In particular, this last form is, in fact, the entropy term adopted by Shannon to describe information. In particular, we require the normalization

$$\sum_{i=1}^{c} p_i = 1. \tag{8.35}$$

Hence, if any individual probability is unity, its contribution is zero, and all other contributions are also zero due to (8.35). Physically, this just means that if we can predict the outcome of an experiment with certainty, then there is no indeterminacy in the outcome. Shannon proved that the information entropy described in this way is unique.

As we have already stated, the ideas of entropy date back to the nineteenth century. They were present in quantum mechanics already in 1927 [24], when von Neumann discussed his ideas on mixtures of quantum states through what we now call the density matrix [25]. In this case, the von Neumann entropy can be defined from the density matrix as

$$S = -\mathrm{Tr}\{\hat{\rho} \ln \hat{\rho}\}, \tag{8.36}$$

where

$$\hat{\rho} = \sum_j \eta_j \mid j\rangle\langle j \mid, \tag{8.37}$$

and η_j is the projection of the total wave function onto the pure states j. If the wave function is written in position space, then the projection onto a position state at x is given by the Wigner function via [8, 26]

$$\eta(\mathbf{x}) = \int d^3\mathbf{p} f_W(\mathbf{x}, \mathbf{p}), \tag{8.38}$$

which gives a true probability for position measurements. This in a sense has carried over to quantum information, when it was shown that the von Neumann entropy may be interpreted as the number of qubits necessary to transmit quantum states emitted by a statistical source [14, 27]. Moreover, it was also shown that the entropy of a subsystem can be greater than the entropy of the entire system only if the system was entangled [28]. So, obviously, the connection between von Neumann's entropy of a quantum system may be taken to be connected to the entanglement of the system. But here we have a problem in that the von Neumann entropy is a projection onto the pure states of the system. In contradiction, the entanglement discussed above is that portion of the system which cannot be separated into its pure state parts [29].

This can be explained a little further following a common approach in which we define a density matrix in terms of tensor product states, as was done in section 8.2. We write this density matrix as [13]

$$\rho = \rho^{(1)} \otimes \rho^{(2)}, \tag{8.39}$$

where the two density matrices on the right represent two subsystems which are uncorrelated. Then, if we compute the density matrix for the general system in terms of this tensor product state, we have a sum of different tensor product states as

$$\rho = \sum_i \eta_i \rho_i^{(1)} \otimes \rho_i^{(2)}, \tag{8.40}$$

with

$$\sum_i p_i = 1, \quad p_i > 0. \tag{8.41}$$

The use of (8.40) generally yields correlated measurements, which are *classical* correlations as the states in (8.40) are separable. Quantum entanglement is described by states which cannot be separated into tensor product states. Such a state is given by (8.1). Thus, if we project out the separable tensor product, or quasi-diagonal, states, as done in (8.18), what is left is a measure of the quantum entanglement of the system. And, following this prescription, if we take away the two Gaussian pulses in figures 8.1, 8.2 and 8.3, what is left is the entanglement. This, by itself, is one of the major reasons for the use of Wigner functions. The connections between entanglement and quantum information, on one hand, and statistical physics and thermodynamics, on the other, is still being explored at the present time [30–32].

Measuring entanglement is a somewhat difficult idea. Perhaps one of the earliest suggestions was to use Bell inequalities as a measure of entanglement [33]. The problem with this lies in the fact that most people assume that a violation of the Bell inequality signals a quantum system. This is not the case. In fact, what is normally recognized as the Bell inequality was introduced by Boole already in 1857 [34], well before quantum mechanics appeared, and it is well known how to violate the inequality classically [35]. More appropriately, Peres demonstrated that one could do a partial transpose of the density matrix for a compound bipartite system, and do this transpose on only one subsystem [36]. The remaining density matrix turned out to be a good test for the existence of entanglement [14]. This is fortunate as it means that measurements of entanglement can be placed on a firmer basis. Now, in general, the non-separable state of (8.1) is used to show the existence of entanglement. But, one normally encounters mixed rather than pure states, and the entanglement of mixed states is no longer just proportional to the simple nonproduct states. In the more complex case of mixed states, one can refer to the system as being entangled if the composite wave function cannot be written as a combination of product states [37]. This is the type of composite state used in the previous section for the coupled harmonic oscillators. There is a huge literature on the manipulation of entanglement [14], but it is not our purpose to review this literature here. Rather, we are interested more in its measurement, as we will discuss in the next several chapters. An important point with respect to this is that entanglement is what it is: it cannot be amplified or increased. The idea of having an axiomatic measure of entanglement was formalized already in 1997 [38], and it has been suggested that this monotonic behavior of entanglement under local operations and classical communications should be the only requirement for entanglement measures [39]. Nevertheless, it can be inferred from the relatively recent dates here, that there are still many entanglement measures that have been proposed, and the youthfulness of the field means that they have not been fully sorted out. Those interested in the great variety of measures, and the properties that they rely upon, should refer to ref. [14], and the book cited in refs [8, 13].

8.4 Some illustrative examples

8.4.1 Photons

Let us first discuss briefly the role of entanglement with photons, primarily because this is often viewed with a Wigner function (but it is also of interest in quantum information, and this is likely to be implemented in the not too distant future). The coupling between a two-level atom and a resonant (quantized) electromagnetic cavity is described by a relatively old model, termed the Jaynes–Cumming model[1] [40]. Generally, the electromagnetic field can interact with the atomic state and lead to Rabi oscillations of the atomic state population. However, something different happens in this particular model, in that the Rabi oscillations collapse and then reappear periodically. The temporal time over which this collapse and reappearance occurs has been termed the revival time [41]. When the field in the cavity is in a coherent state, and the atom is prepared in an excited state, it is found that the atomic and field states become rapidly entangled, and then subsequently disentangle at one half the revival time [42]. That is, the maximum entanglement occurs when the Rabi oscillations are strongest and the minimum entanglement occurs when the Rabi oscillations are minimal. More recently, it has been observed that the photon parity operator also shows similar behavior with one important difference. The peaks of the amplitude of the parity operator seem to occur when the Rabi oscillations are at their minimum. So, the amplitude and decay of this parity operator seems to be shifted by half a period from the Rabi oscillations [43]. In general, the atomic state Rabi oscillations occur with an average frequency [43]

$$\Omega = 2\lambda\sqrt{n+1}, \tag{8.42}$$

where λ is an effective coupling constant between the atom and the electromagnetic cavity, and n is the occupation number of the single mode of the cavity (an equivalent harmonic oscillator mode). This occupation number is given by the creation and annihilation operators of the mode via the normal form

$$n = a^{\dagger}a. \tag{8.43}$$

As the mode occupation builds up, the Rabi frequency increases and the amplitude of these oscillations increases. The photon parity operator is then defined as [43]

$$\Pi_F = (-1)^n e^{in\pi}, \tag{8.44}$$

[1] A more extensive discussion of this model is provided in section 11.1.

where ρ is the density operator. The normal expectation value of this operator may be found to be [42]

$$\langle \Pi_F(t) \rangle = e^{-n} \sum_{s=0}^{\infty} (-1)^s \frac{n!}{s!} \cos(2\lambda t \sqrt{s+1}), \tag{8.45}$$

where the time dependence arises from the time dependence of the effective Rabi frequency. The interesting fact is that for $t = 0$, the cosine is unity, and the expectation value of the parity operator goes asymptotically to zero (the time is scaled with the same coupling factor λ). Hence, when the density peaks and the Rabi oscillations are strongest, the parity operator vanishes, and vice versa. In general, for the initial conditions discussed, the Wigner function of the field splits into a pair of counter-rotating components in phase space [44]. These two components reach a maximum separation at one half the revival time and then recombine at the revival time. Hence, the peak of the parity operator corresponds to the maximum separation of the two parts of the Wigner function. The Wigner function itself can be written as a shifted expectation value of the parity operator, given as [45]

$$W(\alpha) = \frac{2}{\pi} \langle D(\alpha) \Pi_r D^{\dagger}(\alpha) \rangle, \tag{8.46}$$

where D is a displacement operator and α is a complex number that represents a point in phase space, i.e. the real and imaginary parts correspond to position and momentum. The displacement operators work on the two wave functions in the density operator, leading to the displaced versions that appear in the Wigner function description. In figure 8.4, we plot the Wigner function of the parity operator at a time when this operator is near its maximum amplitude and at a positive peak (of the oscillating operator). The x and y axes correspond to the real and imaginary parts of α, and thus to the position and momentum that appear in the Wigner function. While the two major peaks are evident, there are clear interference fringes between them. These fringes themselves oscillate at the Rabi frequency when the parity amplitude is large.

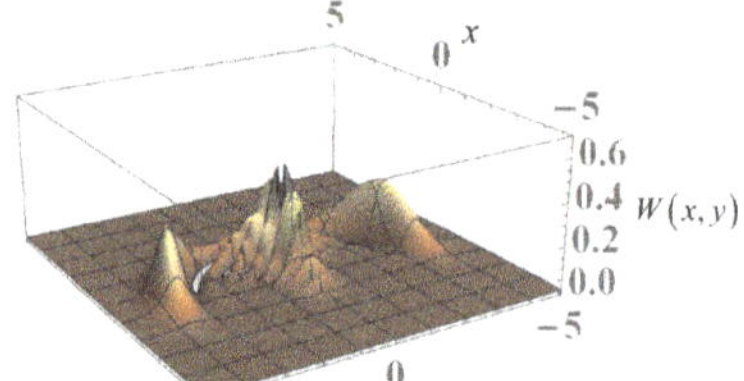

Figure 8.4. The Wigner function for the expectation of the parity operator at the scaled time for which this quantity is a maximum (near one half the revival time) [43]. The x and y axes correspond to the real and imaginary parts of α, as described in the text. This figure is courtesy of Richard Birrittella and Prof. Chris Gerry, and is used with their permission.

Another experiment has shown the ability to actually measure a significant fraction of the Wigner function [46]. Here, two photons are created by spontaneous parametric down conversion of a single high-energy photon. These photons are entangled as the pair of particles created in the EPR experiment discussed above. These two photons are guided into parallel paths and eventually into a beam splitter. Prior to arrival at the beam splitter, one photon undergoes a position translation by an amount 2δ which amounts to the coordinate difference that appears in the wave functions of the Wigner transformation. At the same time, the second photon undergoes a phase shift μ, which corresponds to a momentum shift. After the beam splitter, the combined photons go to a pair of detectors. By varying the position and phase shifts, the authors are able to map out the Wigner function by monitoring the signals at the detectors. Again, the negative-going parts of the Wigner functions are indicators of entanglement in this photon pair. Other nonlinear optics experiments can be used to create non-classical states of the system. One example is the squeezed photon state, which produces non-classical behavior, and the Wigner function can be used to exhibit this non-classical behavior [47]. Others use photons to create what are called Schrödinger cat states, which again are examined with the use of the Wigner function [48]. We will return to this in a later chapter.

8.4.2 Condensed matter systems

An interesting numerical experiment has been carried out with Wigner wave packets [49], in which the splitting of the packet in an Aharonov–Bohm ring [50], in the presence of scattering, is considered. This study is depicted in figure 8.5(a). A waveguide form of an Aharonov–Bohm ring is created using potentials (in red) to define the structure, as shown in the figure. The overall simulation area is 60 nm wide and 65 nm tall. The waveguides (in blue) are 16 nm wide. A Gaussian wave packet in the Wigner function form (such as that in (8.4)) is sent into the structure from the bottom. This packet has a full-width spread of 4 nm ($\sigma = 2$ nm) and an initial momentum of $k_0 \sim 9.4 \times 10^7\ \text{m}^{-1}$. A coherence length, due to the existence of the scattering processes, of 30 nm was assumed for the simulation, which means that the entanglement will be relatively heavily damped by the time the wave packet reaches the output of the A–B ring. The simulation is performed using the signed Monte Carlo particle simulation discussed in section 6.3 [51]. In figure 8.5(b), the two wave packets that have formed from the initial single packet are depicted, at a time of about 45 fs after the start. As these two packets were created from a single input wave packet and remain described by a single Wigner function, they are entangled, and this may be seen by the rapid oscillatory features which exist *outside the ring* (see the arrow in this panel). Some of the penetration into this area could arise from the wave function penetration into the barrier, but this would not be expected to show the transverse oscillation that appears to be here. Figure 8.5(c) depicts the two packets as they are making the turn at the ring corners, at a time of 65 fs. Here, there are additional interesting features that arise as these points are turning points of the classical trajectories. One notes that fringes have formed, much in the manner discussed by Berry [52]. In a sense, this arises from the need for the

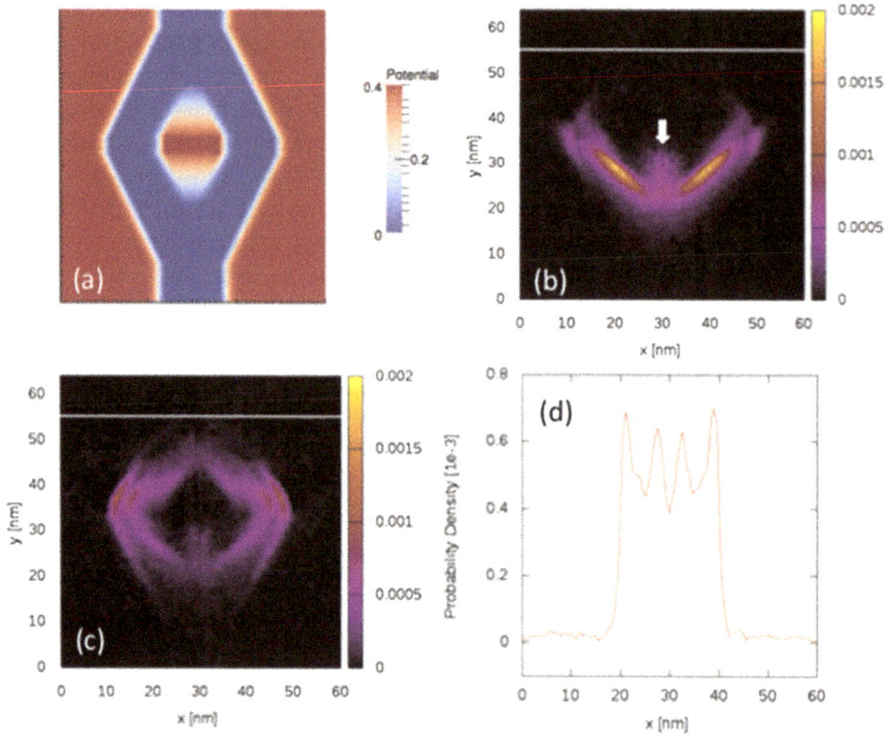

Figure 8.5. Propagation of a localized Wigner wave packet through an Aharonov–Bohm style ring. (a) The waveguides are defined by the potential. (b) At 45 fs, the input packet has clearly divided into two packets, which are entangled (white arrow). (c) At 65 fs, the packets have reached the turning points and developed fringes. (c) When the two packets reconnect (105 fs), they lead to interference fringes. The amplitude is plotted along the white line shown in (b) and (c). Reprinted from P Ellinghaus [49], with permission.

transverse momentum to reverse at these turning points. We also note that the entanglement has still survived, although much weaker and broader due to the dephasing process. Finally, in figure 8.5(d), we see that interference fringes are observed in the plane where the two wave packets reconnect (the plot is across the line indicated in white, 55 nm from the entrance, in panels (b) and (c) of the figure). The number of fringes is limited by the waveguide size, but it is clear that one is seeing the equivalent of the two-slit experiment in this output plane. A magnetic field will shift this interference to left or right and the conductance oscillates in the Aharonov–Bohm effect [53]. Finally, in figure 8.6, we give a video which recreates the entire propagation of the Wigner wave packet through the ring.

The current approach to quantum computing that seems to be the most fruitful is through the use of superconducting flux qubits. As a result, there is considerable interest in the decoherence of these qubits, and how this is affected by the presence of the ensemble of electrons that also exist in the superconducting material. It is possible to cast this problem into one in which the coherence/entanglement can be followed by studying the evolution of the Wigner function of the reduced density matrix of the qubits [54]. Such studies have shown that, under some conditions, the

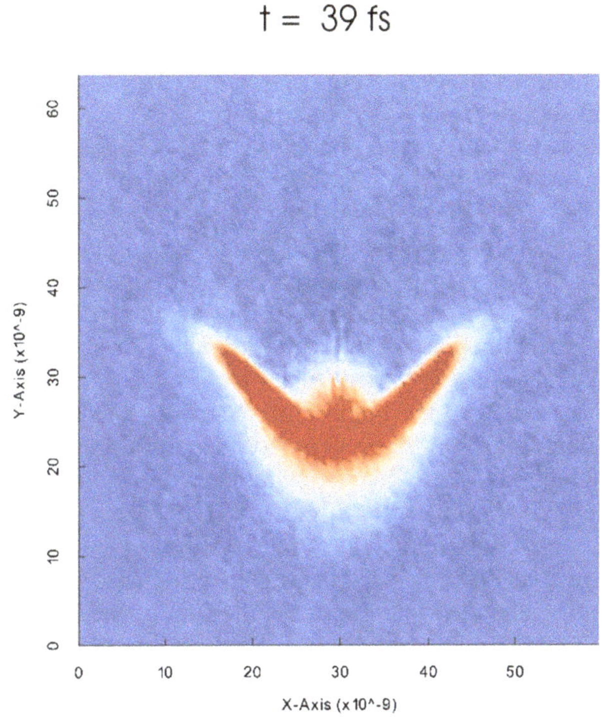

Figure 8.6. A video of the Wigner wave packet propagating through the Aharonov–Bohm ring. Provided by and used from P Ellinghaus [49], with permission. A video is available online at http://iopscience.iop.org/book/978-0-7503-1671-2.

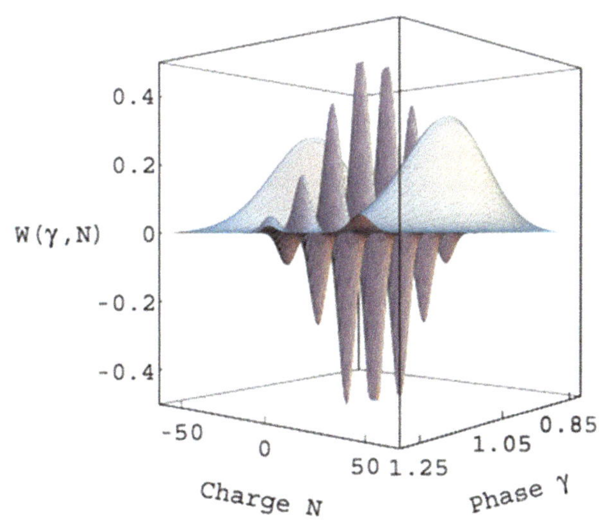

Figure 8.7. Plot of the full Wigner function for the readout Josephson junction. The parameters were taken from [56]. The interference fringes that can be seen between the Gaussians (for the two junctions) correspond to the off-diagonal coherence in the Wigner function.

presence of the electrons can even lengthen the coherence time. In some early experiments, it was shown that a so-called charge-phase qubit would be insensitive to first-order fluctuations in external control parameters, provided the latter were in the neighborhood of an optimal operating point [55]. This experimental work was then studied by a quantum transport simulation incorporating dissipation through a Lindblad-type operator [56, 57]. Here a bias current was introduced into the model of the charge-phase qubit in a manner that acted to provide decoherence of the qubit. In this approach, the current would count the number of electrons passing through the qubit, and such a measurement in quantum mechanics provides decoherence. In this system, a second Josephson junction provides the readout for the qubit. The Wigner function for the readout junction is shown in figure 8.7 [56]. The interference fringes between the two Gaussians are due to the coherence between the two qubit states; the number of such fringes changes with time as the separation of the two Gaussian parts vary their separation in coordinate space.

The Josephson junction is also interesting in its own right, as a nonlinear coupling or control device. For example, if two electromagnetic resonators are coupled in series with a Josephson junction, interesting quantum dynamics can be observed [58]. If the voltage that is applied to the Josephson junction leads to a Josephson frequency across the junction which matches the sum of the frequencies of the two resonators, the Cooper pairs can flow through the entire circuit. This leads to the creation of photons in the two resonators which produce a non-classical, far from equilibrium state. Here, the dynamics can be fruitfully explored by the study of a Wigner function describing the joint dynamics of the two coupled oscillators. The Wigner function for either of the two oscillators can then be found by integrating out the coordinates of the other oscillator. Needless to say, these individual oscillator Wigner functions appear to be Gaussians depending upon the number operator for that oscillator, much like the Gaussians for the individual particles in figure 8.5. Study of the motion of the Wigner function leads to the dynamics of the coupled resonator-junction system, and can even be used to study the quantum fluctuations in the system.

References

[1] Planck M 1900 Protokollbuch https://www.dpg- physik.de/veroeffentlichung/archiv/index.html

[2] Planck M 1950 *Scientific Autobiography and Other Papers* tr. by ed Gaynor F (London: William and Norgate) p 7
cited in Pais A 1991 *Neils Bohr's Times, in Physics, Philosophy, and Polity* (Oxford: Clarendon) p 80

[3] Einstein A, Podolsky B and Rosen N 1935 *Phys. Rev.* **47** 777

[4] Schrödinger E 1935 *Proc. Cambridge Phil. Soc.* **31** 555
Schrödinger E 1936 *Proc. Cambridge Phil. Soc.* **32** 446

[5] Bohm D 1951 *Quantum Theory* (New York: Prentice-Hall)

[6] Bohm D 1952 *Phys. Rev.* **85** 166

[7] Ferry D K *Copenhagen Conspiracy* (Singapore: Pan Stanford) in press

[8] Ozorio de Almeida A M 2009 *Entanglement and Decoherence* ed A Buchleitner, C Viviescas and M Tiersch (Berlin: Springer) pp 157–219
[9] Shifren L and Ferry D K 2002 *J. Comput. Electron.* **1** 55
[10] Kim I and Iafrate G J 2004 *Found. Phys. Lett.* **17** 507
[11] Iafrate G J *private communication.*
[12] Moyal J E 1949 *Proc. Cambridge Phil. Soc.* **45** 99
[13] Mintert F, Viviescas C and Buchleitner A 2009 *Entanglement and Decoherence* ed A Buchleitner, C Viviescas and M Tiersch (Berlin: Springer) pp 61–86
[14] Hododecki R, Horodecki P, Horodecki M and Horodecki K 2009 *Rev. Mod. Phys.* **81** 865
[15] Shannon C 1949 *Proc. IRE* **37** 10
[16] Avery J 2003 *Information Theory and Evolution* (New York: World Scientific)
[17] Feynman R 1965 *The Character of Physical Law* (Cambridge, MA: MIT Press)
[18] Baungratz T, Cramer M and Plenio M 2014 *Phys. Rev. Lett.* **113** 140401
[19] Chan K C, Volkoff T, Kwon H and Jeong H 2017 *Phys. Rev. Lett.* **119** 190405
[20] See also the discussion at L Zyga https://phys.org/news/2017-11-physicists-quantum-coherence-nonclassicality.html
[21] Ellinghaus P, Weinbub J, Nedjalkov M and Selberherr S 2017 *Phys. Status Solidi Rapid Res. Lett.* **11** 1700102
[22] Kreuzer H J 1981 *Nonequilibrium Thermodynamics and its Statistical Foundations* (Oxford: Oxford Univ. Press)
[23] Zubarev D N 1974 Neravnovesnaia Staisticheskaia Termodinamika
Tr. by ed Shepard P J as *Nonequilibrium Statistical Thermodynamics* (New York: Consultants Bureau)
[24] von Neumann J 1927 *Gött. Nach.* **1** 273
[25] von Neumann J 1935 *Mathematische Grundlagen der Quantenmechanik* (Berlin: Springer)
Tr. by ed Beyer R T 1955 *Mathematical Foundations of Quantum Mechanics* (Princeton, NJ: Princeton Univ. Press)
[26] Wigner E P 1932 *Phys. Rev.* **40** 749
[27] Schumacher B 1995 *Phys. Rev.* A **51** 2738
[28] Horodecki P and Horodecki R 1994 *Phys. Lett.* A **194** 147
[29] Werner R F 1989 *Phys. Rev.* A **40** 4277
[30] Popescu S, Short A J and Winter A 2006 *Nat. Phys.* **2** 754
[31] Horodecki M 2008 *Nat. Phys.* **4** 833
[32] Brandão F G S L and Plenio M B 2008 *Nat. Phys.* **4** 873
[33] Popescu S 1995 *Phys. Rev. Lett.* **74** 2619
[34] Boole G 1854 *An Investigation of The Laws of Thought: the Mathematical Theories of Logic and Probability* (London: Walton and Maberly) The book has been reprinted by several publishers, but is available online at archive.org (https://archive.org/details/investigationofl00boolrich) and from Project Gutenberg (www.gutenberg.org/files/15114/15114-pdf.pdf).
[35] Pitowsky I 1994 *Br. J. Phil. Sci.* **45** 95
[36] Peres A 1996 *Phys. Rev.* A **54** 2685
[37] Werner R F 1989 *Phys. Rev.* A **40** 4277
[38] Vedral V M B, Plenio K, Rippin M A and Knight P L 1997 *Phys. Rev. Lett.* **78** 4452
[39] Vidal G 2000 *J. Mod. Opt.* **47** 355
[40] Jaynes E T and Cummings F W 1963 *Proc. IEEE* **55** 89
[41] Norozhny N B, Sanchez-Mondragon I I and Eberly J H 1981 *Phys. Rev.* A **23** 236

[42] Gea-Banacloche J 1990 *Phys. Rev. Lett.* **65** 3385
[43] Birrittella R, Chang K and Gerry C C 2015 *Opt. Commun.* **354** 286
[44] Eiselt J and Risken H 1991 *Phys. Rev.* A **43** 346
[45] Cahill K E and Glauber R J 1969 *Phys. Rev.* **177** 1882
[46] Douce T, Eckstein A, Wallborn S P, Khoury A Z, Ducci S, Keller A, Coudreau T and Milman P 2013 *Sci. Rep.* **3** 3530
[47] See, e.g. Zhang H-L, Yuan H-C, Hu L-Y and Xu X-X 2015 *Opt. Commun.* **356** 223
[48] See, e.g. Seshadreesan K P, Dowling J P and Agarwal G S 2015 *Phys. Scr.* **90** 074029
[49] Ellinghaus P Tech. Univ. Vienna *unpublished.*
[50] Aharonov Y and Bohm D 1959 *Phys. Rev.* **115** 485
[51] Ellinghaus P 2016 *PhD Dissertation* Tech. Univ. Vienna.
[52] Berry M V and Balazs N L 1979 *J. Phys.* A **12** 625
[53] See, e.g. Ren S L, Heremans J H, Gaspe C K, Vijeyaragunathan S, Mishima T D and Santos M B 2013 *J. Phys. Condens. Matter* **28** 435301
[54] See, e.g. Reboiro M, Civitrese O and Tielas D 2015 *Phys. Scr.* **90** 074028
[55] Vion D, Aassime A, Cottet A, Joyez P, Pothier H, Urbina C, Esteve D and Devoret M H 2002 *Science* **296** 886
[56] Hutchinson G D, Holmes C A, Stace T M, Spiller T P, Milburn G J, Barrett S D, Hasko D G and Williams D A 2006 *Phys. Rev.* A **74** 062302
[57] Lindblad G 1976 *Commun. Math. Phys.* **48** 119
[58] Armor A D, Kubala B and Ankerhold J 2015 *Phys. Rev.* A **91** 184508

Chapter 9

Quantum chemistry

In the previous chapters, we have developed the formalism for the Wigner function for a variety of situations. This includes, for example, the equation of motion for a Wigner function whether it be a full distribution or a Gaussian packet. Nevertheless, it cannot have escaped the reader's notice that the applications discussed come mainly from the field of semiconductor transport and semiconductor device simulation. Perhaps this is a result of the authors' bias toward their own field, with many examples readily at hand. Now, however, we want to turn to the usage of the Wigner distribution function and its phase space representation in other fields. The purpose of this chapter and those following is to demonstrate that the use of quantum mechanics in phase space has developed a large and vigorous interest, and that this has led to quite a number of fields of application. We begin this with chemistry, perhaps because of its closeness to physics, but more likely due to the fact that the authors became aware of the use of quantum trajectories in chemistry through the work of Wyatt [1], addressing the tunneling problem with which we were also concerned. In fact, the trajectories in this case were Bohm trajectories, which were studied in the barrier-crossing situation. These were soon applied to reaction rates [2], resonant scattering [3], and multi-mode system dynamics [4]. Soon enough, he began to use the phase space Wigner function to study decoherence [5] and the study of quantum forces, including the nonlocal Wigner force [6]. In truth, however, Wyatt was not the first to bring the Wigner function to use in tunneling in chemistry. To the authors' knowledge, this distinction goes to Tanimura and Wolynes [7], who studied the dynamics of tunneling, resonance and dissipation in the transport across/through a barrier.

From the beginnings mentioned above, the use of the Wigner function and phase space dynamics has spread to a great many sub-fields of science, not the least of which is the barrier-crossing problem that appears in a variety of guises, particularly in chemistry. In this and the following chapters, we discuss these applications in

doi:10.1088/978-0-7503-1671-2ch9

terms of a number of sub-fields, which are chosen by our attempt to classify them and to group the result for common applications.

9.1 Quantum statistics

Even the old quantum mechanics began with the spectra of atoms, and so early work on Wigner functions also focused on this theme. One of the early papers in fact dealt with the scattering from the ^4He atom [8], at low temperatures but above the superfluid regime. The description of the density fluctuations is usually treated by quantum Monte Carlo approaches [9, 10], but Ciccotti *et al* find that the Wigner function gives a good description of the spectrum of these fluctuations, at a lower computational cost. The spectrum they determine gives a proper fit to the observed experimental spectrum that is shifted from the elastic, zero energy exchange peak. Hence, the Wigner function clearly gives the quantum results, although an elastic peak is observed experimentally and is not explained by their approach. Ciccotti *et al* applied their technique to the scattering from ^4He atoms [8]. In this latter work, the authors begin with the expectation value of an operator in terms of the density matrix and then introduce the standard Wigner transform to this equation, and this yields the Wigner form of the expectation value of the operator. They then develop the normal time variation from the equation of motion and path integral, without scattering, discussed in chapter 6. Without the scattering, they use the momentum term in the equation of motion to develop the path integral, as

$$f_W(x, p, t) = e^{iL_0 t} f_W(x, p, 0) + \int_0^t dt' e^{iL_0(t-t')} \int ds\, W(x, s) f_W(x, p-s, t'), \quad (9.1)$$

although they treat a homogeneous system, and

$$L_0 = -\frac{ip}{m}\frac{\partial}{\partial x}. \quad (9.2)$$

The solution is then introduced into the expectation value for an operator corresponding to (3.11). They then show that the correlation function between two observables (and/or their operator forms) to be given by

$$C_{AB}(t, \beta) = \frac{1}{Z}\mathrm{Tr}\{\hat{A}\hat{B}(t)\} = \frac{1}{Z}\mathrm{Tr}\{e^{-\beta\hat{H}}\hat{A}e^{i\hat{H}t/\hbar}\hat{B}e^{-i\hat{H}t/\hbar}\}, \quad (9.3)$$

where β is the inverse temperature, Z is the partition function, and the time variation for B has been introduced by the Heisenberg representation. They then introduce the complex time, usually used with the Matsubara thermal Green's functions (although modified here), and from the similarity to the Matsubara thermal Green's functions, the nonlocal Wigner function depends upon two positions and two momentums. That is, the imaginary time is introduced from the density matrix as

$$e^{-\beta H}e^{iHt/\hbar} \rightarrow e^{-iH\tau_c/\hbar}, \quad \tau_c = t - i\hbar\beta/2. \quad (9.4)$$

This now leads to the quantum Wigner correlation function, for the double phase space as

$$G_{AB}(t, \beta) = \iint \iint dx\, dx' dp\, dp' A_W(x, p) B_W(x', p') \times f_{W2}(x, p, x', p'; t, \beta), \tag{9.5}$$

where the double space Wigner function is normalized in the standard way, but for integration over the double phase space. Statistical averages are determined by summing over a set of configurations of the test system. As we mentioned, this approach is applied to the calculation of the scattering function of ^{4}He, in which one is interested in the time correlation of the density fluctuations. For this, one extends the above formations to the N particle case, from which one is now interested in the Fourier transformed version, as

$$\begin{aligned} F(\mathbf{k}, t) &= \frac{1}{Z} \mathrm{Tr} \left\{ \hat{\rho}_{\mathbf{k}} e^{iH\tau_c^*/\hbar} \hat{\rho}_{-\mathbf{k}} e^{-iH\tau_c/\hbar} \right\} \\ &= \int\int\iint d^N\mathbf{x}\, d^N\mathbf{x}'\, d^N\mathbf{p}\, d^N\mathbf{p}' \sum_{i=1}^{N} e^{i\mathbf{p}\cdot\mathbf{x}/\hbar} \sum_{j=1}^{N} e^{i\mathbf{p}'\cdot\mathbf{x}'/\hbar} \\ &\quad \times W(\mathbf{x}, \mathbf{x}', \mathbf{p}, \mathbf{p}'; 0, \beta). \end{aligned} \tag{9.6}$$

The results are shown in figure 9.1. The dynamic structure factor at a 'position' of 0.164 nm is plotted in the figure. The thick solid line is the Wigner simulation, while the dashed line is the classical one. The dotted line is the experimental data and shows an elastic scattering peak.

We discussed path integrals in chapter 6. These approaches to finding the Wigner function to describe quantum correlation functions have been quite popular in

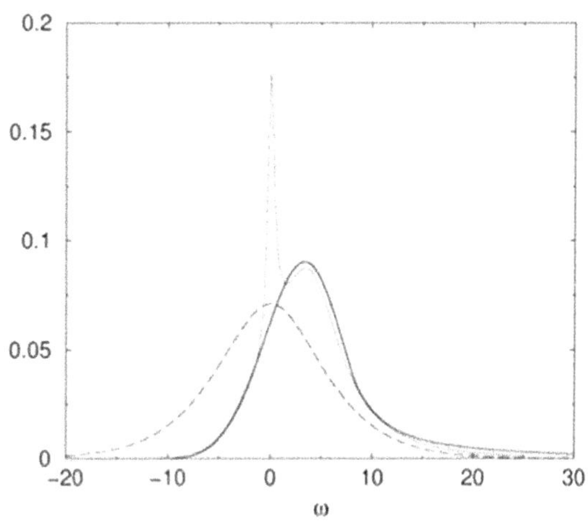

Figure 9.1. The dynamic structure factor at a 'position' of 0.164 nm. The thick solid line is the Wigner simulation, while the dashed line is the classical one. The dotted line is the experimental data and shows a elastic scattering peak. Reprinted from [8], copyright 1999, with permission from Elsevier.

chemistry as well. Poulsen *et al* [11] developed a linearized Feynman–Kleinert variational path integral approach to study the correlation functions for a chain of He atoms as well as to liquid oxygen, obtaining a radial distribution function for the latter that was clearly different than the expected classical result. Another approach to the use of a path integral implemented a forward–backward action behavior to study non-radiative electronic relaxation [12]. In this approach the relevant operators in the correlation functions were Wigner transformed as an approach to study the quantum to classical connections.

The Feynman–Kleinert estimation of the density matrix and the resulting linearized path integral, discussed in chapter 4 as well as chapter 6, has been further developed more recently [13, 14], where the approach was applied to determining the dynamic structure factor in para-hydrogen and ortho-deuterium. The Wigner representation of the density operator has also been applied to study the correlation functions in the spin-boson model [15], where they calculated the equilibrium population difference as a bias was applied. Another approach is to use the Wigner density approximation for real-time correlation functions, but to optimize the latter with what is called a maximum entropy method [16]. The latter authors also studied the momentum autocorrelation function for para-hydrogen.

Another approach to computing symmetrized correlation functions is the phase integration method [17], in which the correlation function is developed from the density matrix and then symmetrized. The Wigner transform is then introduced for the various operators to create a phase space representation of this correlation function. It was then shown that this approach gives a good method for finding the thermal Wigner distribution function [18], and the approach was applied to determine the latter for particles in a double Morse potential for various sets of parameters (variation of the parameters can deform the potential from a single minimum to double minima). More recently, a molecular dynamics approach for determining the microcanonical distribution and connecting it to the Wigner function has been demonstrated [19].

Another method has approached the connection between quantum dynamics and the classical Boltzmann distribution. Here, the exact time-correlation functions for the normal modes of a ring polymer are developed in terms of Feynman paths [20]. Taking the limit of an infinitely long polymer, they find that the lowest mode frequencies take their Matsubara frequency values (as determined from Matsubara Green's functions). Wigner–Moyal transformations of the correlation functions allow connection to the classical phase space dynamics. Tanimura [21] studied the real and imaginary time (as used in the Matsubara Green's functions) for a hierarchical set of Fokker–Planck equations. These were then used to study the phase space Wigner dynamics of a model quantum system coupled to a harmonic oscillator bath.

9.2 Reactions and rates

The discussion of chemical reactions is extensive. Perhaps the earliest work is the diffusion model of Kramers [22]. Since then, the field has expanded in a great many directions. A few of these have introduced the Wigner function as a viable phase space approach to chemical reactions. In this section, we want to briefly discuss some of these phase space approaches.

One of the earliest papers dealt with electron transfer from a donor or acceptor in which there was a lattice relaxation described by a reaction coordinate [23]. In this case, the local state of the lattice around the impurity changes with the charge on the impurity, and this is described by the reaction coordinate. This also changes the energy level, so that optical transitions can change depending upon the charge state; this is sometimes referred to as the Franck–Condon effect. In the study at hand, the interactions can be studied by the spin-boson model, which provides a pragmatic yet realistic formulation for the role of dissipation in the electron transfer [23]. In this work, the electron and reaction coordinates are cast in the Wigner phase space representation; the role of a driving electric field was also considered in the study. Others have also studied electron transfer when coupled to collective boson degree of freedom, including the study of the fluctuations in the system [24]. Again, these latter studies also employed the reaction coordinate for the impurity system.

Let us first describe the approach with the reaction-coordinate model. The idea is about the same whether we deal with a molecule or with condensed matter. For an impurity, say a donor, the upper state corresponds in the condensed-matter system with the conduction band, while the lower state is the actual impurity level (in the molecule, this might actually be an extended state with dispersion). When the electron is excited from the donor impurity, the local potential changes primarily due to the short-range local potential around the impurity. The potential change induces a lattice relaxation around the impurity, and this in turn changes the level of the impurity relative to the conduction band [25], but is usually treated by a change in the conduction band. There are two contributions here. First, the Jahn–Teller theorem tells us that any electronic system with multiple degenerate ground states, such as spin degeneracy in the simple donor impurity case, is unstable against a distortion that removes the degeneracy. This leads to lattice relaxation, which is of course much more prevalent in deep donors as opposed to shallow donors. As a result the atom is actually displaced in position, and this is the reaction coordinate. But there is a second distortion: the change in the band structure between the neutral and the ionized states, as mentioned above, the Franck–Condon effect, which results in a difference between the optical excitation energy and the thermal excitation energy of the impurity [26, 27]. The overall situation, as mentioned in the previous paragraph, is best described by a reaction-coordinate diagram, as shown in figure 9.2. The two potential curves cross at a point x^*, which is the point at which electron transfer occurs. Once the electron is into the second potential V_2, it must gain an energy greater than ε_0 in order to recombine with the donor. In semiconductors, this leads to persistent photoconductivity at low temperatures. Here, the Hamiltonian can be written as [23, 24]

$$H = H_{EL} + H_{RC} + H_B, \tag{9.7}$$

where the bare electronic system is written in terms of a two-state pseudo-spin system as

$$H_{EL} = -\frac{1}{2}[V_1(x) - V_2(x)]\hat{\sigma}_z + \frac{\hbar\Delta}{2}\hat{\sigma}_x \tag{9.8}$$

with the pseudo-spin operators being given by

$$\begin{aligned} \hat{\sigma}_z &= |1\rangle\langle 1| - |2\rangle\langle 2| \\ \hat{\sigma}_x &= |1\rangle\langle 2| + |2\rangle|1\rangle. \end{aligned} \tag{9.9}$$

The coupling term Δ is assumed to be independent of the reaction coordinates. The reaction-coordinate term is given by

$$H_{RC} = \left[\frac{p^2}{2m} + V_1(x) + V_2(x)\right]\begin{bmatrix} 1 & 0 \\ 0 & 1 \end{bmatrix}, \tag{9.10}$$

where the two harmonic potentials in figure 9.2 are given as

$$\begin{aligned} V_1(x) &= \frac{m\omega_0^2}{2}x^2 \\ V_2(x) &= \frac{m\omega_0^2}{2}(x - x_0)^2 - \varepsilon_0, \end{aligned} \tag{9.11}$$

and the bare term is the energy of the non-interacting particles. Here, the so-called reorganization energy is $V_1(x_0)$, and the excitation indicated by the red arrow in figure 9.2 is the sum of this energy and the offset energy ε_0. One notes that the reaction coordinate x is the primary variable in the Hamiltonian and the resulting problem being studied. In both of the approaches in the previous paragraph, the authors cast the problem in terms of the Zusman equation, which is the equation of motion for the density matrix in the above Hamiltonian. The latter is a 2×2 matrix for the pseudo-spin coordinates. The equation is then transformed into one for the Wigner function, and this is used for studying the electron transfer in the molecular system. The important point is that when the electron is transferred from the 'donor'

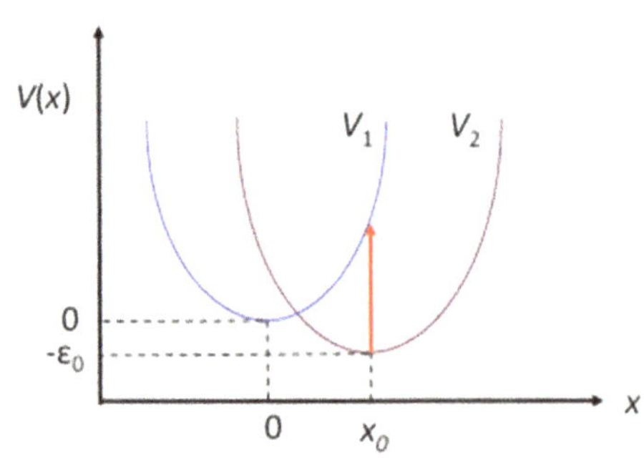

Figure 9.2. The reaction-coordinate diagram for the impurity interaction. The two potentials represent the excitation energies before and after ionization of the impurity. The red arrow describes the sum of the reorganization energy and ε_0 of the structure.

molecule (state $|1\rangle$) to the 'acceptor' molecule (state $2\rangle$), the relaxation process makes this a 'one way' reaction, as the barrier to the back reaction is increased by the molecular relaxation characterized by the reaction coordinate.

The vibrational energy relaxation in a condensed phase system has also been studied with Wigner functions [28]. In this work, the authors use a semi-classical approximation to the quantum-mechanical force–force correlation function, and then Wigner transform the corresponding classical quantities. The result is found to give the exact result at the initial time and to converge to the classical result as h is allowed to go to zero. They use this approach to study the coupling of the vibrational mode to a harmonic boson bath, a diatomic molecule coupled to a linear chain of helium atoms and a expanding/contracting sphere in a monoatomic liquid with Lennard–Jones interactions between the atoms. A similar study of the relaxation of a polyatomic molecule via a radiation-less transition to the vibrational degrees of freedom of the nuclei has also been studied [29].

The activated escape of a particle over a confining barrier is another important problem in physics, biology and chemistry, and the reaction-coordinate approach is one way to deal with it. But there are others. When the system is subjected to an excitation field, the escape rate will vary from the normal Kramers rates due to the non-equilibrium nature of the system. This has been studied using a path integral formulation and subsequent Wigner formulation in the case of a rapidly oscillating external field [30]. In the non-equilibrium state, the fluctuation–dissipation theorem is no longer applicable, as we have noted previously. This requires more care in treating the statistics of the problem, and these authors have proceeded by dividing the time into slow and fast dynamics. This allows them to find time-independent dynamics for the system alone and then to focus on the escape of a particle in the rapid-varying external field. A similar type of problem is the tunneling-based dissociation of the H atom in electronically excited pyrrole [31]. In this latter work, the authors used a trajectory-based approach to calculating the various tunneling probabilities from the phase integrals. The trajectories were limited to straight-line tunneling paths, and it was found that sampling these paths based upon a fixed-energy Wigner distribution gave the best fit to the quantum-mechanical dissociation rates.

The idea of using trajectories based upon the Wigner distribution is somewhat older than the work above, as discussed already in chapter 6. It was shown that this provided a viable approach to the calculations of rate constants in chemical reactions, in which quantum effects are important. Here, weighted classical trajectories are taken from the Wigner distribution that represents the Boltzmann-averaged flux operator [32]. These authors studied two standard potentials, the quartic potential and the symmetric Eckart potential, to illustrate the differences from the simpler parabolic barrier. Another approach for reaction rates used Feynman paths in the presence of an effective potential, as discussed in chapter 4, and this is used then in the Wigner approach to determine correlation functions [33]. These latter authors introduce a characteristic spacing between Feyman paths, which is zero in the classical model, and obtain transmission coefficients for several well-known reaction rate problems. This approach has also been applied to the collinear $H + H_2$ reaction [34].

Reaction rates for condensed phase reactions involving ring polymers have also enjoyed some attention with Wigner dynamics. It has been shown how these dynamics can be adapted to calculate approximate Kubo-transformed flux-side correlation functions to give comparable accuracy to the semi-classical Wigner model [35]. These authors concentrate on a standard model of the reaction that is represented by a double-well potential linearly coupled to a bath of harmonic oscillators. A quite similar approach has been followed more recently for determining the transition states for the ring polymers [36]. In this latter work, the authors show that there is a form of quantum time-correlation function that does not vanish at the initial time and that gives the exact rate which is the same as that of the ring-polymer molecular dynamics approach to the transition-state theory.

9.3 Tunneling

In several previous chapters, tunneling of an electron through a barrier was discussed as an electronic application. It should come as no surprise that tunneling is also a major concern in chemistry. For example, the reaction rates discussed in the previous section dealt with semi-classical dynamics for excitation over a barrier. But the transition can also occur by tunneling through the barrier. As a result, tunneling can be important to the discussion of chemical reactions. Indeed, use of a Wigner function in the chemical reaction process can allow one to study the bridge between classical and quantum treatments in various molecular and chemical systems. Hence there have been a number of papers that consider the tunneling process via the Wigner function. One approach has used the Wigner function to study tunneling via wave packets in both dispersive media and in tunneling well below the barrier height [37]. Here, the authors developed analytical expressions for the Wigner function propagators in terms of the local group velocity and the Wigner phase delay time. As in (9.1), they take the equation of motion and evolve it into an integral evolution equation for the Wigner function

$$f_W(\mathbf{x}, \mathbf{p}, t) = \iint d^3\mathbf{p}'\, d^3\mathbf{x}'\, L_t(\mathbf{x}, \mathbf{p}; \mathbf{x}', \mathbf{p}') f_W(\mathbf{x}', \mathbf{p}', 0), \tag{9.12}$$

where the propagator L_t is given by the Wigner–Weyl transform

$$\begin{aligned} L_t = \left(\frac{4}{h}\right)^3 \iint d^3\mathbf{p}'' d^3\mathbf{x}'' e^{2i(\mathbf{p}\cdot\mathbf{x}''+\mathbf{x}'\cdot\mathbf{p}'')/\hbar} \langle \mathbf{x} + \mathbf{x}'' | \, U \, | \mathbf{p}' + \mathbf{p}'' \rangle \\ \times \langle \mathbf{x} - \mathbf{x}'' | \, U \, | \mathbf{p}' - \mathbf{p}'' \rangle, \end{aligned} \tag{9.13}$$

with U the unitary evolution operator that appears in (7.1) for the Schrödinger equation

$$U = e^{-iHt/\hbar}. \tag{9.14}$$

In dispersive media, this unitary operator for free propagation (absence of a potential) becomes

$$U|\mathbf{p}\rangle = e^{-i\omega(\mathbf{p})t}|\mathbf{p}\rangle \tag{9.15}$$

for a plane wave momentum state. In the simplest dispersive medium, the frequency may be expressed as an analytic function of the momentum, with powers above 2 (the classical harmonic potential). Then the semi-classical approximation gives

$$\omega(\mathbf{p} + \mathbf{p}') - \omega(\mathbf{p} - \mathbf{p}') \sim 2\mathbf{p}' \cdot \frac{\partial \omega(\mathbf{p})}{\partial \mathbf{p}} + \ldots. \tag{9.16}$$

This then gives the result that we used in chapter 8 for the free Gaussian of (8.3), in that

$$f_W(\mathbf{x}, \mathbf{p}, t) = f_W(\mathbf{x} - v(\mathbf{p})t, \mathbf{p}, 0). \tag{9.17}$$

That is, the Gaussian moves as a Gaussian. However, in the normal harmonic potential, this gives a single group velocity for each value of momentum. In the dispersive medium, this is no longer the case, and each momentum contribution has its own velocity.

In figure 9.3 are shown the exact Wigner functions for tunneling through a short-range (delta) function potential

$$V = \frac{\hbar^2 s_0}{2m}\delta(x) \tag{9.18}$$

in one dimension, with $s_0 = 100$, for a free particle (in reduced coordinates). Here, the initial condition is a Gaussian,

$$\psi(x, 0) \sim e^{-i(x-x_0)^2/4\sigma^2 + ip_0 x}, \tag{9.19}$$

with $x_0 = -5$ and $p_0 = 8$, $\sigma = 0.5$, again in reduced units. Here, the dispersive aspect arises from the fact that different momentum states have different tunneling probabilities.

Another approach to the reaction system using a double-well potential has been formulated on the Langevin equation, and tunneling between the two wells

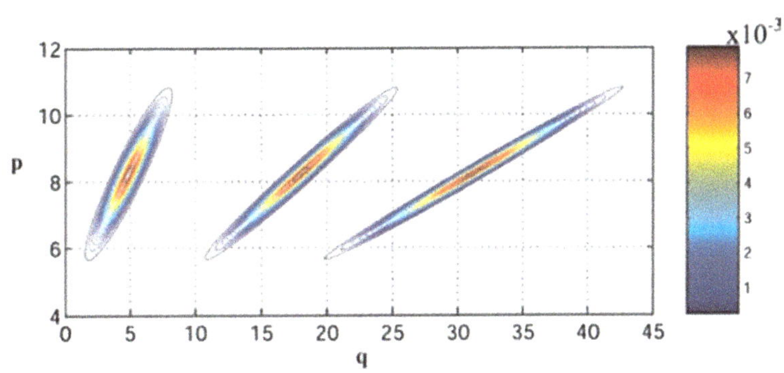

Figure 9.3. The dispersion of the Wigner function for tunneling of a Gaussian wave packet through a local delta function potential at reduced times of 1.2, 2.8, and 4.4. Here, q is the reduced dimension for the position of the wave packet. Reprinted from [37], copyright (2004), with permission from Elsevier.

considered [38]. In this case, the two potential wells are of equal depth, with the potential being described by

$$V(x) = ax^4 - bx^2, \tag{9.20}$$

with a and b parameters that can be adjusted. The two minima are now located at

$$x_0 = \pm\sqrt{\frac{b}{2a}}, \tag{9.21}$$

where $V_0 = -b^2/4a$. Hence adjusting b and a varies the central barrier height by lowering the potential minima, and changes the spacing of these minima. Now, they can use the Hamiltonian

$$H = \frac{\hat{p}^2}{2m} + V(x) + \sum_j \left\{ \frac{\hat{p}_j^2}{2} + \frac{\kappa_j}{2}(\hat{x}_j - \hat{x})^2 \right\}, \tag{9.22}$$

where the last term describes a set of harmonic oscillators to which the particle motion is coupled. Here, the harmonic oscillators are described via a frequency ω_j and the coupling constant κ_j. The authors then eliminate the reservoir degrees of freedom 'in the usual way'[1] [39] to obtain the retarded Langevin equation for the particle [38]:

$$\frac{d^2\hat{x}}{dt^2} + \int_0^2 dt'\gamma(t - t')\hat{x}(t') + \frac{dV(\hat{x})}{dx} = F(t), \tag{9.23}$$

where the random force noise operator and the memory kernel are given by

$$F(t) = \sum_j \left[\{\hat{x}_j(0) - \hat{x}(0)\}\kappa_j \cos(\omega_j t) + \hat{p}_j(0)\sqrt{\kappa_j} \sin(\omega_j t) \right] \tag{9.24}$$

and

$$\gamma(t) = \sum_j \kappa_j \cos(\omega_j t) \to \int_0^\infty \kappa(\omega)\rho(\omega)\cos(\omega t), \tag{9.25}$$

respectively, where the last term is the classical continuum limit for the memory kernel. The authors then assume that the noise is represented by a canonical thermal Wigner distribution representing the distribution of harmonic oscillator coordinates. They can then take a thermal average of the response for the system to the noise sources. This leads to the result that the important aspect of the noise source is its force–force correlation function, which they finally write as [38]

$$c_F(t - t') = \sum_i \frac{D_i}{\tau_i} \exp\left(-\frac{|t - t'|}{\tau_i} \right), \tag{9.26}$$

[1] The phrase 'in the usual way' can refer to a variety of approaches. The most common (used here) is the introduction of projection operators to trace out the reservoir degrees of freedom.

which is a summation over a set of exponentials for which the parameters D_i and τ_i are assumed to be known. This type of noise source is commonly used in condensed-matter physics to represent a well-known low-frequency noise called $1/f$ noise that appears in most electronic systems. This is different from the normal thermal noise in such systems, which is characterized more by the temperature. Finally, these authors use this approach to study the time-dependent transmission coefficient from one well to the other.

The ubiquitous double-barrier potential has been discussed in several of the previous chapters (for example, figures 4.2, 6.1, 8.2, and several figures in chapter 6 all deal with this structure). Typically, these are one-dimensional structures and exceedingly convenient for the study of Wigner function tunneling. For a single barrier, only energies near the top of the barrier are presented with an easier path for tunneling. If we take two dimensions, however, things can become more complicated. One approach is to take a one-dimensional potential and add a magnetic field normal to that one dimension [40]. More recently, a more extensive study of magneto-tunneling through a single barrier in two dimensions, $V(x, y)$, has been presented [41]. Here, the potential is taken to be flat-topped just as in the one-dimensional case, and the magnetic field is oriented normal to the plane of the potential. In this case the wave function decays in the barrier region as

$$\psi(x, y) \sim \exp\left(-\frac{Cx^2}{l_B^2}\right), \tag{9.27}$$

where C is a constant and the magnetic length is

$$l_B = \sqrt{\frac{\hbar}{eB}}, \tag{9.28}$$

with B the magnetic field. In the presence of the magnetic field in the Schrödinger equation, the vector potential is chosen in the Landau gauge (along the y-axis), so that boundaries in the y-direction become important. Hence, it is assumed that the potential has infinite walls at $y = \pm a$. These authors find that when the magnetic field is near a particular value, termed B_R, the decay of the wave function is no longer exponential, and this point is called a Euclidean resonance. This magnetic field is given by [41]

$$B_R = \frac{\sqrt{2m|E|}}{ea}\alpha, \quad \alpha \sim 1.66. \tag{9.29}$$

Here, E is the energy of the particle. In this case, the Euclidean resonance produces behavior quite like normal Wigner tunneling through a double barrier; that is, the resonance produces an effective potential with wells introduced into the normal barrier. The effect has some connection to a similar case studied in the double-well barrier with a magnetic field, where the tunneling direction was x and the field was oriented in the x–z plane [42]. Here, it was found that chaotic dynamics could be observed in the transport in the neighborhood of the resonant tunneling energy.

One of the most discussed questions in the consideration of tunneling is the so-called tunneling time. As one may guess, there are almost as many suggestions for this time as there are researchers studying tunneling. A nice review of the topic has been given by Jonson [43]. One of the oldest suggestions comes from Landau and Lifshitz [44], who discussed the decay of a quantum state due to *outgoing* waves when this decay comes from tunneling. This bears certain similarities to the considerations of Leggett of dissipation from a two-state system in the presence of dissipation [45, 46], which again has considerable implications for the reaction-coordinate discussions above. In particular, the tunneling time is related to the lifetime of the metastable state from which the tunneling occurs, which leads to

$$\Delta\tau^{\varphi} = 4\hbar/\Gamma, \tag{9.30}$$

where Γ is the spectral linewidth of the state. Another typical suggestion arises from the so-called phase time (see, for example, the Wigner–Smith phase delay in section 6.2), where an incoming wave scatters from a central potential. The outgoing wave is then determined in a spherical harmonic expansion in a manner that conserves angular momentum [44]. One then estimates the passage of the incoming wave at x_1, t_1 to the outgoing wave passing into the asymptotic regime at x_2, t_2. Here, one can then define a diagonal element of the scattering matrix as $S_k = e^{2i\delta_k}$, where δ_k is an obvious phase shift. Then if we take the magnitude of the two positions as the same radius from the scattering potential, we can write the phase delay as

$$\tau_k^{\varphi}(k) = t_2 - t_1 = \frac{1}{v_k}\left(2|x_i| + 2\frac{d\delta_k}{dk}\right), \quad v_k = \frac{1}{\hbar}\frac{dE}{dk}, \tag{9.31}$$

where the last expression defines the group velocity of the wave packet. This form has obvious similarities to the phase delay introduced by Bohm [47] and by Wigner [48]. It is obvious that this connects directly to the Wigner–Smith phase time discussed in section 6.2, and the phase delay time discussed there is only slightly more elegant than the above expression. Another suggestion was given by Smith [49], that being the dwell time of the wave packet in the barrier region, but his formula is precisely the phase delay time of section 6.2. Because the ideas of the tunneling time are important in the world of devices as well as in the world of chemical reactions, much time has been spent on arguing the nature of the tunneling time. It is probably as interesting to note that there has also been many suggestions on how to measure the tunneling time [50, 51]. While suggestive, though, each method is beset with the question of just what each method actually measures, particularly in light of the various suggestions for a tunneling time.

The problem that has appeared in the study of chemical reactions is that some studies have given tunneling times that vanish [52–54], or that are independent of the barrier thickness [55]. These results might suggest that non-relativistic quantum mechanics is violated. Recent work [56] has suggested that one should use a transition time similar to that of (9.31) or in other cases a tunneling flight time, but show that their results tend to agree with estimates using a Wigner tunneling packet. Tunneling flight time is very closely related to the idea of the transversal time of a Wigner packet introduced by Barker [57], and used extensively in device considerations. Such a transversal time is not likely to give a vanishing result.

9.4 Spectroscopy

One of the advanced methods for the study of the dynamics of a chemical system is through femtosecond pump-probe spectroscopy. Needless to say, this is another window into subjects such as reaction kinetics, and provides another method of studying the quantum coherence in the chemical system when it is coupled to a dissipative environment. A molecule or a condensed-matter system is bound when it sits at the minimum of a potential structure that balances the attractive and repulsive forces acting upon the individual atoms. In linear pump-probe spectroscopy, one applies first a femtosecond laser pulse to the system. A smaller part of this initial pulse is separated from it by a beam-splitter and sent through an adjustable delay apparatus; the latter is just a system to cause the second, smaller pulse to be delayed relative to the first, before it too is applied to the system. The first pulse excites the system, and the second pulse then is used to study the changes that occur in the optical response due to the exciting pulse. Some of the earliest work in this area studied the carrier excitation in Ge [58, 59]. One of the interesting aspects of the direct transition in Ge is that the small effective mass would normally lead to state filling (which would cut off the optical absorption), but this is counterbalanced by the large density excitation leading to many-body effects which narrow the band gap slightly [60]. As the conduction band moves downward, or the valence band moves upward, the optical excitation is continuously focused upon a new set of states. As time has passed from the early studies, more optical processes have been probed by the pump-pulse method, including Raman scattering [61] and the optical Kerr effect [62], which is a change in the index of refraction of the system under excitation (a very nonlinear optical effect). In particular, the use of short-pulse Raman scattering has allowed one to directly measure the velocity overshoot and non-equilibrium carrier distribution in semiconductors [63]. Raman scattering has also allowed one to probe the non-equilibrium optical phonons generated by the relaxation of the carriers [64].

Once the experimental processes became almost commonplace, it was only natural that theoretical studies began to appear. It then became possible to directly integrate the equations of motion for the atoms in the molecule or solid, through a determination of the proper wave function in the presence of the optical fields. This then could be used to determine the density matrix and/or the Wigner function and to solve an equivalent quantum Fokker–Planck equation for the motion, if or when the potential in which the atoms sit was determined. Given the density matrix or the Wigner function for the atomic position and motion, one can then calculate the nonlinear optical response that gives rise to the experimental results from the pump-probe studies. One of the earliest papers here was the work of Tanimura and Maruyama [65], in which the confining potential was taken to be a displaced Morse potential. The effects of the dissipative bath are introduced via a damping operator based upon Gaussian white noise fluctuations. The particular atomic motion was described by wave packets in the Wigner representation, and this motion was followed as a function of time after excitation by the laser pulse.

We should not forget, in the study of fast processes, that there is also a large field of continuous wave excitation of molecules and condensed material. There still remains a significant problem in determining the spectroscopy as the system is a difficult quantum many-body problem. Nevertheless, one can study the spectroscopy using the Wigner transform of the classical Boltzmann distribution to calculate the correlation functions for incident and scattered optical dynamics [66]. Here, the authors specifically calculated the van Hove spectrum of liquid ^{4}He at low temperature (but well above the superfluid transition). Their approach used the Feynman–Leinert path integral approach discussed in chapter 4. Another use of the Wigner function was to specifically study the presence of quantum effects in the dynamic motion of molecules and whether or not these effects could be seen in the spectroscopy [67]. In this latter work, they studied the molecule as an anharmonic oscillator coupled to a harmonic heat bath. Such a model includes the essence of vibrational relaxation and dissipation via the heat bath. A real-time path integral approach was used to determine the temporal response and the optical properties, through the use of the Wigner phase space representation. Finally, the dependence upon the initial condition for evolution of the photodynamics was studied for the pyrrole molecule, with the quantum situation being described via a Wigner distribution [68]. Here it was found that the use of the quantum distribution obtained from the initial conditions fit to the evolving photodynamics much better than a classical thermal distribution.

References

[1] Lopreore C L and Wyatt R E 1999 *Phys. Rev. Lett.* **82** 5190
[2] Wyatt R E 1999 *J. Chem. Phys.* **111** 4406
[3] Na K and Wyatt R E 2001 *Int. J. Quantum Chem.* **81** 206
[4] Wyatt R E and Na K 2001 *Phys. Rev.* E **65** 016702
[5] Na K and Wyatt R E 2003 *Phys. Scr.* **67** 169
[6] Rowland B A and Wyatt R E 2006 *Chem. Phys. Lett.* **426** 209
[7] Tanimura Y and Wolynes P G 1992 *J. Chem. Phys.* **96** 8485
[8] Ciccotti G, Pierleoni C, Capuani F and Filinov V S 1999 *Comput. Phys. Commun.* **121-2** 452
[9] Jarrel M and Gubernatis J E 1996 *Phys. Rep.* **269** 133
[10] Boninsegni M and Ceperly D M 1996 *J. Low Temp. Phys.* **104** 339
[11] Poulsen J A, Nyman G and Rossky P J 2003 *J. Chem. Phys.* **119** 12179
[12] Shi Q and Geva E 2004 *J. Phys. Chem.* A **108** 6109
[13] Smith K K G, Poulsen J A, Nyman G and Rossky P J 2015 *J. Chem. Phys.* **142** 244112
[14] Smith K K G, Poulsen J A, Nyman G and Rossky P J 2015 *J. Chem. Phys.* **142** 244113
[15] Montoya-Castillo A and Reichman D R 2017 *J. Chem. Phys.* **146** 024107
[16] Liu J and Miller W H 2008 *J. Chem. Phys.* **128** 124111
[17] Monteferrante M, Bonella S and Ciccotti G 2011 *Mol. Phys.* **109** 3015
[18] Beutler J, Borgis D, Vuilleumier R and Bonella S 2014 *J. Chem. Phys.* **141** 084102
[19] Orr L, Hernández de la Peña L and Roy P-N 2017 *J. Chem. Phys.* **146** 214116
[20] Hele T J H, Willatt M J, Muolo A and Althorpe S C 2015 *J. Chem. Phys.* **142** 134103

[21] Tanimura Y 2015 *J. Chem. Phys.* **142** 144110
[22] Kramers H A 1940 *Physica* **7** 284
[23] Goychuk I, Hartmann L and Hänggi P 2001 *Chem. Phys.* **268** 151
[24] Ankerhold J and Lehle H 2004 *J. Chem. Phys.* **120** 1436
[25] Kittel C 1966 *Introduction to Solid State Physics* 7th edn (New York: Wiley)
[26] Jaros M 1982 *Deep Levels in Semiconductors* (Bristol: Adam Hilger)
[27] Baranowski J M, Grynberg M and Porowski S 1982 *Handbook on Semiconductors* vol 1 ed W Paul (Amsterdam: North-Holland) ch 6
[28] Shi Q and Geva E 2003 *J. Phys. Chem.* A **107** 9059
[29] Sergeev A V and Segev B 2002 *J. Phys.* A **35** 1769
[30] Shit A, Chattopadhyay S and Chaudhuri J R 2014 *Chem. Phys.* **431-2** 26
[31] Xie W, Domcke W, Farantos S C and Grebenshchikov S Yu 2016 *J. Chem. Phys.* **144** 104105
[32] Smedarchina Z and Fernández-Ramos A 2002 *J. Chem. Phys.* **117** 6022
[33] Poulsen J A, Li H and Nyman G 2009 *J. Chem. Phys.* **131** 024117
[34] Li H, Poulsen J A and Nyman G 2011 *J. Phys. Chem.* **115** 7338
[35] Craig I R and Manolopoulos D E 2005 *J. Chem. Phys.* **122** 084106
[36] Hale T J H and Althorpe S C 2013 *J. Chem. Phys.* **138** 084108
[37] Kallush S, Tannenbaum E and Segev B 2004 *Chem. Phys. Lett.* **396** 261
[38] Bark D, Bag B C and Ray D S 2003 *J. Chem. Phys.* **119** 12973
[39] Kreuzer H J 1981 *Nonequilibrium Thermodynamics and its Statistical Foundations* (Oxford: Clarendon)
[40] Kluksdahl N C, Kriman A M and Ferry D K 1989 *High Magnetic Fields in Semiconductors II (Springer Series in Solid-State Science)* vol 87 ed G Landwehr (Berlin: Springer) p 335
[41] Ivlev B 2007 *Phys. Rev.* A **76** 022108
[42] Fromhold T M, Eaves L, Sheard F W, Leadbeater M L, Foster T J and Main P C 1994 *Phys. Rev. Lett.* **72** 2608
[43] Jonson M 1992 *Quantum Transport in Semiconductors* ed D K Ferry and C Jacoboni (New York: Plenum) pp 193–238
[44] Landau L D and Lifshitz E M 1958 *Quantum Mechanics* (Oxford: Pergamon)
[45] Caldeira A O and Leggett A J 1981 *Phys. Rev. Lett.* **46** 211
[46] Leggett A J 1984 *Phys. Rev.* B **30** 1208
[47] Bohm D 1951 *Quantum Theory* (Englewood Cliffs, NJ: Prentice-Hall) pp 257–63
[48] Wigner E P 1955 *Phys. Rev.* **98** 145
[49] Smith F T 1960 *Phys. Rev.* **118** 349
[50] Büttiker M and Landauer R 1982 *Phys. Rev. Lett.* **49** 1739
[51] Baz A I 1960 *Sov. J. Nucl. Phys.* **4** 182
[52] Terlina L *et al* 2011 *Nat. Phys.* **11** 503
[53] Low F E and Mende P F 1991 *Ann. Phys.* **210** 380
[54] Ivanov I A and Kim K T 2015 *Phys. Rev.* A **92** 053418
[55] Hartman T E 1962 *J. Appl. Phys.* **33** 3427
[56] Petersen J and Pollak E 2017 *J. Phys. Chem. Lett.* **8** 4017
[57] Barker J R 1985 *Physica* B **134** 22
[58] Auston D H and Shank C V 1974 *Phys. Rev. Lett.* **32** 1120
[59] Kennedy C J, Matter J C, Smirl A L, Wieche H, Hopf F A, Pappus S V and Scully M O 1974 *Phys. Rev. Lett.* **33** 419

[60] Ferry D K 1978 *Phys. Rev.* B **18** 7033
[61] May P G and Sibbett W 1983 *Appl. Phys. Lett.* **43** 624
[62] Etchepare J, Grillon G, Thornazeau I, Migus A and Antonetti A 1985 *J. Opt. Soc. Am.* B **2** 649
[63] Grann E D, Tsen K T, Sankey O F, Ferry D K, Salvador A, Botcharev A and Morkoç H 1995 *Appl. Phys. Lett.* **67** 1760
[64] Tsen K T, Klem J and Morkoç H 1986 *Sol. State Commun.* **59** 537
[65] Tanimura Y and Maruyama Y 1997 *J. Chem. Phys.* **107** 1779
[66] Poulsen J A, Nyman G and Rossky P J 2004 *J. Chem. Phys.* A **108** 8743
[67] Sakurai A and Tanimura Y 2011 *J. Phys. Chem.* A **115** 4009
[68] Barbatti M and Sen K 2016 *Int. J. Quantum Chem.* **116** 762

IOP Publishing

The Wigner Function in Science and Technology

David K Ferry and Mihail Nedjalkov

Chapter 10

Signal processing

The previous chapter, dealing with the use of the Wigner function in quantum chemistry, was littered with phrases such as 'semi-classical' and 'quasi-classical'. Yet we consider the use of the Wigner function to be a quantum approach that happens to be based upon phase space considerations. But, then, it appears that the Wigner function has been applied in many areas where we would not normally think of quantum mechanics. Is the propagation of an electromagnetic wave through turbulent air a quantum system? One would not normally think so. Yet, in this chapter we will explore the use of Wigner functions in many areas that we think are classical, and perhaps classical to the extreme. So, perhaps the above phrases have more importance to the topic than we might expect.

As an example, suppose you are interested in market economics, in the movement of supply and demand curves, if for no other reason than an interest in both prices and profits. One mathematical technique to utilize is classical game theory. Although there were earlier discussions of such games of strategy, modern game theory arose from John von Neumann [1], one of our pioneers of quantum theory. Although one might surely think of this as a classical field, in fact modern approaches have begun to refer to quantum game theory [2–4]. The example we wish to take is said to be a new approach to this topic that is more applicable to the economic world [5]. Just as we might expect, various quantum strategies are expressed as vectors in a Hilbert space, and these can be defined as superpositions of trading decisions (scattering processes?). In this approach supply and demand are conjugate operators in the Hamiltonian which is called the risk inclination operator. Naturally, there is a counterpart to the Planck constant, termed the economical Planck constant h_{E}. The wave function basis set is said to be the defining part of a players behavior and strategy, and this is then Wigner transformed into the phase space according to the above operators via

$$f_W(x, p) = \frac{1}{h_E}\int dx' e^{-ix'p/h_E}\frac{\langle x + x'/2|\psi\rangle\langle\psi|x - x'/2\rangle}{\langle\psi|\psi\rangle}. \tag{10.1}$$

Here, x represents the demand and p represents the supply in this approach. Interestingly enough, they find that the prices are represented as a diffusion process, and this can be related to a classical diffusion picture. In this approach, they find a model of price movement that is inspired by and related to the quantum-mechanical evolution of physical particles. The novelty in their approach lies in the wave function which can be complex quantities subject to normal quantum descriptions. The picture that results is intriguing because it goes right to the heart of how quantum mechanics affects our everyday life, and whether or not quantum phenomena exist beyond the world of sub-atomic processes. It seems clear that quantum phenomena are far more ubiquitous than the founders thought.

In this chapter, we want to explore the use of the Wigner function in the areas of signal processing and propagation, areas which are widespread throughout science, especially as we envision such science to affect our everyday world. The above example fits within this scheme as one often thinks, while reading the financial section of the paper, 'What is the market trying to tell me?' This makes a very personal view of signal processing.

10.1 Signal propagation

The study of propagation of either acoustic or electromagnetic waves through the environment has been a subject of study almost since the work of Maxwell. This is because the environment is inhomogeneous, whether it is the atmosphere, the solid earth, or a water system. The inhomogeneity can be uniform in the directionality or it can introduce birefringence by which waves split into different directions depending upon their polarizations. Indeed, in the electromagnetic situation, one can often use the ray approach to characterize the medium of interest [6]. But many other approaches have been used. The type of inhomogeneity varies, and there are two basic considerations that have to be made. The first is the (relatively) slow spatial variation of the parameters of the medium, while the second is the fluctuations of these parameters. For example, the atmosphere has regular gradients whether by humidity, dust, or altitude. Fluctuations can arise by irregularities in the interface between atmosphere and, for example, sea surface. Radar signals at sea have always had a problem with random sea scatter of the signals, and most early attempts focused upon trying to account for these instead of thinking about how they could be used. Similarly, acoustic wave analysis of the Earth's inhomogeneities seek to find regular variations, such as those arising from oil deposits, in the presence of the entirely random nature of the Earth's composition. Modern developments (defined as within the last half century), such as ground-penetrating radar and sonar imaging, have improved the situation considerably, but have not eliminated the need to study in detail the nature of the inhomogeneous media through which the waves are propagated.

So it is clear that wave processes and their processing methods can give insight into the nature of the media through which the signals pass. Indeed, as in other fields, the analysis of the statistical data can provide a great deal of information on the scattering fields that are finally detected. It has long been realized that random potentials that appear in the Schrödinger equation have their analogy in the random nature of the dielectric function in electromagnetic wave propagation. This analogy has been pursued by many, but is sometimes limited by the fact that the classical approach does not have the uncertainty principle in its evaluation [7]. An example discussed in this latter paper addresses the ray approach mentioned above. Normally, one has to deal with the fact that both the slope and the position of the ray are specified at each spatial point. Quantum mechanically, these two would be conjugate operators. The approach in the latter work is to consider only the straight trajectories in the analysis, and to expand these trajectories into a set of eigenfunctions. Here, the amplitude of the field u is represented by a plane wave expansion [7]

$$\begin{aligned} u(\mathbf{x}, \sigma_0) &= \left(\frac{k}{2\pi}\right)^2 \int_{-\infty}^{\infty} \int d^2\bar{\kappa}_0 \varphi(\bar{\kappa}_0, \sigma_0) e^{ik\bar{\kappa}_0 \cdot \mathbf{x}} \\ &= \int_{-\infty}^{\infty} \int d^2\bar{\kappa}_0 \varphi(\bar{\kappa}_0, \sigma_0) g(\mathbf{x}, \sigma_0) \end{aligned}, \tag{10.2}$$

where $\bar{\kappa}_0$ is a unit vector (slope) in the direction of propagation of the wave and σ_0 is the mean value of the fluctuations. In the second line, g is the spectral function of the waves. From the wave field, one then defines a two-point random function of the field via the center-of-mass coordinates (2.100)

$$\Gamma(\mathbf{R}, \mathbf{s}, \sigma) = u\left(\mathbf{R} + \frac{\mathbf{s}}{2}, \sigma\right) u^*\left(\mathbf{R} - \frac{\mathbf{s}}{2}, \sigma\right). \tag{10.3}$$

As one might expect, this two-point function is then Fourier transformed into the Wigner function for the wave

$$f_W(\mathbf{R}, \mathbf{p}, \sigma) = \int d^3\mathbf{s} \Gamma(\mathbf{R}, \mathbf{s}, \sigma) e^{ik\mathbf{p}\cdot\mathbf{s}}. \tag{10.4}$$

With this approach, the author is able to demonstrate that ray uncertainty plays an important role in generating approximate forms for the field propagators in random media. A similar approach using two times as well as two spatial points in the random function has also been developed [8]. This now provides a two-time Wigner function transform that is basically the full two-time, two-space Green's function discussed in chapter 2. This function is shown to satisfy a Fokker–Planck type of equation, in order to find the precise profile of the space-frequency correlation on a scale below that of the transport mean free path. More recently, the polarization of the waves through random media has been added to the transport [9]. Here, a mode approach is adopted, and intermode scattering introduces the randomness of the medium. They can then study the loss of polarization that arises from the intermode scattering, with the use of the Wigner function to describe the energy density of the wave. Again, introduction of the Wigner function leads to a set of stochastic equations which describe the propagation and the role of the inhomogeneities. Numerical solutions of one-dimensional waves has also been studied through discretization of the space [10].

While we have discussed approaches using the Wigner function, the study of propagation in random media is much older, so that we should not give the impression it began with the first paper listed above. The more general approach, which has been around for a long time, is the radiative transfer approach [11]. The specific entity of interest is the phase space quantity known as the intensity **I** of the wave. This is given at a position **x** and direction **s** by the equation [11]

$$\mathbf{s} \cdot \nabla I(\mathbf{x}, \mathbf{s}) + \gamma I(\mathbf{x}, \mathbf{s}) = \int d\Omega' \overline{\mathbf{P}}(\mathbf{s}, \mathbf{s}') \cdot I(\mathbf{x}, \mathbf{s}'), \tag{10.5}$$

where $\overline{\mathbf{P}}$ is the phase space scattering matrix (tensor) that describes scattering from one point to another in direction, and Ω is the spherical solid angle subtended by the two direction vectors. Recently, this approach has been studied using a more statistical approach which utilizes the Wigner function (in a two-time representation, so that they are more related to the Green's functions) to describe spatial correlation functions for the fields [12]. With this approach, the author is able to better define the intrinsic assumptions and approximations that exist in traditional radiation transfer theory.

Most of the above approaches are applied to propagation *through* random media. But there are also important problems in which the propagation is *over* random media. For radars that track various movements, contamination can arise by random variations in the ionosphere above the region of propagation. This can act like a random surface, especially for waves propagating in the atmospheric 'waveguide', so-called as the waves are reflected from above by the ionosphere and from below by the Earth's surface. The unstable and temporally varying properties of the ionosphere can cause an effective frequency modulation of a high-frequency radar system [13]. The transition to a phase space representation, using what is called the Wigner–Ville distribution, allows an estimation of the extent of the ionospheric disturbance to the signal. Reflection of an over-the-horizon radar from the sea surface can be described in terms of a Bragg scattered signal that provides Doppler-shifted lines corresponding to the positive and negative first Bragg peaks. These are usually small, and the Bragg frequency shifts are often described as

$$f_B \sim \pm\, 0.1\sqrt{f_0}\, \cos\beta, \tag{10.6}$$

where f_0 is the radar frequency and β is the grazing angle of the radar waves with the sea surface. Then the ratio of the energy in the positive and negative Bragg peaks can be written as

$$r = [\tan(\vartheta/2)]^s, \tag{10.7}$$

where $s \sim 2$–6, and θ is the angle between the radar beam and the wind direction at the surface. The noise that is generated provides a compression in the time domain to each range cell of the (digitized) radar, which may be expressed as [13]

$$s_0(m) = a_{B+}e^{i2\pi f_{B+}mT} + a_{B-}e^{i2\pi f_{B-}mT} + \sum_{j=1}^{p} a_j e^{i2\pi f_j mT} + g_0(mT). \tag{10.8}$$

Here, T is the frequency-sweep period, m is the number of radar pulses in a coherent integration time, g_0 is the background noise, f_{B^+} and f_{B^-} are the positive and negative first-order Bragg frequencies and a_{B^+} and a_{B^-} are their amplitudes. The radar echo signal is a set of discrete peaks, the number of which is the maximum value of m, given as M. This may be divided into an upper and lower frequency domain, corresponding to the positive and negative Bragg peaks, and the set of each of these may be then Fourier transformed into time-domain signals $s_{r^\pm}(m)$, where the upper sign is for the positive Bragg peaks and the negative sign is for the negative Bragg peaks. These time-domain signals are now used to generated the Wigner–Ville distributions, one for each peak, as

$$f_{W-V,\, r^+}(m,\, k) = 2 \sum_{j=-M/2+1}^{M/2-1} h_1(j) s_{r^+}(m + j) s_{r^+}(m - j) e^{-i(4\pi k/M)}, \qquad (10.9)$$

and here m and j denote the discrete time and frequency in the discretized spectrum and signal, and k is the discrete frequencies in the Fourier domain. Here, h_1 is a sliding Hamming window to filter the ranges used in the summation. From such processing as this, it is possible to then estimate the amount of noise contamination that exists in the signal, and this estimate can be used to enhance the radar return signal.

A quite similar approach can be used in sonar signals, where the contamination now comes from the ocean floor in (relatively) shallow water [14]. In this case, one is not affected by the time variations of the sea floor, and geological and geophysical information can be used to generate the nature of the contamination signal. In principle, detailed sea floor mapping can provide the necessary information. In the absence of such information, the noise generated on the sonar signal from the sea floor can be reverse processed to provide an estimate of the nature of the sea floor; in essence this is the inverse mapping approach. The approach uses two trains of pulses transmitted from various directions and distances from the detector. For example, the sources can be placed below the sea surface. The Wigner transform of the signal is determined for each signal as a two-dimensional function of time and instantaneous frequency as

$$f_W(t, f) = \int_{-\infty}^{\infty} x(t + \tau/2) x(t - \tau/2) e^{-2\pi i f \tau} d\tau, \qquad (10.10)$$

where x is the time domain signal of interest. The advantage of using the Wigner transform lies in the fact that a linear frequency modulation appears just as an harmonic frequency in this transform plane. However, there are interactions between different coherent modes which do require some additional processing. Nevertheless, this approach has important advantages and extends the frequency range over which normal waves can be used, and increases the resolution and accuracy of each set of mode parameters in the analysis.

10.2 Wavelets

Wavelets in the classical world are quite similar to what we call wave packets in the quantum world. They represent a wave-like oscillation that lasts only a few periods and in which the envelope function is similar to a Gaussian or Lorentzian in time. Obviously, a Fourier transform of the wavelet has a limited frequency spectrum, so that convolving the wavelet with a data signal will pick a certain range of frequencies and can be used to devolve weak signals from noise. In the original usage, wavelets referred to groups of plane waves [15], but we know that these produce localized wave packets. These wavelets can then be used to study optical phenomena. The wavelet description has been applied to a wide variety of time-limited signals, such as gastric [16] and seismic disturbances [17]. It is thus quite natural that the mathematics of these wavelets has been greatly expanded and they have become useful in a wide range of applications, in particular signal processing.

The basic ideas today are closely related to quantum mechanics. That is, we develop a Hilbert space of localized wave packets, typically described by

$$\psi_{jk}(x) = 2^{j/2}\psi(2^j x - k), \tag{10.11}$$

where the factor of 2 is obviously governed by modern digital processing. Of course, these functions must be orthonormal, with

$$\int_{-\infty}^{\infty} \psi_{lm}^*(x)\psi_{jk}(x)\, dx = \delta_{lj}\delta_{mk}. \tag{10.12}$$

While this is written in one dimension, it is obviously extendible to multiple dimensions, and the use of wavelets in data compression is quite common today. For example JPEG 2000 uses this compression method to reduce the size of image files. Naturally, any function of space can be expanded in a series of these orthonormal functions.

The wavelet transform enters signal processing when the spatial variable is taken over to a temporal variable. And, in this endeavor, the Wigner–Ville transform describes a short-time Fourier transform that presents the signal in a mixed phase space of time and frequency. For example, the Wigner–Ville distribution describes a more confined region of the time–frequency plane when treating epileptic signals in ECG analysis, although there are problems from cross-term interactions [18]. In this guise, the Wigner–Ville distribution of a time-dependent signal is given by (10.10), and wavelets are used to describe the various functions $x(t)$ that appear in the equation [19–23]. In these approaches, the wavelet used for the wavelet transform differs from (10.11) and is described by

$$\psi(\tau, a) = \frac{1}{\sqrt{a}}\psi\left(\frac{t - \tau}{a}\right), \tag{10.13}$$

where, again, the main wave function is a localized wave packet. The two parameters shift and expand it by various amounts. In some areas, the Wigner–Ville approach is considered to be an alternative to the continuous wavelet transform rather than an addition to the processing [24].

References

[1] von Neumann J 1928 On the theory of games of strategy *Math. Ann.* ed J Von Neumann (Princeton, NJ: Princeton Univ. Press) pp 13–42
[2] Meyer D 1999 *Phys. Rev. Lett.* **82** 1052
[3] Eisert J, Wilkens M and Lewenstein M 1999 *Phys. Rev. Lett.* **83** 3077
[4] Piotowski E W and Sladkowski J 2002 *Progress in Mathematical Physics Research* ed C V Benton (New York: Nova Science)
[5] Piotowski E W and Sladkowski J 2005 *Physica* A **345** 185
[6] Kline M and Kay I W 1965 *Electromagnetic Theory and Geometrical Optics* (New York: Wiley)
[7] Mazar R 1998 *Comput. Struct.* **67** 119
[8] Fannjiang A C 2007 *Europhys. Lett.* **80** 14005
[9] Borcea L and Garnier J 2016 *Wave Motion* **63** 179
[10] Marica A and Zuazua E 2015 *Found. Comput. Math.* **15** 1571
[11] Chandrasekar S 1960 *Radiative Transfer* (New York: Dover)
[12] Mudaliar S 2013 *Radio Sci.* **48** 535
[13] Chengyu H, Tian X and Qian B 2014 *IET Radar Sonar Navig.* **8** 742
[14] Belov A I and Kuznetsov G N 2014 *Acoust. Phys.* **60** 191
[15] Stoney G J 1903 *Phil. Mag.* **5** 264
[16] Alvarez W C and Zimmermann A 1926 *Am. J. Physiol.* **78** 405
[17] Ricker N 1955 *J. Acous. Soc. Am.* **27** 199
[18] Gutiérrez J, Alcántara R and Medina V 2001 *Med. Eng. Phys.* **23** 623
[19] Narasimhan S V, Haripriya A R and Shreyamsha Kumar B K 2008 *Signal Process.* **88** 1
[20] Qian S and Chen D 1999 *IEEE Sign. Process. Mag.* **16** 52
[21] Tang B, Liu W and Song T 2010 *Renew. Energy* **35** 2862
[22] Yu X, Ji H, Chen S, Liu X and Zeng Q 2013 *Adv. Mater. Res.* **805-6** 1962
[23] Xu C, Wang C and Liu W 2016 *J. Vib. Acoustics* **138** 051009
[24] Gramatikov B 2013 *Biomed. Eng. Online* **12** 41

IOP Publishing

The Wigner Function in Science and Technology

David K Ferry and Mihail Nedjalkov

Chapter 11

Quantum optics

In many aspects, the consideration of optical signals is a mere extension of the classical approach to the propagation of waves. And, indeed, the first section below will deal with just that aspect, particularly the propagation of optical waves through disordered regions. But optical signals also hold the promise of quantum coherent waves, which takes us well beyond the classical propagation of waves. The majority of this chapter will deal with that aspect, as it seems that it is here that the Wigner function shows its almost normal application as a method of illustrating the coherence and entanglement of the waves. An example of this lies in the world of clocks and the international measurements for the standard value of the second, the international unit for time.

For a long time, the definition of the second relied upon the rotation of the Earth. This changed dramatically with the arrival of quantum mechanics and atomic clocks. In 1967, the standard second was defined in terms of the difference frequency between two hyperfine levels of cesium atoms. The accuracy of such clocks has improved considerably over time. The drawback, however, is that the cesium frequencies are in the microwave range, and it has been noted that clocks using optical frequencies should provide higher degrees of accuracy, even though optical frequencies are more difficult to measure. The development of lasers whose spectrum consists of discrete equally spaced frequencies, the so-called frequency *comb*, has made such measurements easier [1, 2]. The comb is generated most commonly by a pulsed mode-locked laser whose pulse repetition rate is governed by the round trip pulse time of the optical cavity, and the usefulness of the comb lies in the fact that it is useful for mapping the optical frequencies into a range of lower frequencies in which the frequency measurement is more tenable. This has led to the development of single trapped ions for the purpose of making frequency standards, and new approaches based upon Sr and Yb have appeared [3]. Now, optical clocks based upon these atoms have matured considerably and have surpassed the cesium clocks in the accuracy of time measurements [4]. People are now looking at ultracold spin-

doi:10.1088/978-0-7503-1671-2ch11

polarized fermion systems, with the hope of achieving longer atom-light coherence times. But this requires more understanding of the dynamics of various dissipative processes within the atoms, because the overall ‘clock’ represents an open quantum system, subject to a range of interactions even at ultracold temperatures. One recent approach provides a unified theoretical treatment, beyond simple mean-field theory, which treats the full many-body interactions of a nuclear spin-polarized alkaline earth atom during the light–atom interaction [5]. By developing the various quantum equations of motion, they are able to use a truncated Wigner approximation to obtain the relevant solutions to these equations; so we find Wigner approaches within the world of modern atomic clocks.

11.1 Propagation

Much as with classical electromagnetic waves in the previous chapter, the propagation of optical waves is of great interest, especially as these waves can be coherent quantum systems on their own. We discussed the use of coherent states already in chapter 3, and one can easily consider these as quantum propagating states. Samson [6] has considered the propagation of these types of states in generalized media. A path integral for the coherent states is regularized in that it is shown to be a set of averages over polygonal paths with a given set of vertices for each polygon. Samson shows that the distribution of the path centroid tends to the Wigner function, which is defined as the joint distribution of the path operators $P(n_i)$, where n_i is the number of vertices in the i'*th* polygon. In addition, it is shown that the Wigner function will be non-positive if the dominant paths for a path centroid in a given region have Berry phases close to multiples of π. In essence, this work demonstrates a comment made by Feynman about trying to connect classical and quantum physics [7]. In the latter, Feynman was concerned with negative probabilities, a concept we are by now quite familiar with in relation to Wigner functions. As Samson points out, if we try to apply classical concepts to quantum systems, a cost is incurred to do so. Consider a function f of commuting variables, for which we can describe an expectation value for the equivalent operator function in quantum mechanics as [6]

$$\langle \hat{f} \rangle = \mathrm{Tr}\{\hat{f}\hat{\rho}\} = \int_{\Gamma} dx\, f(x) W(x). \tag{11.1}$$

Here, x takes values in some space Γ, and $W(x)$ is a normalized distribution that depends only upon the density matrix $\hat{\rho}$ [8]. It is this property that Samson investigates via coherent wave packets, and demonstrates that $W(x)$ is the Wigner function, complete with its negative probabilities. He goes on to conclude that this gives insight into the quantum Monte Carlo sign problem, discussed in chapter 6. If the variables of interest that are sampled in the Monte Carlo simulation are non-commuting operators, then he shows that the distribution of their time averages will be a Wigner function and the negative excursions relate directly to the non-commuting property, as discussed already in chapter 3.

In a similar manner, there have been studies of the propagation of optical waves and the formation of images. It has been shown that the intensity distribution of waves, whose rays lie along different paths in the two-dimensional image space, can be described in terms of a Wigner function [9]. These authors suggest that using such an analysis can be a good approach to analyzing systems suffering from spherical aberrations. Indeed, it has been shown that a ray-based approach can be made to mimic the Wigner function itself for these optical beams [10], and this approach can handle both Fresnel and Fraunhofer diffraction around edges. Indeed, the Wigner function is an excellent tool for modeling the optical field for light propagation, whether for the stationary field or for the propagation of pulses, such as coherent packets [11]. The latter author discusses the usefulness of the Wigner function in various transforms that are common in optics, more recently the effect of partially coherent but decentered annular beams, on the skewness and sharpness of the wave propagation [12]. Among the various transforms mentioned by Alonso [11], one that has become quite useful is the fractional Fourier transform [13, 14]. The fractional Fourier transform, of order θ, can be defined, for the quantum case, as [11]

$$f_\theta(\mathbf{s}) = \left[\frac{\tan\theta - i}{2\pi\hbar\sigma\tan\theta}\right]^{N/2} \int d^N\mathbf{x}\, f(\mathbf{x}) \exp\left[i\frac{(x^2 + \sigma^2 s^2)\cos\theta - 2\sigma\mathbf{x}\cdot\mathbf{s}}{2\hbar\sigma^2\sin\theta}\right], \tag{11.2}$$

where $\mathbf{x}$ is an N-dimensional (position) vector and σ is a scaling constant. The fractional Fourier transform is a unitary operation. This transform is a type of linear canonical transformation and is particularly useful in optics and simple quantum systems. This transform has the property that its squared modulus gives the Radon–Wigner transform, which is a basic tool in generating the Wigner distribution function [15]. Let us consider a complex function $f(x)$, in one dimension, so that the Wigner transform is given as

$$f_W(x, \xi) = \int_{-\infty}^{\infty} dx' f\left(x + \frac{x'}{2}\right) f^*\left(x - \frac{x'}{2}\right) e^{-2\pi i \xi x'}. \tag{11.3}$$

The Radon transform is defined in terms of two orthogonal dimensions, which can be taken as x and y. If we rotate these coordinates by an angle θ, the new coordinates may be written as x_0 and y_0, and the Radon transform is defined as (note that the angle exists in the relation of y to y_0) [16]

$$R_g(x_0, \theta) = \int_{-\infty}^{\infty} g(x, y)\, dy_0. \tag{11.4}$$

Obviously, then, the Radon–Wigner transform is the Radon transform of the Wigner function (11.3).

As with classical electromagnetic signals, the passage of optical signals through random media is of considerable interest. Here, the random media is more associated with small targets or embedded obstacles than with variations in the boundary conditions. One problem, of course, is the analysis of the back-scattered signal, often referred to as the double-passage problem. This has been studied by looking at the correlation between the forward and backward waves through the use of a paired field measure, which is a two-point random function. Its spectral transforms produce the Wigner function and the ambiguity function [17, 18]. Here, the phase variation along the main propagation direction is given by

$$U(\mathbf{x}, \sigma) = u(\mathbf{x}, \sigma)e^{ik\sigma}, \tag{11.5}$$

where σ is the range coordinate along the main propagation direction and $\mathbf{x}$ is a two-dimension variable in the plane normal to this direction of propagation. Then, the paired field measure is the Wigner representation for two such fields:

$$\Gamma(\mathbf{R}, \mathbf{s}, \sigma) = u\left(\mathbf{R} + \frac{\mathbf{s}}{2}, \sigma\right)u^*\left(\mathbf{R} - \frac{\mathbf{s}}{2}, \sigma\right), \tag{11.6}$$

where the coordinates have been combined in the usual way (see (3.1)):

$$\mathbf{R} = \frac{\mathbf{x}_1 + \mathbf{x}_2}{2}, \quad \mathbf{s} = \mathbf{x}_1 - \mathbf{x}_2. \tag{11.7}$$

The Fourier transform in the difference variable of course leads to the Wigner function, which can be used to evaluate the role of the randomness of the media. The Wigner functions themselves have been measured by the use of a heterodyne detection technique [19]. More recently, studies have focused upon the higher-order moments of the partially coherent light beams as they propagate through atmospheric turbulence [20].

11.2 The Jaynes–Cummings model

The coupling between a two-level atom and a resonant (quantized) electromagnetic cavity is described by a relatively old model, termed the Jaynes–Cummings model [21], which we have already discussed briefly in chapter 8, where a short derivation of the interaction was presented. It turns out that the process of coupling a coherent optical beam to such an atom is of great interest in quantum optics, so we want to discuss the model in a little more detail. The model treats the atom as a two-level system, and describes the interaction of this atom as it interacts with a quantized mode of an optical cavity. The principle of the model is analogous to the problem of two coupled classical pendulums that are connected to produce an interaction[1]. This interaction leads to a coupling of the two modes in a way in which all of the energy oscillates between the two. At a given instant of time, pendulum 1 may be static while all the energy is in the oscillation of pendulum 2, and at a later time the process is in the opposite state. In the quantum case, one pendulum is the harmonic oscillator mode of the electromagnetic cavity, while the second is the atom, which is described as an oscillator by its Rabi frequency. Hence, the field can interact with the atomic state and lead to Rabi oscillations of the atomic state population. As with the pendulums, the Rabi oscillation is not static in amplitude. Something different happens, in that the Rabi oscillations collapse and then reappear periodically, just as the energy moves from one pendulum to the other. The temporal time over which this collapse and

[1] One can find many videos illustrating this effect on YouTube.com. One such video which clearly indicates the energy transfer can be found at https://www.youtube.com/watch?v=8JhDbR7tDbg.

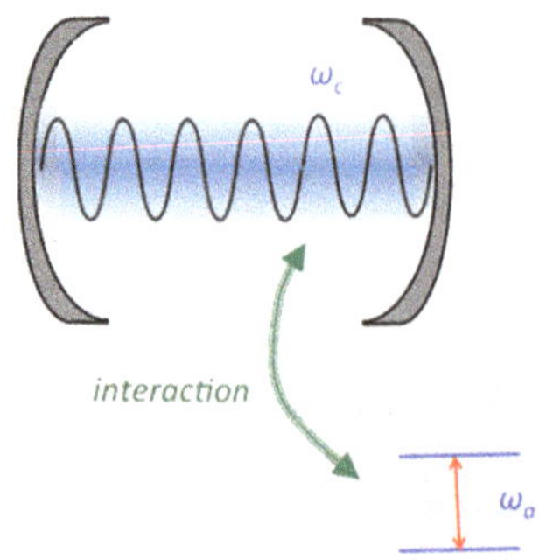

Figure 11.1. In the Jaynes–Cummings model, an electromagnetic wave, depicted in the cavity at the top, couples to a two-level atom, depicted at the bottom. This has some similarities to a classical set of two pendulums which are coupled to one another.

reappearance occurs has been termed the revival time [22]. When the field in the cavity is in a coherent state, and the atom is prepared in an excited state, it is found that the atomic and field states become rapidly entangled, and then subsequently disentangle at one half the revival time [23]; that is, the maximum entanglement of the atomic states occurs when the Rabi oscillations are strongest and the minimum entanglement occurs when the Rabi oscillations are minimal. More recently, it has been observed that the photon parity operator also shows similar behavior with one important difference. The peaks of the amplitude of the parity operator seem to occur when the Rabi oscillations are at their minimum. So, the amplitude and decay of this parity operator seems to be shifted by half a period from the Rabi oscillations [24]. Because the parity operator is proportional to the number density of the photon field, this latter is also in agreement with the two-pendulum concept, as the Rabi oscillation is a minimum when the majority of the energy is in the photon oscillator. The Jaynes–Cummings model was revived in this century [25] and applied explicitly to the problem of a trapped ion interacting with a laser field.

The Hamiltonian can be described relatively simply in terms of the various components as discussed above, and is shown in figure 11.1. First, the electromagnetic field or the laser field is quantized as usual in terms of a series of harmonic oscillators that describe each of the modes that can exist in the cavity or wave [26]. Since we are interested in a single mode, we write the wave energy as

$$H_{EM} = \hbar\omega_c \hat{a}^\dagger \hat{a}, \tag{11.8}$$

where the operators are the normal creation and annihilation operators for the oscillator field, and ω_c is the frequency of this mode. The atomic levels are just the two levels of the atom, which we describe as a pseudo-spin index with the spin up state as the upper level and the spin down state as the lower level, so that we can write the Hamiltonian as

$$H_a = \hbar\omega_a \frac{\hat{\sigma}_z}{2}. \tag{11.9}$$

Hence, the atomic frequency describes the separation of the two states. The state of the atom is described by its polarization,

$$\hat{S} = \hat{\sigma}_+ + \hat{\sigma}_-, \tag{11.10}$$

where $\hat{\sigma}_+$ and $\hat{\sigma}_-$ are the raising and lowering operators of the two-level atom [27]. Hence, the polarization is equivalent to the position operator in a normal harmonic oscillator, with the various parameters being folded into a pre-factor discussed shortly. The interaction between the electromagnetic field and the atom occurs through the electric field of the wave. Normally, the potential would be proportional to $eE_z z$, so here we take the position as the polarization and write the interaction energy as

$$H_{\text{int}} = \frac{\hbar\Omega}{2}\hat{E}\hat{S} = \frac{\hbar\Omega}{2}E_z(\hat{a}^\dagger + \hat{a})(\hat{\sigma}_+ + \hat{\sigma}_-). \tag{11.11}$$

The parameter Ω incorporates all the various constants arising from the connection between the operators (from the harmonic oscillators) that appear in (11.11) and their corresponding position operators. The creation and annihilation operators each have their time variation determined by their respective frequencies, so the four cross terms will have slow variations due to the sum of the frequencies and fast variations due to the difference in the two frequencies. Generally, the fast components are ignored and only the slow components are retained in what is called the rotating wave approximation. The Hamiltonian is generally separated into two commuting parts, which are described by

$$\begin{aligned} H_1 &= \hbar\omega_c\left(\hat{a}^\dagger\hat{a} + \frac{\hat{\sigma}_z}{2}\right) \\ H_2 &= \hbar\delta\frac{\hat{\sigma}_z}{2} + \frac{\hbar\Omega}{2}(\hat{a}\hat{\sigma}_+ + \hat{a}^\dagger\hat{\sigma}_-), \end{aligned} \tag{11.12}$$

where

$$\delta = \omega_a - \omega_c \tag{11.13}$$

is the so-called detuning of the frequency between the field and the atom. Now, this creates a proper 2×2 Hamiltonian, whose eigenvalues are given in terms of the number operator (first term in the parentheses of line one of (11.12)) for the field as

$$E_\pm = \hbar\omega_c\left(n + \frac{1}{2}\right) \pm \frac{\hbar}{2}\Omega_n(\delta), \tag{11.14}$$

where

$$\Omega_n = \sqrt{\delta^2 + \Omega^2(n+1)} \tag{11.15}$$

is the Rabi frequency for the specific detuning parameter. If the detuning is small, the atomic state Rabi oscillations occur with an average frequency $\Omega_c = 2\Omega\sqrt{n+1}$.

As the mode occupation builds up, the Rabi frequency increases and the amplitude of these oscillations increases. The photon parity operator is then defined as

$$\Pi_F = (-1)^n = e^{i\pi n}. \tag{11.16}$$

The normal expectation value of this operator is found to be [24]

$$\langle \Pi_F(t) \rangle = e^{-n} \sum_{s=0}^{\infty} (-1)^s \frac{n^s}{s!} \cos(2\Omega t \sqrt{s+1}), \tag{11.17}$$

where the time dependence arises from the time dependence of the effective Rabi frequency. The interesting fact is that for $t = 0$, the cosine is unity, and the expectation value of the parity operator goes asymptotically to zero (the time is scaled with the same coupling factor λ). Hence, when the density peaks and the Rabi oscillations are strongest, the parity operator vanishes, and vice versa. In general, for the initial conditions discussed, the Wigner function of the field splits into a pair of counter-rotating components in phase space [28]. These two components reach a maximum separation at one half the revival time and then recombine at the revival time. Hence, the peak of the parity operator corresponds to the maximum separation of the two parts of the Wigner function. The Wigner function itself can be written as a shifted expectation value of the parity operator, given as [29]

$$W(\alpha) = \frac{2}{\pi} \langle D(\alpha) \Pi_F D^{\dagger}(\alpha) \rangle, \tag{11.18}$$

where D is a displacement operator and α is a complex number which represents a point in phase space, i.e. the real and imaginary parts correspond to position and momentum. The displacement operators work on the two wave functions in the density operator, leading to the displaced versions that appear generating the Wigner function. In figure 11.2, we plot the Wigner function of the parity operator at a time when this operator is near its maximum amplitude and at a positive peak (of the oscillating operator). The x and y axes correspond to the real and imaginary parts of α, and thus to the position and momentum of the function. While the two

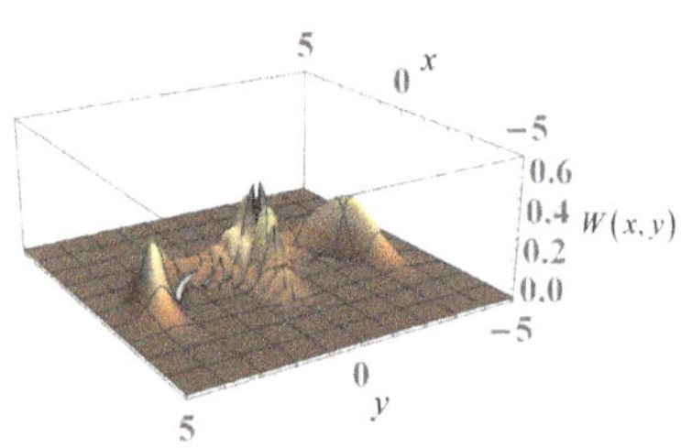

Figure 11.2. The Wigner function for the expectation of the parity operator at the scaled time for which this quantity is a maximum (near one half the revival time [24]). The x and y axes correspond to the real and imaginary parts of α, as described in the text. This figure is courtesy of Richard Birrittella and Prof. Chris Gerry, and is used with their permission.

major peaks are evident, there are clear interference fringes between them. These fringes themselves oscillate at the Rabi frequency when the parity amplitude is large.

Many variations on the Jaynes–Cummings model have appeared over time. First, the model has been generalized through the development of a Markovian master equation approach, rather than sticking with the Hamiltonian form [30, 31]. Then, the model has been extended to a double Jaynes–Cummings model by incorporating two two-level atoms within the structure [32, 33], as well as to a pair of three-level atoms [34]. An anti-Jaynes–Cummings model used the interchange of the two field operators in the second term of H_2 in (11.12) [35]. By reconstructing the Wigner function, these latter authors were able to reconstruct the wave fields themselves. The Jaynes–Cummings model has been studied in a strong coupling regime (very small detuning) as a method of investigating broken inversion symmetry [36]. Finally, Miranowicz *et al* replaced the electromagnetic field with the phonon field in the study of a nanomechanical resonator [37].

11.3 Squeezed states

One of the most ubiquitous objects in quantum mechanics is the Gaussian wave packet, because it can be used to create the minimum uncertainty wave packet. We have used this in several earlier chapters, and now want to consider it further in connection with squeezed states. Let us start with the wave function from (8.3) as

$$\Psi(x) = \sqrt{\frac{1}{\sqrt{2\pi}\sigma}}\,e^{-x^2/4\sigma^2}. \tag{11.19}$$

It is easy to show that the uncertainty in the position of this wave packet is simply [38]

$$\langle(\Delta x)^2\rangle = \sigma^2. \tag{11.20}$$

If we now Fourier transform this into a momentum wave function, we obtain

$$\varphi(k) = \frac{1}{\sqrt{2\pi}}\int_{-\infty}^{\infty}\psi(x)e^{-ikx}dx = \left(\frac{2\sigma^2}{\pi}\right)^{1/4}e^{-\sigma^2k^2}, \tag{11.21}$$

and the uncertainty in momentum is

$$\langle(\Delta k)^2\rangle = \frac{1}{4\sigma^2}. \tag{11.22}$$

Hence, this wave packet satisfies

$$\Delta x\Delta p = \frac{\hbar}{2}, \tag{11.23}$$

which is the minimum uncertainty obtainable in quantum mechanics [39].

Now, the argument goes that this minimum uncertainty wave packet can be used to create the coherent state, by adding a propagator as

$$\Psi(x) = \sqrt{\frac{1}{\sqrt{2\pi\sigma}}}e^{-x^2/4\sigma^2+ip_0x/\hbar}, \tag{11.24}$$

which is a propagating minimum uncertainty packet, and often used to describe a single photon in quantum optics:

$$\varphi(k) = \left(\frac{2\sigma^2}{\pi}\right)^{1/4} e^{-\sigma^2(k-k_0)^2}, \quad k_0 = p_0/\hbar. \tag{11.25}$$

That is, the wave packet remains a minimum uncertainty wave packet as it propagates. On the other hand, we can also construct the Wigner function from the wave packet (11.24) through the use of (3.2) to obtain

$$f_W(x, k) = \frac{1}{h}e^{-x^2/2\sigma^2-2\sigma^2(k-k_0)^2}. \tag{11.26}$$

Hence, the coherent state (11.24) gives us a standard Wigner function in phase space that moves with the average momentum of the wave packet. We note that this Wigner function is positive semi-definite (it can take the values of zero at infinite distances from the center in either position or momentum). So, in quantum optics, the idea of a single photon is represented by a minimum uncertainty wave packet.

The idea of the minimum wave packet lies in just this position and momentum spreading of the packet. One problem, of course, is that the photon is supposedly traveling at the speed of light. This gives a great difficulty in describing just what the spread in momentum represents. On the other hand, if the photon represents a particular mode in a resonator, such as in the Jaynes–Cummings experiments of the previous section, then it has a mode momentum that differs from the speed of light. We have to remark that in both cases there is a difference between the expectation value of the momentum and the speed of light, if for no other reason than that the photon has zero mass. Hence, momentum is described by the propagation constant of the wave, which is linear in the frequency (in free space) as

$$k = \frac{\omega}{c} = \frac{p}{\hbar}. \tag{11.27}$$

So, there is no problem of talking about a spread of momentum as it arises from a spread of frequencies required to make the concentrated wave packet. Now, the natural units of our so-called harmonic oscillator mode, say of the cavity, is such that we have

$$\Delta x = \Delta k = \frac{1}{2}. \tag{11.28}$$

The normal convention is to think about the wave amplitude, the electric field, as being X (with 0 phase) and the node value Y (at phase of $\pi/2$) being the out-of-phase component (such as the magnetic field). This convention then leads to define the normal oscillator as

$$\Delta X \Delta Y \geqslant \frac{1}{4}, \tag{11.29}$$

so that we can talk about the usual case as having $\Delta X \geqslant 1/2$. Hence, these two values create a different phase space, which is still described via a minimum uncertainty wave packet. The idea of a *squeezed state* is to reduce the uncertainty in the amplitude of the wave [40, 41], by utilizing nonlinear optics. In the first case [40], four-wave mixing was used, while in the latter case [41] parametric down conversion was applied to the optical wave. In figure 11.3, we illustrate the differences in the Wigner function for the normal wave packet and the squeezed wave packet, which has been squeezed in amplitude X.

It should be noted that the above techniques both involved nonlinear optics. This is because in a normal electromagnetic wave the electric and magnetic fields are related in magnitude by the wave impedance of the medium in which the wave is traveling. Since this impedance is an intrinsic property of the linear medium, one cannot generally vary the ratio of the two fields with linear operations. Hence, one needs to use nonlinear processing to gain the squeezed photon state. As observed in figure 11.3, the Wigner function is positive semi-definite for the squeezed pure state. However, if a mixed state is considered, then the Wigner function develops the well-known oscillations. Originally, the squeezed state was proposed as a method to reduce the noise in an optical interferometer [42–44]. We write the squeezed state as a set of operations on the vacuum state [42, 44]

$$| \, \alpha, \zeta \rangle = D(\alpha) S(\zeta) | \, 0 \rangle, \tag{11.30}$$

where the displacement operator

$$D(\alpha) = e^{-|\alpha|^2/2} e^{\alpha \hat{a}^\dagger} e^{\alpha * \hat{a}} \tag{11.31}$$

generates the coherent state from the vacuum state, and

$$S(\zeta) = \exp\left(\frac{\zeta *}{2} \hat{a}^2 - \frac{\zeta}{2} \hat{a}^{\dagger 2} \right) \tag{11.32}$$

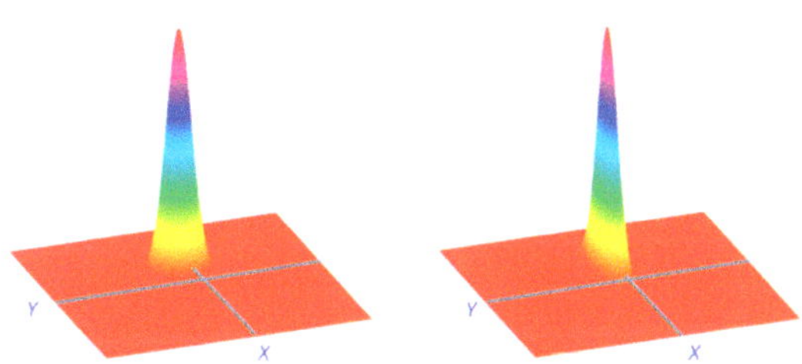

(a) (b)

Figure 11.3. (a) The standard minimum uncertainty wave packet (11.26) shown in the wave coordinates. (b) The squeezed packet has reduced spread in X and enlarged spread in Y.

is the squeezing operator. Here, the complex mode amplitude is given by its two quadrature components from above as the operator

$$a = X + iY, \tag{11.33}$$

with the expectation value

$$a = \langle a \rangle = \langle X + iY \rangle. \tag{11.34}$$

In addition, the squeeze parameter

$$\zeta = re^{i\vartheta} \tag{11.35}$$

is an arbitrary complex number. This squeezing takes the normally circular distribution in phase space into an elliptical distribution, as shown in figure 11.3(b). When the state is a superposition of several modes, the Wigner distribution develops negative excursions, as mentioned above [45].

In later work, the squeezed state was used to investigate the spatial extent of single photons using a two-photon coincidence experiment [46]. Here, the single photon was passed through a double slit, and the coincidence technique used to reconstruct the single-photon Wigner function after passage through the slit. With the modified Wigner function, the authors feel that this allows them to fully characterize the scattering behavior of the two slits. In other work, a squeezed state was used to do phase estimation on an interferometer via the Wigner function for the squeezed state [47]. The phase estimation is carried out by using parity detection via the probe state.

In recent years, there has been a growing investigation both for better creation of squeezed states as well as for applications of these states. A more general treatment of the role of nonlinearity in creating squeezed states has appeared [48], and a new time evolution operator to describe the propagation of the squeezed state has appeared [49]. It has been demonstrated that the squeezed state can be generated by a parametric amplifier process [50]. The squeezed state has been generated from a normal harmonic oscillator mode [51], and from two modes from an optomechanical system [52]. In addition, the concept of fractional squeezing has been introduced [53].

The use of a squeezed state to reduce the decoherence of a coherent optical pulse has also been demonstrated [54], although one might argue that this was the reason squeezed states were developed in the first place. The difference between the quantum and classical nature of the optical pulse has been discussed by using a distance measure to determine the difference between the Wigner and the Husumi distribution, and this is claimed to represent a roughness measure for the propagation of the squeezed state [55].

11.4 Coherence I

As might be expected, the use of specially created photon states, such as the squeezed states discussed above, has been of considerable interest in the world of quantum information in the optical sciences [56]. The generation of the squeezed state is of interest, as some examples were discussed in the previous section. In recent years, there has been interest in using the photon addition and subtraction process to create these squeezed states [57–59]. Then, this process has been used to create optical vortex states, which may be described by the wave function [60]

$$\psi_{\sigma, m}(x, y) = A(x - iy)^m \exp\left(-\frac{x^2 + y^2}{2\sigma^2}\right). \tag{11.36}$$

This is said to be a vortex state because the circulation (defined as the line integral around the path of the gradient of the argument of the wave function) of the argument for a closed contour that incorporates the origin is $2\pi m$. This circulation is closely related to the Einstein–Brillouin–Keller quantization condition [61–63] and, of course, the Berry phase [64]. Originally, the vortex was studied in classical fields [65], but has become of interest in quantum optics in connection with squeezed states. The idea of vortex states was extended to a more complicated structure [66]:

$$\psi_{\sigma, m} = A\left[(x - x_0) + e^{i\vartheta}(y - y_0)\right]^m \exp\left(-\frac{(x - x_0)^2}{2\sigma_x^2} - \frac{(y - y_0)^2}{2\sigma_y^2}\right), \tag{11.37}$$

where it was shown that these states produce a squeezed optical state that is also a vortex state. The vortex state has also been created by a nonlocal photon subtraction process [67, 68]. In this latter work, the authors created a photonic chip to continuously generate and manipulate entangled states of light. The entanglement is created by nonlocal photon absorption of two separable states by means of directional couplers with high transmissivity. The resulting delocalized photon is then manipulated by a reconfigurable interferometer, which produces the desired state after photon counting. The final quantum state has the appeal of being both squeezed light and a single photon.

Multimode photon addition and subtraction of photons arising from Gaussian states can lead to non-Gaussian states, such as squeezed states, which are then useful in quantum computation [69]. Such states have shown usefulness for entanglement distillation [57, 70, 71]. It has been shown that a theoretical framework for such multimode photon-added and -subtracted states can be developed which yields a general Wigner function for the states [69]. Quite generally, quantum computing can outperform classical machines only when entanglement, and the presence of negative components in the Wigner function, exist [72]. The generation of such a state by homodyne detection, which is basically a subtractive process, has been shown to give such a Wigner function [73]. Such a reconstructed Wigner function for photons at 1.55 micron is illustrated in figure 11.4. Here, the data are not corrected for any kind of loss, and the peak negative value reaches down to -0.063 ± 0.004, which is about 20% of the strongest negativity possible (which would correspond to a perfect measurement of the pure single-photon Fock state). The degree of non-classicality in such added and subtracted states used to create squeezed states has been characterized recently, as well [74, 75].

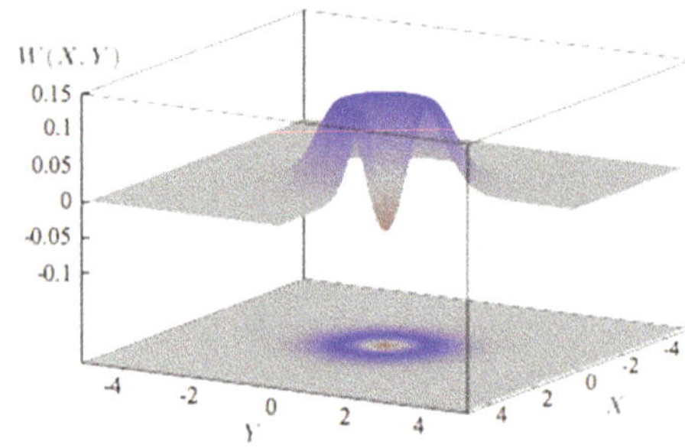

Figure 11.4. Reconstructed Wigner function of the single-photon-subtracted, phase-randomized weakly squeezed vacuum state. Reprinted with permission from C Baune, J Fiurásek, and R Schnabel, *Phys. Rev.* A **95** 061802 (2017). Copyright 2017 by the American Physical Society.

Finally, let us turn to the detection of the squeezed state, usually through some form of reconstruction of the Wigner function. One recent approach used point-by-point sampling of the Wigner function via an ultrafast parametric down-conversion process. A loss-tolerant, time-multiplexed detector, based upon a fiber-optical cavity and a pair of photon-number-resolving avalanche photodiodes were used. By proper data processing and pattern tomography, the properties of the light states could be determined with outstanding accuracy [76]. Another approach used parity and phase detection within an interferometer to study the two-mode squeezed state and its resulting Wigner function [77]. As may be guessed, the use of quantum tomography is a relatively standard approach to reconstructing the Wigner function, and this is often done with homodyne detection. This often includes a phase-sensitive amplifier to amplify the quadrature component of the light. Recently, it has been shown that a travelling wave parametric amplifier based upon an optical nonlinear crystal can be used to enhance the quality of the Wigner function reconstruction [78]. Here, they generate the squeezed-single-photon states with the Wigner function

$$f_W(q, p) = \frac{2q^2/s + sp^2 - 1}{\pi} \exp(-q^2/s - sp^2), \tag{11.38}$$

where s is

$$s = e^{2r_1}, \tag{11.39}$$

with r_1 the parametric gain. This Wigner function contains a negative-valued area along one quadrature and is squeezed along the other, as illustrated in figure 11.5.

Other approaches to the detection and observation of non-classical states have appeared, as well, in recent years, such as the Knill–Laflamme–Milburn type interferometer, which is composed of three waveguides coupled via three beam splitters [79] to generate non-classical states from a coherent initial state. This type of interferometer has been used to create a controlled-NOT gate with optical elements [80]. Recent work has used the interferometer to create non-classical states as measured by the negativity of the Wigner function [81]. In other work, a new experimental method of recognizing non-Gaussian multi-photon states has been proposed [82]. They recognize that this is a property of Fock states, which are quantum states for a fixed

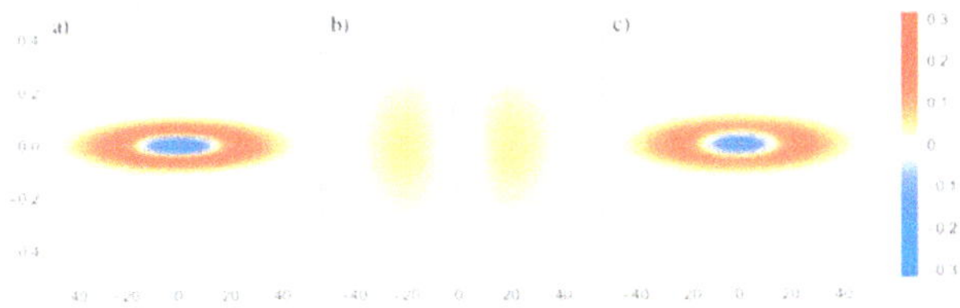

Figure 11.5. (a) The Wigner function of a squeezed-single-photon state, given by (11.38) with $r_1 = 3$ (26 dB of squeezing). (b) The Wigner function reconstruction without the amplification, but with 95% detection efficiency. (c) The Wigner function reconstruction with 95% detection efficiency and 20 db of pre-amplification. Black ellipses encircle the negative-valued area. The axes are scaled disproportionally. Reprinted from Knyazev *et al* [78], under the Creative Commons attribution 3.0 unported license. Copyright 2018 by the Institute of Physics Publishing.

number of particles. In this latter work, the authors show that their proposed approach will give an experimental method of creating the Fock states from normal multi-photon states, and provide the increased negativity of the Wigner function.

11.5 Coherence II

The introduction of the quantum process in the previous section makes a nice transition into the discussion of optical qubits. One approach has treated qubit-oscillator states. In its most basic form, this is just another variation of the Jaynes–Cummings model in which the two-state atom is replaced by the qubit. In this case, the Hamiltonian becomes [83]

$$H = \hbar\omega_c\hat{a}^\dagger\hat{a} + \frac{\hbar\Omega}{2}\hat{\sigma}_x + \hbar\lambda\hat{\sigma}_z(\hat{a}^\dagger + \hat{a}), \tag{11.40}$$

which differs from (11.12) in the choice of spinors for the last two terms. The change introduces the complex nature of the qubit which replaces the two-state atom. This has been studied again, more recently, within the rotating wave approximation [84, 85]. This latter work has used the results to examine the differences among various information-theoretic measures. And, as we discussed earlier, the qubit-oscillator system has been studied for broken inversion symmetry [36].

Squeezed states have also been used for representing qubits. Measuring these states relies upon the existence of non-demolition detectors that can detect a single photon without destroying it in an absorption process. Such detectors often rely upon the phase shift produced by the photon traveling through a nonlinear medium [86]. Recently, two-mode qubit-like entangled squeezed states have been compared with entangled coherent states, through the use of the Wigner functions for the two types of states [87]. The Wigner function and its negative excursion have been used to signify the entanglement of as many as 3000 atoms by a single photon [88]. The role of entanglement for generation of the negative portions of the Wigner function has also been studied for multi-qubit GHz-squeezed states [89].

The qubit-oscillator system has also been used to create what are called Schrödinger kitten states [84, 85, 90], so-called because it is a microscopic superposition of two coherent states with opposite phases [91]. By passing these states

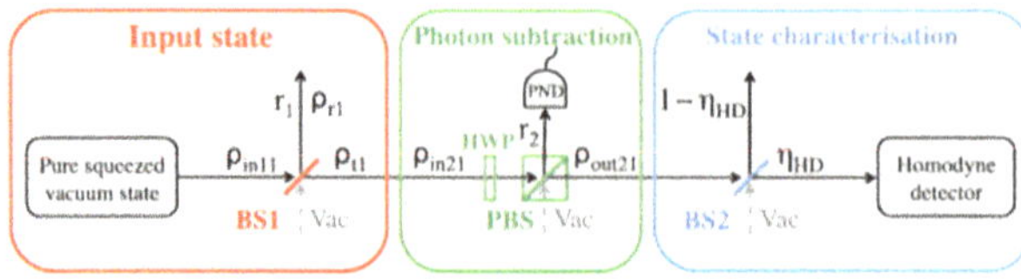

Figure 11.6. Theoretical model for Schrödinger kitten state generation with experimental imperfections. $\rho_{in21} = \rho_{t1}$, r_1 is the reflectivity of BS1, t_1 is the transmission of BS1, r_1 is the reflectivity of PBS, ρ_{in11} is the input state density matrix of BS1, ρ_{t1} is the transmitted state density matrix of BS1, ρ_{r1} is the reflected state density matrix of BS1, ρ_{out21} is the output state density matrix of PBS, η_{HD} is the homodyne detection efficiency, HWP is a half-wave plate, BS1 is a beam splitter for modeling the impurity of the squeezed vacuum state, PBS is a polarization beam splitter, PND is a photon-number detector, and BS2 is a beam splitter for modeling the inefficiency of the homodyne detector. Reprinted from Song *et al* [93], under the Creative Commons attribution 3.0 unported license. Copyright 2018 by the Institute of Physics Publishing.

through a Kerr medium, the gain by such a process can exceed the normal optical losses and lead to squeezed kitten states. Such kitten states, as well as single-photon states, also have been generated from the steady state output of a continuously driven optical parametric oscillator [92]. In principle, the kitten state can be generated by subtraction of photons from a squeezed vacuum state [93]; the process of doing so is shown in figure 11.6. The model includes three sections, each outlined by a box in the figure. The red box creates the input state. The green box is the photon subtraction portion, which involves a 'magic' reflector that is arbitrarily tunable via a half-wave plate and a polarization beam splitter. Finally, the third part is the state characterization. The pure squeezed state in the initial state is considered to have been produced from parametric down conversion, and this state is said to consist of a photon-number distribution with only even numbers of photons (impurities in production would lead to odd numbers of photons and is handled by the impure beam splitters).

11.6 Bell states

As pointed out above, the use of specially created photon states, such as the squeezed states discussed above, and more particularly entangled states, has been of considerable interest in the world of quantum information in the optical sciences [56]. The idea of entangled states in quantum mechanics arises as early as the Einstein–Podolsky–Rosen work [94] and the subsequent introduction of the phrase 'entanglement' by Schrödinger [95]. In fact, Schrödinger called entanglement the most important aspect of quantum mechanics. The creation of so-called Bell states became more interesting after the publication of the Bell inequality [96], although the inequality was known already more than a century earlier [97]. Since the inequality was known well before Bell, it can hardly be used to distinguish between classical and quantum mechanics. Nevertheless, the use of these Bell states are quite useful to describe entangled states which are necessary for the world of quantum information.

Let us first look at the idea of the nonlocal character of the entangled state, and how it appears in optics. It was earlier shown that one can produce the optical analog of an EPR, or Bell state, in the process of non-degenerate-optical parametric amplification (NOPA) in the strong squeezing limit [98]. The NOPA is a nonlinear interaction of two squeezed modes in a nonlinear medium in the presence of a strong classical optical pump. Then, the interaction Hamiltonian is given by [99]

$$H = i\chi(\hat{a}^\dagger\hat{b}^\dagger - \hat{a}\hat{b}), \tag{11.41}$$

where the two sets of operators $\hat{a}$, $\hat{a}^\dagger$ and $\hat{b}$, $\hat{b}^\dagger$ refer to the two squeezed states and χ is a parameter which measures the strength of the second-order susceptibility of the nonlinear medium in the presence of the pump. The initial state of the system consists of the two vacuum modes the NOPA generates [99]

$$| NOPA\rangle = e^{r(\hat{a}^\dagger\hat{b}^\dagger - \hat{a}\hat{b})} | 0, 0\rangle, \tag{11.42}$$

where $r = \chi t$ is a dimensionless parameter that characterizes the interaction and the interaction time. One can decompose the evolution operator in terms of the number states as [99]

$$e^{r(\hat{a}^\dagger\hat{b}^\dagger - \hat{a}\hat{b})} = e^{\tanh(r\hat{a}^\dagger\hat{b}^\dagger)}\left(\frac{1}{\cosh(r)}\right)^{(\hat{a}^\dagger\hat{a}+\hat{b}^\dagger\hat{b}+1)} e^{-\tanh(r\hat{a}\hat{b})}. \tag{11.43}$$

This leads to a diagonal decomposition in terms of the number states of the two modes as [99]

$$| NOPA\rangle = \frac{1}{\cosh(r)}\sum_{n=0}^{\infty}[\tanh(r)]^n \; | n, n\rangle. \tag{11.44}$$

This latter equation can be rewritten in the form of Bell states as

$$| NOPA\rangle = \frac{1}{\cosh(r)}\sum_{n=0}^{\infty}[\tanh(r)]^n \int dq \int dq' \; | q, q'\rangle\langle q, q'|n, n\rangle. \tag{11.45}$$

The scalar products can be written in terms of the Hermite polynomials that are the basis states for the assumed oscillator properties of the modes, in dimensionless units,

$$\langle q|n\rangle = \frac{1}{2^n\sqrt{\pi}n!}H_n(q)e^{-q^2/2}, \tag{11.46}$$

and the following summation formula [99]:

$$\sum_{n=0}^{\infty}\lambda^n\langle q|n\rangle\langle n|q'\rangle = \frac{1}{\sqrt{\pi(1-\lambda^2)}}\exp\left(-\frac{q^2+q'^2-2\lambda qq'}{2(1-\lambda^2)}\right), \tag{11.47}$$

where λ is an arbitry constant < 1, to yield the final result

$$| NOPA\rangle = \frac{1}{\sqrt{\pi}}\int dq \int dq' \exp\left(-\frac{q^2+q'^2-2qq'\tanh(r)}{2[1-\tanh^2(r)]}\right) | q, q'\rangle. \tag{11.48}$$

This result is a regularized version of the EPR or Bell state, which includes a Gaussian smoothing function. Finally, in the long time limit ($r \to \infty$), one can write the Bell state as

$$|EPR\rangle = \lim_{r\to\infty} |NOPA\rangle \sim |\,0, 0\rangle + |\,1, 1\rangle + \cdots \tag{11.49}$$

From the state (11.48), we can develop the Wigner function for the entangled state as

$$\begin{aligned} f_W(\alpha, \beta) &= \int \frac{d^2\alpha'}{\pi^2} \int \frac{d^2\beta'}{\pi^2} e^{\alpha\alpha'^* - \alpha^*\alpha' + \beta\beta'^* - \beta^*\beta'} \langle \hat{D}_a(\alpha')\hat{D}_b(\beta')\rangle \\ &= \frac{4}{\pi} \exp[-2\cosh(2r)(|\,\alpha\,|^2 + |\,\beta\,|^2) + 2 \\ &\quad \times \sinh(2\pi)(\alpha\beta + \alpha^*\beta^*)], \end{aligned} \tag{11.50}$$

where the Ds are the displacement operators in the Wigner definition and the expectation value is carried out with the NOPA wave function (11.48). This Wigner function is positive semi-definite, as it has no negative parts.

In light of the discussion in the previous sections, this Wigner function does not display the entanglement that is expected from properly entangled states. In a sense, then, using this state to probe the Bell inequality is almost certain to generate classical results, even if it shows a violation of the inequality. As we remarked above, such a violation is no proof of the quantum character of the system, and that would be supported by the non-negative Wigner function found here. As a result, interest has passed to states in which the entangled pair pass along different paths, and the use of, for example, a path-entangled state such as [100]

$$|\,\psi\rangle = \frac{1}{\sqrt{2}}(|\,N\rangle_a\,|\,0\rangle_b + e^{i\varphi}\,|\,0\rangle_a\,|\,N\rangle_b), \tag{11.51}$$

where there are N photons in one mode and the second mode is a vacuum state. Here, φ is an arbitrary phase shift. The wave function is not separable and hence is entangled, as suggested by Bohm and Aharonov [101]. Such states are locally correlated for any finite N [100]. From the individual states that make us the wave function, it is possible to define operators which characterize parity measurements, which for a given state may be written as

$$\begin{aligned} \hat{\Pi}^+(\alpha) &= \hat{D}(\alpha)\sum_{k=0}^{\infty} |\,2k\rangle\langle 2k\,|\,\hat{D}^\dagger(\alpha) \\ \hat{\Pi}^-(\alpha) &= \hat{D}(\alpha)\sum_{k=0}^{\infty} |\,2k+1\rangle\langle 2k+1\,|\,\hat{D}^\dagger(\alpha), \end{aligned} \tag{11.52}$$

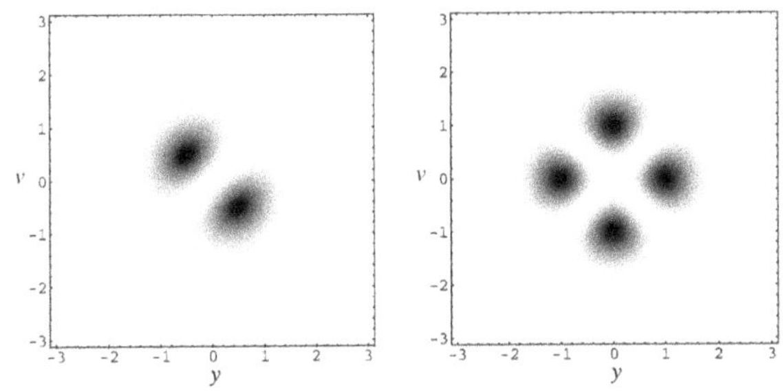

Figure 11.7. (a) The marginal Wigner function for $N = 1$. (b) The marginal Wigner function for $N = 3$. Reprinted with permission from Wildfeuer *et al* [100]. Copyright 2007 by the American Physical Society.

where, again, the operators D are the normal displacement operators. Then, the corresponding operator for the correlated measurement of the parity on mode a and b, where the modes correspond to those in (11.51), can be defined by

$$\hat{\Pi}(\alpha, \beta) = [\hat{\Pi}_a^+(\alpha) - \hat{\Pi}_a^-(\alpha)] \otimes [\hat{\Pi}_b^+(\beta) - \hat{\Pi}_b^-(\beta)], \tag{11.53}$$

which up to a numerical factor is equivalent to the Wigner function for this correlated measurement [102, 103]. By using the dimensionless complex amplitudes of the two modes as $\alpha = x + iy$ and $\beta = u + iv$, and then integrating over x and y, one produces a marginal Wigner function illustrating the dependence on the second mode. In figure 11.7, we plot this marginal Wigner function for the case of $N = 1$ and $N = 3$.

A different definition of the entangled Bell states varies somewhat from the previous definitions. In the latter case, we may define the states as [104]

$$\begin{aligned} |\varphi^{\pm}(z)\rangle &= \frac{1}{\sqrt{2(1 \pm e^{-4|\alpha|^2})}}(|\alpha, \alpha\rangle \pm |-\alpha, -\alpha\rangle) \\ |\psi^{\pm}(z)\rangle &= \frac{1}{\sqrt{2(1 \pm e^{-4|\alpha|^2})}}(|\alpha, -\alpha\rangle \pm |-\alpha, \alpha\rangle), \end{aligned} \tag{11.54}$$

which are supposed to correspond to the normal type of Bell states:

$$\begin{aligned} |\varphi^{\pm}\rangle &= \frac{1}{\sqrt{2}}(|00\rangle \pm |11\rangle) \\ |\psi^{\pm}\rangle &= \frac{1}{\sqrt{2}}(|01\rangle \pm |10\rangle). \end{aligned} \tag{11.55}$$

In this case, the states are proper tensor product states, with the individual states being defined by displacement operators as [105]

$$|\alpha\rangle = D(\alpha)|0\rangle = \exp\left(-\frac{|\alpha|^2}{2} - \alpha\hat{a}^{\dagger}\right)|0\rangle = e^{-|\alpha|^2}\sum_{n=0}^{\infty}\frac{\alpha^n}{\sqrt{n!}}|n\rangle. \tag{11.56}$$

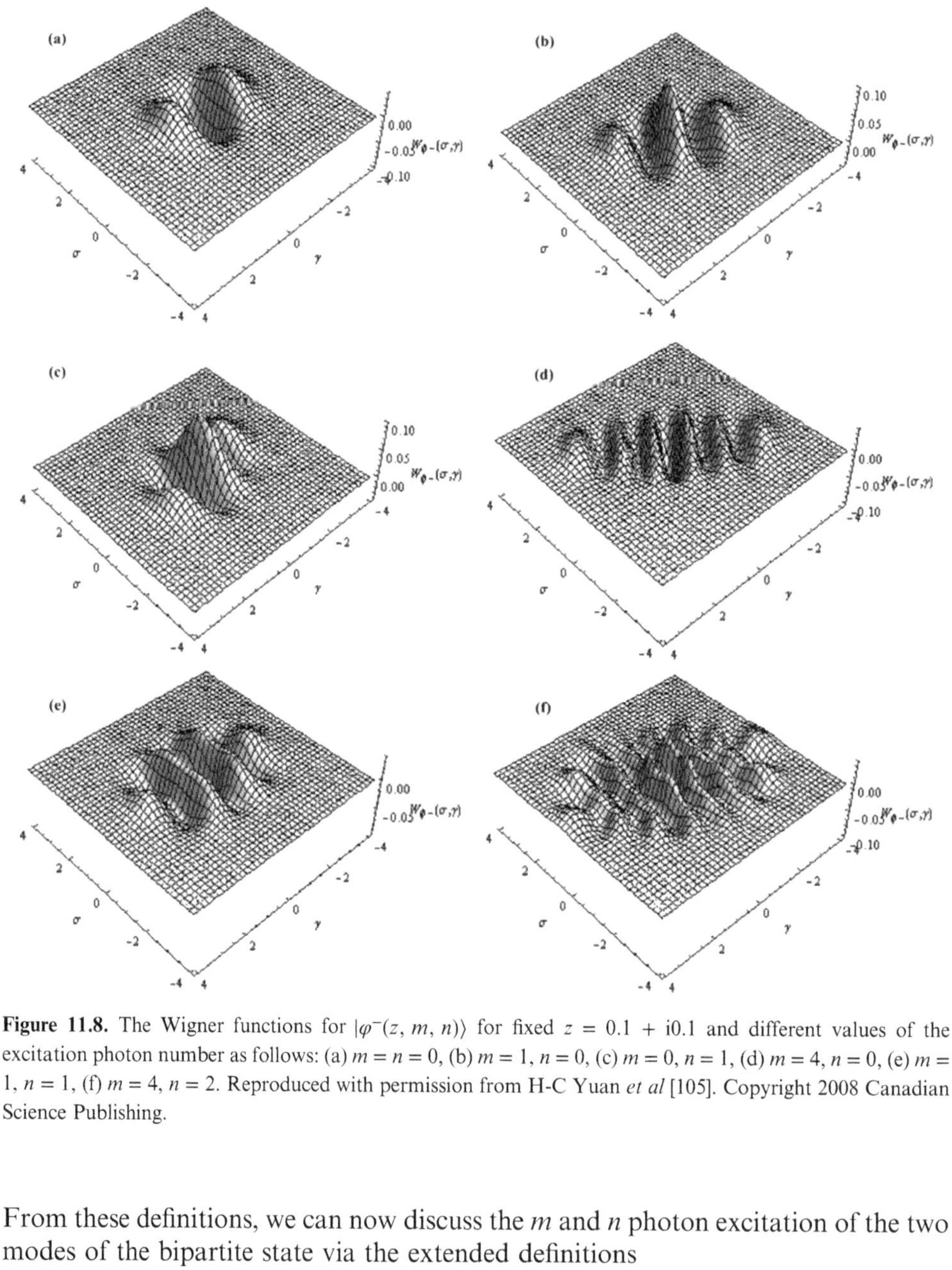

Figure 11.8. The Wigner functions for $|\varphi^{-}(z, m, n)\rangle$ for fixed $z = 0.1 + i0.1$ and different values of the excitation photon number as follows: (a) $m = n = 0$, (b) $m = 1, n = 0$, (c) $m = 0, n = 1$, (d) $m = 4, n = 0$, (e) $m = 1, n = 1$, (f) $m = 4, n = 2$. Reproduced with permission from H-C Yuan *et al* [105]. Copyright 2008 Canadian Science Publishing.

From these definitions, we can now discuss the m and n photon excitation of the two modes of the bipartite state via the extended definitions

$$\begin{aligned} |\,\varphi^{\pm}(z, m, n)\rangle &= \frac{1}{\sqrt{2(1 \pm e^{-4|\alpha|^2})}} \hat{a}^{\dagger m} \hat{b}^{\dagger n} (|\,\alpha, \alpha\rangle \pm |-\alpha, -\alpha\rangle) \\ |\,\psi^{\pm}(z, m, n)\rangle &= \frac{1}{\sqrt{2(1 \pm e^{-4|\alpha|^2})}} \hat{a}^{\dagger m} \hat{b}^{\dagger n} (|\,\alpha, -\alpha\rangle \pm |-\alpha, \alpha\rangle). \end{aligned} \tag{11.57}$$

If we introduce the position and momentum of the two modes as

$$z = \frac{1}{\sqrt{2}}(x_1 + ip_1), \quad z' = \frac{1}{\sqrt{2}}(x_2 + ip_2)$$
$$\gamma = z + z'^*, \quad \sigma = z - z'^*, \tag{11.58}$$

the two-mode Wigner function can be introduced as

$$f_W(\sigma, \gamma) = \int \frac{d^2\eta}{\pi^3} \mid \sigma - \eta\rangle\langle\sigma + \eta \mid e^{\eta\gamma^* - \eta^*\gamma}, \tag{11.59}$$

with the individual two-mode state being defined by the application of the above equations to yield

$$\mid \eta\rangle = \exp\left[-\frac{\mid \eta \mid^2}{2} + \eta\hat{a}^\dagger - \eta^*\hat{b}^\dagger + \hat{a}^\dagger\hat{b}^\dagger\right] \mid 00\rangle, \quad \eta = \eta_1 + i\eta_2. \tag{11.60}$$

In figure 11.8, the Wigner function for several different combination of mode excitations is shown, and it is clear that for other than the ground state, there are extended negative excursions which clearly show the entanglement of the two modes. Similar-type entanglement has been achieved with vortex beams [106] and the states have been used for teleportation [107].

In optical communications, one also has to worry about synchronization of qubits when they are flying through various media. Recently, it has been possible to control the interference between two nearly pure photons that emerge from two independent quantum memories [108]. Controlled storage times of 1.8 μs were achieved with sufficient purity so that the Wigner function showed sufficient negative excursions which could be confirmed with homodyne detection. In other work, control of the entanglement dynamics has been achieved via a Kerr nonlinearity that is mediated by cavity detuning in a two-photon process [109]. Finally, the typical EPR or Bell state has been studied with pseudo-spin-type measurements [110].

References

[1] Hall J L 2006 *Rev. Mod. Phys.* **78** 1279

[2] Hansch T W 2006 *Rev. Mod. Phys.* **78** 1297

[3] Gill P, Barwood G P, Klein H A, Huang G, Webster S A, Blythe P J, Hosaka K, Lea S N and Margolis H S 2003 *Meas. Sci. Technol.* **14** 1174

[4] Bloom B J, Nicholson T L, Williams J R, Campbell S L, Bishof M, Zhang X, Zhang W, Bromley S L and Ye J 2014 *Nature* **506** 71

[5] Rey A M *et al* 2014 *Ann. Phys.* **340** 311

[6] Samson J H 2000 *J. Phys.* A **33** 5219

[7] Feynman R P (ed) 1987 *Quantum Implications* ed B J Hiley and F D Peat (London: Routledge) p 235

[8] Stratonovich R L 1956 *Zh. Eksp. Teor. Fiz.* **31** 1012 trans in R. L. Stratonovich 1957 *Sov. Phys. JETP* **4** 891

[9] Saavedra G, Furlan W D, Silvestre E and Sicre E E 1997 *Opt. Commun.* **139** 11

[10] Oh S B, Kashyap S, Garg R, Chandran S and Rasker R (ed) 2010 *Eurographics* ed T Akenine-Möller and M Zwicker (Blackwell: Oxford)
[11] Alonso M A 2011 *Adv. Opt. Photon.* **3** 272
[12] Yang T, Ji X, Li X, Zhang H and Wang T 2016 *Opt. Commun.* **359** 146
[13] Condon E U 1937 *Proc. Nat. Acad. Sci.* **23** 185
[14] Namias V 1980 *J. Inst. Math. Appl.* **25** 241
[15] Weimann S *et al* 2016 *Nat. Commun.* **7** 11027
[16] Furlan W D and Saavedra G 2009 *Phase Space Optices: Fundamentals and Applications* ed M E Testorf, B Hennelly and J Ojeda-Castañeda (New York: McGraw-Hill)
[17] Mazar R, Kodner L and Samelsohn G 1997 *J. Opt. Soc. Am.* A **14** 2809
[18] Cheng C-C and Raymer M G 2000 *Phys. Rev.* A **62** 023811
[19] Wax A and Thomas J E 1998 *J. Opt. Soc. Am.* **15** 1896
[20] Li X, Ji X, Wang T and Zhu J 2013 *J. Opt.* **15** 125720
[21] Jaynes E T and Cummings F W 1963 *Proc. IEEE* **51** 89
[22] Norozhny N B, Sanchez-Mondragon I I and Eberly J H 1981 *Phys. Rev.* A **23** 236
[23] Gea-Banacloche J 1990 *Phys. Rev. Lett.* **65** 3385
[24] Birrittella R, Chang K and Gerry C C 2015 *Opt. Commun.* **354** 286
[25] Hessian H A and Mohamed A-B A 2008 *Laser Phys.* **18** 1217
[26] Shore B W and Knight P L 1993 *J. Mod. Opt.* **40** 1195
[27] See, e.g. ed Merzbacher E 1970 *Quantum Mechanics* 2nd edn (New York: Wiley) ch 13; also Feynman R P, Leighton R B and Sands M 1965 *The Feynman Lectures on Physics* vol III (Reading, MA: Addison-Wesley) sec 11
[28] Eiselt J and Risken H 1991 *Phys. Rev.* A **43** 346
[29] Cahill K E and Glauber R J 1969 *Phys. Rev.* **177** 1882
[30] Ashrafi S M and Bazrafkan M R 2014 *Chin. Phys.* **23** 090303
[31] de los Santos-Sánchez O, Récamier J and Jáuregui R 2015 *Phys. Scr.* **90** 074018
[32] Ghorbani M, Faghihi M J and Safari H 2017 *J. Opt. Soc. Am.* B **34** 1884
[33] Pandit M, Das S, Roy S S, Dhar H S and Sen U 2018 *J. Phys.* B **51** 045501
[34] Faraji E and Tavassoly M K 2015 *Opt. Commun.* **354** 333
[35] Lv D, An S, Um M, Zhang J, Zhang J-N, Kim M S and Kim K 2017 *Phys. Rev.* A **95** 043813
[36] Shen L-T, Yang Z-B, Wu H-Z and Zheng S-B 2016 *Phys. Rev.* A **93** 063837
[37] Miranowicz A, Bajer J, Lambert N, Liu Y-X and Nori F 2016 *Phys. Rev.* A **93** 013808
[38] Ferry D K 2001 *Quantum Mechanics* 2nd edn (Bristol: Institute of Physics Publishing) p 16
[39] Heisenberg W 1927 *Z. Phys.* **43** 172
[40] Susher R E, Holberg L W, Yorke B, Mertz J C and Valley J F 1985 *Phys. Rev. Lett.* **55** 2409
[41] Wu L-A, Kimble H J, Hall J L and Wu H 1986 *Phys. Rev. Lett.* **57** 2520
[42] Caves C M 1981 *Phys. Rev.* D **23** 1693
[43] Mandel L 1982 *Phys. Rev. Lett.* **49** 136
[44] Walls D F 1983 *Nature* **306** 141
[45] Zayed E M E, Daoud A S, Al-Laithy M A and Naseem E N 2004 *Chaos Solitons Fractals* **24** 967
[46] Kang Y, Cho K, Noh J, Vitullo D L P, Leary C and Raymer M G 2010 *Opt. Exp.* **18** 1217
[47] Xu X-X and Yuan H-C 2015 *Quantum Inf. Process.* **14** 411
[48] Albarelli F, Ferraro A, Paternostro M and Paris M G A 2016 *Phys. Rev.* A **93** 032112

[49] Ren G, Du J-M, Yu H-J and Zhang W-H 2016 *Optik* **127** 3828
[50] Xu S, Xu X-X, Liu C-J, Zhang H-L and Hu L-Y 2017 *Optik* **144** 664
[51] Ramirez R and Reboiro M 2016 *Phys. Lett.* A **380** 1117
[52] Shakeri S, Mahmoudi Z, Zandi M H and Bahrampour A R 2016 *Opt. Commun.* **370** 55
[53] Lv C-H, Cai Y-X and Wang Y-W 2016 *Optik* **127** 4057
[54] Le Jeannic H, Cavallés A, Huang K, Filip R and Laurat J 2018 *Phys. Rev. Lett.* **120** 073603
[55] Lemos H C F, Almeida A C L, Amaral B and Oliveira A C 2018 *Phys. Lett.* A **382** 823
[56] Bouwmeester D and Zeilinger A 2000 *The Physics of Quantum Information* ed D Bouwmeester, A K Eikert and A Zeilinger (Berlin: Springer)
[57] Ourjoumtsev A, Dantan A, Tualie-Brouri R and Grangier P 2007 *Phys. Rev. Lett.* **98** 030502
[58] Kim M S 2008 *J. Phys.* B **41** 133001
[59] Lee C-W, Lee J, Nha H and Jeong H 2012 *Phys. Rev.* A **85** 063815
[60] Agarwal G S, Puri R R and Singh R P 1997 *Phys. Rev.* A **56** 4207
[61] Einstein A 1917 *Werh. Deutsch. Phys. Gell* **19** 82
[62] Brillouin L 1926 *J. Phys. Radium* **7** 353
[63] Keller J B 1958 *Ann. Phys.* **4** 180
[64] Berry M V and Balazs N L 1979 *J. Phys.* A **12** 625
[65] Berry M V 1981 *Physics of Defects* ed J-P Poirier, M Kleman and R Balian (North Holland: Amsterdam)
[66] Li Y-Z, Jia F, Zhang H-L, Huang J-H and Hu L-Y 2015 *Laser Phys. Lett.* **12** 115203
[67] Barral D, Liñares J and Balado D 2016 *J. Opt. Soc. Am.* B **33** 2225
[68] Barral D, Balado D and Liñares J 2017 *Photonics* **4** 2
[69] Walschaers M, Fabre C, Parigi V and Treps N 2017 *Phys. Rev.* A **96** 053835
[70] Takahashi H, Neergaard-Nielsen J S, Takeuchi M, Takeoka M, Hayasaka K, Furusawa A and Sasaki M 2010 *Nat. Photon.* **4** 178
[71] Navarrete-Beniloch C, Carcia-Patrón R, Shapiro J H and Cerf N J 2012 *Phys. Rev.* A **86** 012328
[72] Mari A and Eisert J 2012 *Phys. Rev. Lett.* **109** 230503
[73] Baune C, Fiudrásek J and Schnabel R 2017 *Phys. Rev.* A **95** 061802
[74] Thapliyal K, Samantray N L, Banerji J and Pathak A 2017 *Phys. Lett.* A **381** 3178
[75] Lu D-M 2017 *Int. J. Theor. Phys.* **56** 3514
[76] Harder G, Silberhorn C, Rehacek J, Hradil Z, Motka L, Stoklasa B and Sánchez-Soto L L 2016 *Phys. Rev. Lett.* **116** 133601
[77] Li H-M, Xu X-X, Yuan H-C and Wang Z 2016 *Chin. Phys.* B **25** 104203
[78] Knyazev E, Spasibko K Y, Chekhova M V and Khalil F Y 2018 *New J. Phys.* **20** 013005
[79] Krnill E, Laflamme R and Milburn G J 2001 *Nature* **409** 46
[80] Ralph T C, White A G, Munro W M and Milburn G J 2001 *Phys. Rev.* A **65** 012314
[81] Xu X-X, Yuan H-C and Ma S-J 2016 *J. Opt. Soc. Am.* B **33** 1322
[82] Straka I, Lachman L, Hlousek J, Miková M, Micuda M, Jezek M and Filip R 2018 *NPJ Quantum Inform.* **4** 4
[83] Irish E K, Gea-Banacloche J, Maritn I and Schwab K C 2005 *Phys. Rev.* B **72** 195410
[84] Chakrabarti R and Yogesh V 2016 *J. Phys.* B **49** 075502
[85] Balamurugan M, Chakrabarti R and Virgin Jenisha B 2017 *Physica* A **473** 428
[86] Yamamoto Y, Imoto N and Machida S 1986 *Phys. Rev.* A **33** 3243
[87] Najarbashi G and Mirzaei S 2016 *Opt. Commun.* **377** 33

[88] Mcconnell R, Zhang H, Hu J, Cuk S and Vuletic V 2015 *Nature* **519** 439
[89] Siyouri F-Z 2017 *Commun. Theor. Phys.* **68** 729
[90] Chakrabarti R and Yogesh V 2018 *Physica* A **490** 886
[91] Paris M G A 1999 *J. Opt.* B **1** 662
[92] Mølmer K 2006 *Phys. Rev.* A **73** 063804
[93] Song H, Kuntz K B and Huntington E H 2013 *New J. Phys.* **15** 023042
[94] Einstein A, Podolsky B and Rosen N 1935 *Phys. Rev.* **47** 777
[95] Schrödinger E 1935 *Naturwiss.* **23** 823–44 trans by Trimmer J D 1980 *Proc. Am. Phil. Soc.* **124** 323
[96] Bell J 1964 *Physics* **1** 195
[97] Boole G 1854 *An Investigation of the Laws of Thought* (London: Walton and Maberly) The book has been reprinted by several publishers, but is available online at archive.org (https://archive.org/details/investigationofl00boolrich) and from Project Gutenberg (https://www.gutenberg.org/files/15114/14114-pdf.pdf)
[98] Reid M D and Drummond P D 1988 *Phys. Rev. Lett.* **60** 2731
[99] Banaszek K and Wódkiewicz K 1999 *Acta Phys. Slovaca* **49** 491
[100] Wildfeuer C F, Land A P and Dowling J P 2007 *Phys. Rev.* A **76** 052101
[101] Bohm D and Aharonov Y 1957 *Phys. Rev.* **108** 1070
[102] Royer A 1977 *Phys. Rev.* A **15** 449
[103] Moya-Cessa H and Knight P L 1993 *Phys. Rev.* A **48** 2479
[104] Munhoz P P, Semião F L, Vidiella-Barranco A and Roversi J A 2008 *Phys. Lett.* A **372** 3580
[105] Yuan H-C, Li H-M and Fan H-Y 2009 *Can. J. Phys.* **87** 1233
[106] Stoklasa B, Motka L, Rehacek J, Hradil Z, Sánchez-Soto L L and Agarawal G S 2015 *New J. Phys.* **17** 113046
[107] Seshadreesan K P, Dowling J P and Agarwal G S 2015 *Phys. Scr.* **90** 074029
[108] Makino K, Hashimoto Y, Yoshikawa J-I, Ohdan H, Toyama T, van Loock P and Furusawa A 2016 *Sci. Adv.* **2** e1501772
[109] Ateto M S 2017 *Quantum Inf. Process.* **16** 267
[110] Xiang Y, Xu B, Mista Jr L, Tufarelli T, He Q and Adesso G 2017 *Phys. Rev.* A **96** 042326

IOP Publishing

David K Ferry and Mihail Nedjalkov

Chapter 12

Quantum physics

It is important to return to physics for this last chapter. After all, the use of the Wigner function began in physics, with some saying it is a third route to achieve quantization [1], following the work of Heisenberg [2] and Schrödinger [3]. It is doubtful that Wigner claimed such a distinction. Rather, he simply pointed out that quantum mechanics, as far as the first two routes were concerned, did not allow for both position and momentum to be described simultaneously as a result of quantum probabilities. Instead, he introduced, in 1932, a method of transforming the wave function to produce a classical-like distribution function of the two sets of variables [4], while pointing out that the function had been developed at some earlier time by Szilard and himself. In fact, what he claimed is that this new function provided a method for determining quantum corrections for the thermal equilibrium system. That is, he dealt with the thermodynamics of the system in a manner which neither of the two predecessors had done.

The importance of the Wigner function, and the equation of motion that was easily derived for this function, lay in its clear correspondence to the standard Boltzmann equation for statistical physics. But there was an important difference that was often overlooked. In Boltzmann's approach, it was clear that classical statistical physics depended upon the fields derived from potentials via Hamilton's equations of motion. In quantum mechanics, however, it is the potentials themselves which are important in any equations of motion that devolve from the Liouville equation's various quantities, such as the density matrix or the Schrödinger equation itself. This difference led to a radical difference in philosophy between the Wigner function equation of motion and the classical Boltzmann equation.

A secondary difference was also noted, which arose from the fact that the Wigner function was non-positive definite in any state other than the ground state. That is, it had negative excursions over phase space regions comparable in size to the range of the uncertainty principle. For quite some time after Wigner's introduction of his function, this negative excursion was of interest only for the questions it raised about

doi:10.1088/978-0-7503-1671-2ch12

the interpretation of probabilities [5]. It became much more interesting in the latter half of the twentieth century when the world became interested in entanglement [6], and it was recognized that the negative excursions of the Wigner function clearly related to this quantity. We have already discussed this important attribute of the Wigner function in each of the last several chapters in which the various applications of the Wigner function in science have been discussed. Now, we want to return to the roots of the Wigner function, and discuss its role in modern physics and the study of quantum effects in physics.

12.1 The harmonic oscillator

One of the earliest treatments of the harmonic oscillator in terms of the Wigner function was that of Groenewold [7]. Here, he reviewed the quantum treatment of the harmonic oscillator in terms of the Hermite polynomials, then showed that a Wigner function composed from two different (or same) eigenfunctions would lead to a solution defined in terms of associated Laguerre polynomials, which on their own formed a complete set for the treatment of the oscillator. Moreover, he demonstrated that the position and momentum variables of the Wigner function would move in phase space exactly as those in a classical oscillator. Groenewold's work is significant if for no other reason than the fact that the harmonic oscillator is ubiquitous in quantum physics: it represents one of the few exactly solvable systems using the Schrödinger equation. Consequently, it is used to describe the quantization of a variety of different fields, whether of electromagnetic origin or of mechanical motion, such as phonons in condensed matter. In these applications, the field is expanded in a number of modes, each of which is described by an harmonic oscillator. We observed this already in the electron–phonon interaction in chapter 7 and in the Jaynes–Cummings model of chapter 11, where each cavity mode was described as an harmonic oscillator. It is therefore not unexpected that an enormous amount of work has been devoted to further understanding of this basic system. The importance of phase space has been re-emphasized by a treatment of the harmonic oscillator in more recent times [8–10]. This simple system has also been considered by the time dependence of the Wigner function in phase space by a Monte Carlo technique [11].

The description of the field as an array of harmonic oscillators means that it has become natural to describe the background 'bath' as an ensemble of harmonic oscillators, and this can be used to study the interaction of other quantum systems with such a bath. In this type of interaction, it has been shown that one has a generalized set of uncertainty relations and that the bath does introduce forces akin to Brownian motion [12, 13]. The development of the quantum phase space formulation with the Wigner function for the harmonic oscillator has been treated in the realm of shape invariance common to the supersymmetric version of quantum mechanics [14], and coupling a single harmonic oscillator to a bath of other harmonic oscillators can produce interesting effects [15, 16].

The harmonic oscillator is also known for other applications in science through its natural connection to the pendulum. It is known that a driven pendulum or coupled pendulums can give rise to chaotic dynamics. Consequently, it becomes a useful system to study the transitions to quantum chaos [17]. In this regard, the driven harmonic oscillator, and its Wigner function behavior, have also been studied [18] by describing the interaction representation for this approach. Another approach is to look at the harmonic oscillator in a confining hard-wall potential, for example the equivalent of a particle in a box [9, 19]. Naturally, such confinement modifies the wave functions, but the basic shapes of the Wigner functions for the modes are not changed much at low energies, although they certainly go over to the box modes at higher energies. It has also been shown that the presence of a driving electric field distorts the Gaussian correlation functions that arise in treating the Wigner functions of the oscillator [20], which is somewhat surprising as the presence of a constant electric field (an exactly solvable model) generally only shifts the harmonic oscillator central position and its energy levels.

12.1.1 The driven oscillator

Some of the earliest studies of the parametrically driven harmonic oscillator studied the presence of damping [21, 22]. A normal harmonic oscillator is described by the Hamiltonian [22]

$$H(t) = \frac{p^2}{2m} + \frac{k(t)}{2}x^2, \tag{12.1}$$

where p is the normal momentum operator and $k(t) = m\omega^2$. Of course, in phase space, (12.1) is a standard classical Hamiltonian, and one particular example of a driven system is the Mathieu oscillator, where

$$k(t) = m[\omega^2 + F\cos(\omega_0 t)]. \tag{12.2}$$

The behavior of the oscillator depends crucially upon the amplitude F of the driving force, and the presence of any dissipation. However, when we add dissipation, the quantum system becomes more complicated, and the approach is to couple the driven oscillator to a bath of other oscillators, as mentioned above. Now, we can write the total Hamiltonian as [21]

$$H_T(t) = H(t) + H_B + H_{\text{int}}, \tag{12.3}$$

where the bath is typically

$$H_B = \sum_{\nu=1}^{N}\left(\frac{p_\nu^2}{2m_\nu} + \frac{m_\nu\omega_\nu^2}{2}x_\nu^2\right) \tag{12.4}$$

and the interaction is given as

$$H_{\text{int}} = -x\sum_{\nu=1}^{N}\gamma_\nu x_\nu + x^2\sum_{\nu=1}^{N}\frac{\gamma_\nu^2}{2m_\nu\omega_\nu^2}, \tag{12.5}$$

where x is the position operator of the driven oscillator (12.1) and γ_ν is the coupling strength between this oscillator and the bath. Standard approaches of quantum mechanics can then be used to show that various Markovian equations can be found for the forced motion of the harmonic oscillator [22]. A different approach to the interaction that leads to the damping of the harmonic oscillator was considered by Ankerhold [23]. In this case, the properties are considered with a Feynman path integral using the influence functional approach to describe the interaction with the bath [24, 25]. In this latter case, the influential functional is governed by the damping kernel

$$K(\vartheta) = \int_0^\infty \frac{d\omega}{\pi} I(\omega) \frac{\cosh(\omega\hbar\beta/2 - i\vartheta)}{\sinh(\omega\hbar\beta/2)}, \tag{12.6}$$

where $\vartheta = s - i\tau, 0 \leqslant s \leqslant t, 0 \leqslant \tau \leqslant \hbar\beta$ are the parameters for the thermal Green's functions upon which the path integral is based, and the spectral density $I(\omega)$ is related to the damping through

$$\gamma(s) = \frac{2}{m}\int_0^\infty \frac{d\omega}{\pi}\frac{I(\omega)}{\omega}\cos(\omega s). \tag{12.7}$$

This enabled Ankerhold to write a Fokker–Planck equation for the Wigner distribution in the strong damping limit, in what is known as the over-damped case [23]. More recently, the nature of the spectral density for the bath has been investigated for its role in affecting the quantum dynamics of the harmonic oscillator [26].

A harmonic oscillator coupled to a heat bath, as above, has also been studied for the case in which the external force is a 'kicked' impulsive force [27]. Here, the harmonic oscillator is described by the normal Hamiltonian (12.1), and the kick term is

$$H_K(t) = K\cos(\mu x)\sum_n \delta(t - nT_K), \tag{12.8}$$

where μ is the wave number of the kicked potential, K is the strength of the potential and T_K is the period of the kicks. The ratio of the oscillator period and the kick period is defined to be $q = 2\pi/\omega T_K$, and the normalized damping is given as $\beta = 2\gamma/\omega$. By introducing the parameters $\eta = \mu/\sqrt{2m\omega/\hbar}$ and $\kappa = K\mu^2/\sqrt{2}m\omega$, it is possible to then derive a master equation first achieved by Caldeira and Leggett [21]. The kicked oscillator classically is known to produce chaos under the proper conditions, and the quantum case is shown in terms of the Wigner functions in figure 12.1. It is clear that the existence of a resonance, where the number of kicks per oscillator period is an integer, differs from the non-resonant case, and significantly differs from the chaotic case. Finally, the nature of the quantumness of the oscillator as it undergoes decoherence has been investigated [28], which shows that it gradually becomes classical in its behavior, although it has also been demonstrated that the phase space trajectories can retain non-classical behavior [29]. In this latter work, the interaction term (12.5) leads to the damping kernel (12.6) to give the interaction

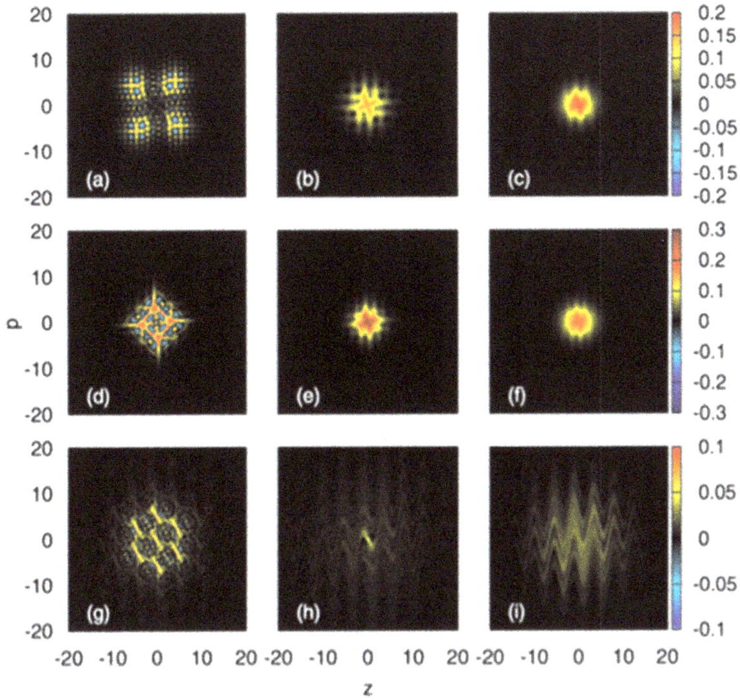

Figure 12.1. Wigner functions after the 36th kick for different coupling constants: $\beta = 0.001$ (first column), 0.01 (second column), and 0.1 (third column) for the resonant case (first row) with $q = 4$, $\kappa = -0.8$, and $\eta^2 = \pi$; the non-resonant case (second row) with $q = 4$, $\kappa = -0.8$, and $\eta^2 = (1 + \sqrt{5})\pi/2$; and the chaotic case (third row) with $q = 6$, $\kappa = -4.5$, and $\eta^2 = 1$. Reprinted figure from Prado Reynoso *et al* [27], copyright 2017 by the American Physical Society.

between the oscillator and the heat bath. In figure 12.2, we illustrate how this quantumness is retained in some features, by plotting two trajectories, referred to as $\mathbf{r}_\pm$, that are trajectories in phase space and related to the position and momentum as $\mathbf{r}_\pm = q_\pm + ip_\pm$, where the $q_\pm$ are the two coordinates that naturally arise in the density matrix $\rho(q_+, q_-) = \langle q_+|\hat{\rho}|q_-\rangle$. These two variables are then combined in the standard Wigner forms to determine the Wigner function and the momenta in the system. While the two trajectories shown in figure 12.2 begin at the same point in phase space, they eventually appear on opposite sides of phase diagram, which leads to the damped classical response shown also in the figure.

The dynamics and phase space properties of a confined harmonic oscillator have also been computed for the situation in which the potential is modified to include the spin–orbit interaction potential and the Zeeman terms [30]. Here, the authors extract the regular behavior as well as the chaotic behavior of the harmonic oscillator in this potential. Indeed, the statistical distribution characterizing the energy spectrum depends crucially upon the parameters of the potential model, as one might expect for a chaotic system.

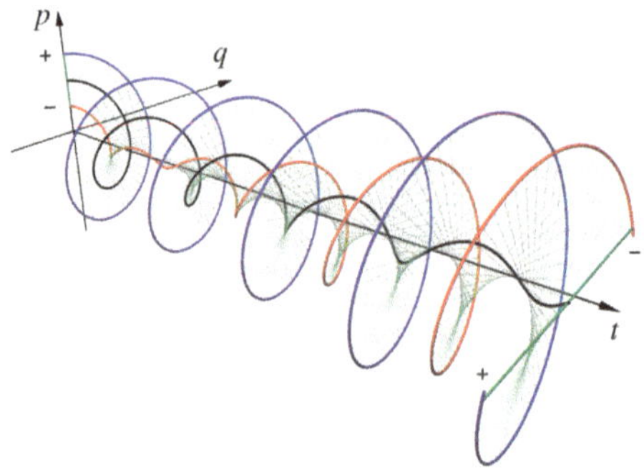

Figure 12.2. The time evolution of a pair of phase space trajectories $\mathbf{r}_{\pm}$, where the positive-signed trajectory is in blue and the negative-signed trajectory is in red. The corresponding classical trajectory is in black and decays with time, while the quantum trajectories actually grow with time. Reprinted from Pachón *et al* [29], with permission from Elsevier.

12.1.2 Qubits

In recent years, the quantum bit, or qubit, has become of great interest due to the overall interest in quantum computing. In this area, the coupling of a harmonic oscillator to the qubit has attracted some interest. When a harmonic oscillator is influenced by linear damping and parametric gain from an external force, its coherent states are generally characterized by a Gaussian Wigner function, as we describe several times throughout this book. Now, qubit states can be encoded in the $|0\rangle$ and $|1\rangle$ states of the harmonic oscillator [31]. One important question arises over when the system is affected by normal changes of the variables of the damping and driving forces: whether or not the qubit changes in the same manner. This question was addressed by these latter authors with the result that, if coherent states are used to drive the oscillator system, the fidelity of the qubit can be established by measurements of the first and second moments of the physical variables involved. Other work on quantum information has addressed similar problems via the Wigner function, by looking at a system based upon the Coulomb interaction between two trapped ions that produces nonlinear coupling between the normal (oscillator) modes of motion at the single photon level [32]. This latter work demonstrated that the system acts like a parametric oscillator which holds promise for providing continuous variables in a quantum computation scheme. Here, the Wigner function is closely related to the parity operator in quantum optics as discussed in chapter 11. Clear distinctions between the Wigner functions for different oscillator states are found in simulations.

As discussed repeatedly, however, the need for entanglement translates into a need for non-Gaussian states in the total Wigner function of the system. Detection of the non-Gaussian part of the Wigner function becomes a criterion for observations of the entanglement. This becomes of particular importance in the mixed states that are likely to be encountered in quantum information. A set of criteria which are able to detect the non-Gaussian property for single-mode quantum states of a harmonic oscillator has been developed recently [33]. The authors extend this to find several bounds on the values of the Wigner function for convex mixtures of Gaussian states, and these in turn define a class of sufficient criteria for the existence of quantum non-Gaussian parts of the Wigner function. More recently, the presence of non-Gaussian pure states generated in by an anharmonic oscillator and approximations to them using a truncated Wigner approximation have been studied [34]. The truncation method is to remove the third- and higher-order derivatives that arise from the expansion of the Wigner potential. Since this modifies the corresponding terms in the resultant Liouville equation, or the Wigner equation of motion, there is no guarantee that any non-Gaussian correlations predicted by the truncated Wigner function would be accurate. However, these latter authors have shown that such non-Gaussian states can be reliably predicted, even with the truncation mentioned.

In nuclear physics, there is an interesting concept termed the scissors mode, which is a counter revolution of the protons against the neutrons in deformed nuclei. While there was some doubt about this mode, it has been explored using time-dependent Hartree–Fock equations for a harmonic oscillator model in the presence of spin–orbit interaction. A Wigner transform is then used to study the properties of this state as well as the presence of quadrupole–quadrupole interactions [35, 36]. Later, a study of quadrupole–octupole interaction used similar methods [37]. The presence of the scissors mode is thought to represent the fact that there are hidden angular momenta in the model [38]. In this work, it is clear that the use of the Wigner transforms helps understanding considerably.

12.2 Quantum physics

Some of the earliest work employing the Wigner function in basic quantum mechanics lies in trying to ascertain the capability of extending quantum mechanics to a macroscopic domain, in contradistinction to the standard interpretation of Copenhagen. The question arises as to how decoherence appears and transforms the quantum system into the classical equivalent [39]. In this latter approach, a consistent histories approach via a decoherence functional shows that the quantum trajectories evolve in a manner such that the probability distribution peaks around the classical histories of the system, especially when the initial positions and momenta are given by a smeared version of the Wigner function. Further work by others also leads to the result that using a Wigner transform for quantum operators leads to a classical result, especially for spectral line shapes [40]. At about the same time, the features of the time-dependent Wigner function were reviewed [41]. This connection to classical initial values and the quantum propagator were studied further in later work [42]. Decoherence in a one-dimensional attractive Bose gas due to the presence of random walks in momentum space was also studied [43], and when these are caused by phonons, it is known to definitively introduce decoherence [44]. Other work has looked at the quantum-classical transition in open quantum dots [45]. Hence, it is fair to say that the decoherence problem appears in many guises and is still a problem of some interest in quantum physics. The Wigner function seems to be an especially good method for attacking this problem.

In condensed matter physics, the presence of the polar optical mode phonon can hybridize with the carriers to form polarons. This formation of the polaron was studied in general transport theory using the Wigner function in its momentum and frequency dependent form [46], where the quantum dynamics could be followed in time-dependent manner via an ensemble Monte Carlo technique. This represents a more in-depth analysis of the electron–phonon dynamics than, for example, the general phase space formulation that has been known since Wigner [47].

12.2.1 Parity again

In the previous section above, the truncated Wigner approach was mentioned. This approach was used to examine commuting operators and their time-dependent averages for the quantum response problem [48], where it was found that connecting the behavior of time-symmetric operators with conventional symmetric orderings of operators was a non-trivial task. Because of the phase space representation of the Wigner function, it also appears to be uniquely suited to the study of general statistical mechanics [49, 50]. It has also been a viable approach to study the basic ideas of quantization and commutators [51, 52]. More recent work has shown the advantages of the Wigner function for arbitrary quantum systems [53] and for study of various imaging elements in a scanning transmission electron microscope (STEM) [54]. In Tilma *et al* [53], a development of the Wigner function for a two-level spin system is based upon the self-conjugate Stratonovich–Weyl correspondence, which leads to the expression

$$\hat{\Delta}^{(2)}(\vartheta, \varphi) = \frac{1}{2}\left[\hat{1}^{(2)} - \sqrt{3}\,\hat{U}_2^{(2)}\sigma_z\left(\hat{U}_2^{(2)}\right)^{\dagger}\right], \tag{12.9}$$

where the Euler angles (ϑ, φ) parameterizing the representation space are set by the rotation operator

$$\hat{U}_2^{(2)}(\vartheta, \varphi, \Phi) = e^{i\sigma_x\varphi}e^{i\sigma_y\vartheta}e^{i\sigma_z\Phi}. \tag{12.10}$$

Using the invariance of the 2×2 identity $\hat{1}^{(2)}$ matrix under this operation, we can define the operator

$$\hat{\Pi}^{(2)} = \frac{1}{2}[\hat{\mathbf{1}} + \sqrt{3}\sigma_z] \tag{12.11}$$

so that (12.9) becomes

$$\hat{\Delta}^{(2)}(\vartheta, \varphi) = \frac{1}{2}\left[\hat{U}_2^{(2)}\hat{\Pi}^{(2)}\left(\hat{U}_2^{(2)}\right)^{\dagger}\right]. \tag{12.12}$$

In a high spin space, the formulas become more complicated, with (12.10) becoming

$$\hat{U}_2^{(2j+1)}(\vartheta, \varphi, \Phi) = e^{iJ_1\varphi}e^{iJ_2\vartheta}e^{iJ_3\Phi}, \tag{12.13}$$

where the J_i are the generators of the $(2j+1)$ dimensional representation of SU(2). Then, (12.11) and (12.12) become

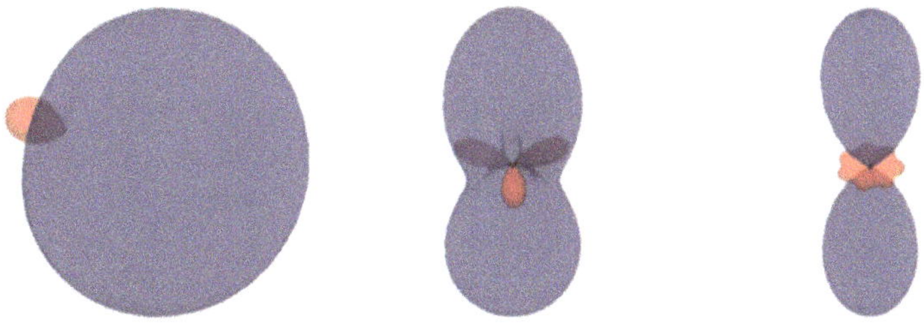

Figure 12.3. Polar plot of the Wigner function for (12.14) for high spin states of dimension (a) 1/2, (b) 3/2, and (c) 7/2. Here, the authors have used as examples normalized states of the form $|i, j\rangle + (j, -j)$. Note that there are $2j$ interference terms and the various images are not to the same scale. For these plots, red is negative and blue is positive. Reprinted with permission from T Tilma *et al* [53]. Copyright 2016 by the American Physical Society.

$$\hat{\Pi}^{(2j+1)} = \hat{1}^{(2j+1)} - N(2j+1)\hat{\Lambda}_{(2j+1)^2-1}$$
$$\hat{\Delta}^{(2j+1)}(\vartheta, \varphi) = \frac{1}{2j+1}\hat{U}_2^{(2j+1)}\hat{\Pi}^{(2j+1)}\left(\hat{U}_2^{(2j+1)}\right)^{\dagger}, \qquad (12.14)$$

where

$$N(2j+1) = \sqrt{(2j+2)(2j+1)(2j)/2} \qquad (12.15)$$

and $\hat{\Lambda}_{(2j+1)^2-1}$ is a $(2j+1) \times (2j+1)$ diagonal matrix in which the diagonal terms are given by $\sqrt{2/(2j+1)(2j)}$, with the exception of the last element which is $-\sqrt{2(2j)/(2j+1)}$. In figure 12.3, the second line of (12.14) is plotted for three different superposition spaces of 1/2, 3/2, and 7/2. While the shapes of the Wigner functions here are quantitatively different from the normal form used, these functions do exhibit the interference in a common manner, in that there are $2j$ interference terms (red is negative and blue is positive).

12.2.2 Quantum Hall effect

While there have been two Nobel prizes in studies of the quantum Hall effect, notably for the effect itself and for the fractional quantum Hall effect, there remains significant interesting physics in these systems. In particular, the existence of single quasi-particle excitations has been compared to quantum optics in many regards. For example, the Hanbury–Brown–Twist [55] and Hong–Ou–Mandel experiments [56] suggest interesting electron optics. But there are differences between true optics and electron systems, in that the latter are fermions subject to the Coulomb interaction between the electrons. In addition to the real parts of this interaction, imaginary parts are attributed to scattering, which can cause decoherence in quantum effects. In recent work, a study was made of the time-resolved properties of two different single-electron-like excitations in a quantum Hall edge state [57]. The two excitations are Landau quasi-particles that correspond to a Lorentzian wave packet emitted from a quantum dot, and a Levitov quasi-particle which corresponds to applying a Lorentzian time-dependent potential with quantized flux [58]. Then, the excitations can be studied effectively utilizing the Wigner function to illustrate their decoherence [57]. This is achieved by using the frequency–time version of the Wigner phase space, which demonstrates that the energy-resolved excitation disappears into a wave of electron–hole excitations. The use of the Wigner function allows them to gain access to both the time variation and the energy content of the single-electron excitation. More recently, levitons have been examined as excitations in the fractional quantum Hall effect [59] and a form of crystallization of these excitations has been shown to occur. Such excitations have also been studied in two-dimensional topological insulators where three electron collisions can be found [60]. These excitations also have been demonstrated to have a use in two-particle interferometry in quantum edge channels, a process which may have an impact in some approaches to quantum computing [61].

12.2.3 Qubits

With the above discussion, we have come back to qubits and quantum computing and/or quantum information theory. It is precisely in this field that the Wigner function provides ready visualization of the entanglement within the system. For example, in figure 12.4 we show the Wigner function for a Schrödinger cat state [62]. One should compare this figure with figures 8.1 and 8.2, because the present one verifies the statements made in chapter 8 about the observability of the entanglement. In particular, the latter authors point out that this cat state is very similar to those presented in the situation where non-classical states of light can be made in quantum optics [63]. It has been proposed that the Wigner function for any quantum system can written in terms of a displaced or rotated generalized parity operator (12.14), for an arbitrary number of states, or qubits, with the Wigner function being given by (12.12) [62]. These authors then examine several cases, one of which is a Wigner function whose kernel comprises a tensor product of one-qubit kernels. In the case of a single qubit, the parity operator is given as (12.11), while the two-qubit operator involves four states and appears as

$$\hat{\Pi}^{(4)} = \frac{1}{4}[\hat{1} \otimes \hat{1} + \sqrt{5}\hat{1} \otimes \sigma_z + \sqrt{5}\sigma_z \otimes \hat{1} + \sqrt{5}\sigma_z \otimes \sigma_z]. \tag{12.16}$$

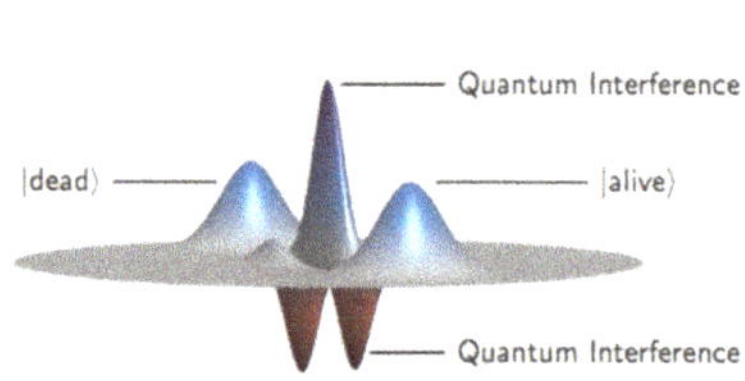

Figure 12.4. The iconic textbook example of a Wigner function for a Schrödinger cat state. The bell shapes represent the 'dead' and 'alive' possible states and the oscillations between them indicate the quantum interference (entanglement) between these states. A similar Wigner function without these interference terms would represent the classical states, such as for a coin toss, as discussed in chapter 8. Reprinted with permission from R P Rundle *et al* [62], under the Creative Commons attribute 3.0 license from the American Physical Society.

Combining this definition of the extended parity operator with the rotation kernel from (12.13), one reaches the extended two-qubit Wigner function, which is expressed as [62]

$$W_{SU(4)}(\{\vartheta_i, \varphi_i\}) = \mathrm{Tr}\{\hat{\rho}\, U_2 \hat{\Pi}^{(4)} U_2^{\dagger}\}. \tag{12.17}$$

Here, each qubit brings with it two degrees of freedom which are expressed by the two Euler angles, so we wind up with a four-dimensional Wigner function. Such four-dimensional functions are obviously not easy to visualize, but one can take slices of the functions. In figure 12.5, we show some examples of Wigner function slices for two Bell states. Specifically, panels (a) and (b) show the equal angle slice for $W_{SU(4)}(\vartheta, \varphi, \vartheta, \varphi)$, while panels (c) and (d) show the slice $W_{SU(4)}(\vartheta_1, 0, \vartheta_2, 0)$.

In order to demonstrate that this function is indeed easy to construct, the authors [62] have used the five-qubit processor made available on the web through IBM's *Quantum Experience* project (since taken down to be replaced with a larger number of qubits in the processor). This project employed five qubits, in which a central

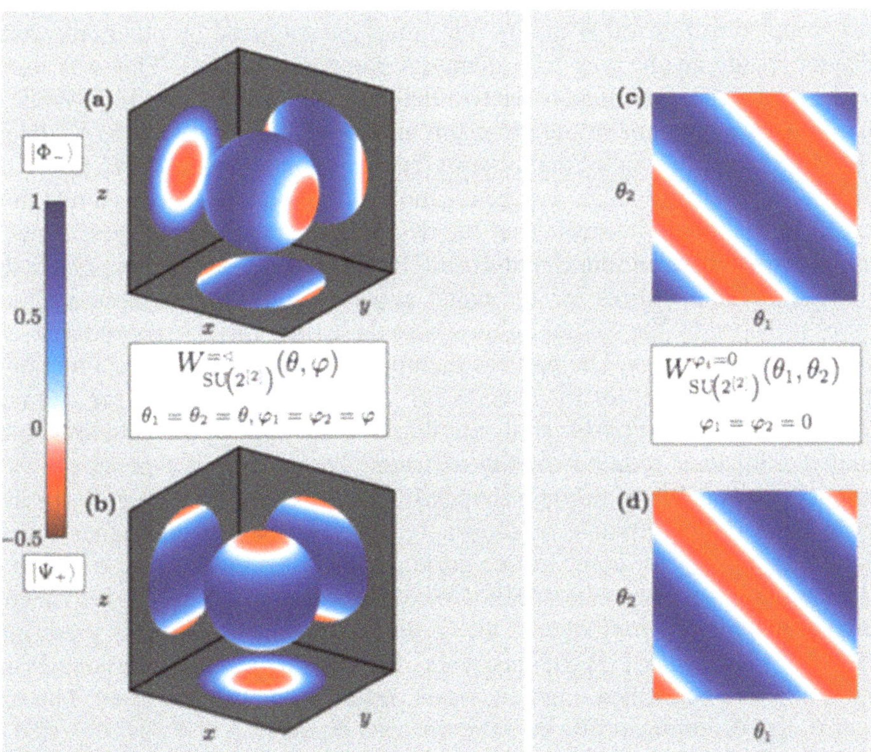

Figure 12.5. Slices from the four-dimensional Wigner function of two qubits given by (12.17) (the two Wigner functions are identified in the figure as well as the text), representing maximally entangled Bell states. The three-dimensional plots in panels (a) and (b) are the projections of the Bloch sphere of these two states. Panels (c) and (d) show the projection onto the two Euler angles for each qubit. Reprinted with permission from R P Rundle *et al* [62], under the Creative Commons attribute 3.0 license from the American Physical Society.

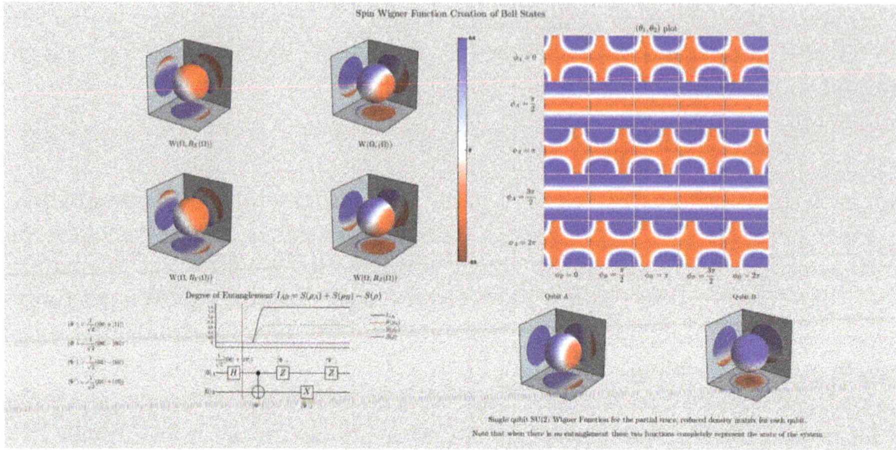

Figure 12.6. Time evolution of the Bell state as seen through the Wigner functions of figure 12.5. Reprinted with permission from R P Rundle *et al* [62], under the Creative Commons attribute 3.0 license from the American Physical Society.

qubit is coupled to four other qubits. This machine was used by others to produce interesting results on the way to a quantum computer [64, 65]. This was used to construct the Wigner function for the two Bell states shown in figure 12.5. While this use of slices conveys some of the information, more is found in looking at the time development of the Bell states that arise in the two-qubit formulation. In figure 12.6, the Wigner function dynamics for the creation of the four Bell states (shown in the lower left of the figure) illustrate that the two Wigner functions for these states are rotations of each other in four dimensions. The Bloch spheres in the upper left show four two-dimensional slices for the equal angle case as well as three additional rotations provided by the second qubit, where these rotations correspond to each of the three coordinate axes. The plots at the upper right are the Wigner functions of the polar angles for each of the spheres for various fixed values of the azimuthal angles. The bottom right panels show the Wigner functions for the individual qubits calculated from their reduced density matrices. The bottom left panel shows the progress of the simulation through the algorithm illustrated in the figure. In figure 12.7, an equivalent animation is shown for the tensor product Wigner function version of figure 12.6 for comparative purposes [62].

Experimental measurements which allow reconstruction of the Wigner functions for discrete atomic systems [66] have also been developed, following the prescription outlined by Tilma *et al* [53]. In this work, the complete and continuous Wigner function of a well-controlled single two-level cesium atom was measured. This atom was controlled deterministically in a micron-sized dipole trap and was subjected to a nearly pure dephasing process in the measurement. This work demonstrated that, for an arbitrary pure state of a qubit, the Wigner function always possesses a negative region that vanishes if the purity of an arbitrary mixed state is less than two-thirds. We will return again to qubits in the next section where we discuss a major implementation approach to qubit systems.

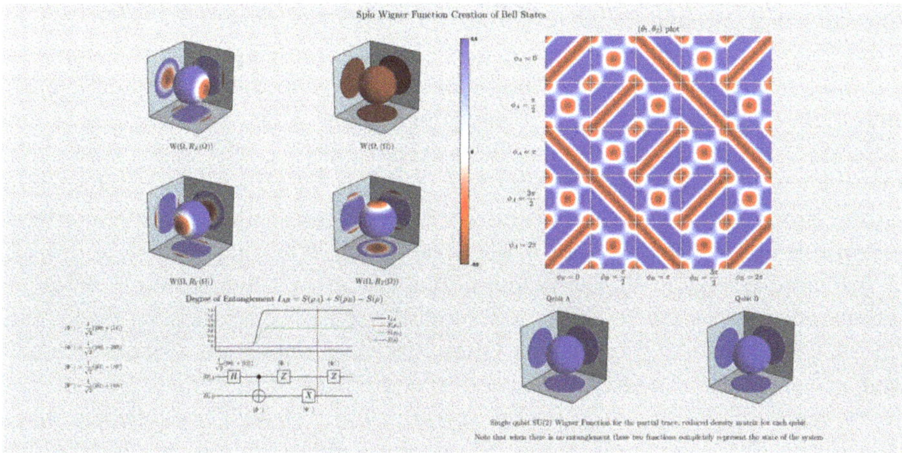

Figure 12.7. Time evolution of the tensor product Wigner function for the Bell state shown in figures 12.5 and 12.6. Reprinted with permission from R P Rundle *et al* [62], under the Creative Commons attribute 3.0 license from the American Physical Society. A video is available online at http://iopscience.iop.org/book/978-0-7503-1671-2.

12.3 Superconductivity

Our ultimate interest in superconductivity lies both in the use of Josephson junctions and in the use of circuits with these devices for qubits. In many of these applications, Wigner functions have become one of the standard methodologies for illustrating the entanglement and phase coherence of the circuits. The Josephson junction is a tunnel junction in which the materials on either side of the insulator (or non-superconducting material) are superconductors. As it is a tunnel junction, it is natural that the Wigner function comes into play just as for semiconductor tunnel junctions. One of the earliest uses of the Wigner function for this device studied the small junction limit in which single-electron behavior could be observed [67] (in superconductors, for temperatures above absolute zero, the carriers are a mix of Cooper pairs and normal electrons above the gap). Normally, with Josephson junctions, the tunneling is carried out by Cooper pairs, but these authors have generated a master equation that describes both the Cooper pair tunneling and the single-electron tunneling. But it is also known that the Josephson junction can be a microwave source, and it has been observed that one can see a so-called Riedel peak in the response of the junction [68, 69]. The Josephson frequency is given by $f_J = 2eV/h$, and the Riedel peak is observed at $f_J = 4\Delta/h$; one finds this peak at a bias voltage of $2\Delta/e$ [69], clearly being related to the superconducting gap Δ (the factor of 2 appears as the normal energy to raise a Cooper pair above the gap is twice the energy gap due to the $2e$ charge). In the modeling of the single-electron tunneling, it is found that the Riedel peak and the gap for the quasi-particle-pair interference are shifted to $2\Delta/e - e/2C$, while the quasi-particle tunneling gap shifts to $2\Delta/e + e/2C$ [67]. Here, the term $e/2C$ is the Coulomb charging energy for single-electron effects, where C is the capacitance of the small junction. Hence, there is a hybridization between the superconducting gap and the Coulomb gap. In this latter paper, the master equation was developed for the Wigner function describing the charging effect on Cooper pairs and quasi-particle pairs, as well as the single-electron tunneling in the junction. More recently, the situation in which a superconducting ring, incorporating a single mesoscopic Josephson junction, has been considered [70]. The Wigner function in number phase representation shows that the state of the system evolves into a quantum superposition of two coherent states which clearly demonstrate interference and negative values for the Wigner function. On the other hand, these authors find that the number phase Wigner function evolves in a classical manner. This ring-junction system was studied further to look at the role of dissipation on the evolution of the Wigner function [71]. In this latter case, the dissipation is incorporated by coupling the system to a reservoir, and this expanded system is then projected back onto the reduced density matrix for the ring-junction system. They conclude that the two coherent states survive even in the presence of dissipation, at least for weak dissipation. The number phase representation in the Wigner space has been studied more recently [72], although it is not a new concept [73, 74].

Vortices can appear in a dual Josephson junction (which is the case for a 2+1 dimensional non-superconducting material is placed between two 3+1 dimensional superconductors and constitutes a model of localization for a four-dimensional electromagnetic field [75]), when these junctions are incorporated in a ring, under the conditions that the system lies in an insulating state. There is a duality between electrons in the superconducting state and vortices in the insulating state [76]. Such vortices have been studied in the presence of an illuminating microwave field in order to assess the role of dissipation [77]. As above, the system becomes insulating when the Coulomb charging energy $E_C = 2e/C$ is greater than the Josephson coupling energy E_J ($=\Phi_0 I_c/2\pi$, where $\Phi_0 = h/e$ is the flux quantum and I_c is the critical current in the junction). Here, the vortex mass is extremely low, being proportional to the Coulomb charging energy, and the vortices behave as quantum particles. In this case, the vortices circulate around the ring with high mobility and tunnel through the dual Josephson junction. The presence of non-classical microwaves (at very high frequencies) affects the vortex current through the quantum noise of the microwaves. As before, the two coherent states can become entangled, and this is illustrated by using the Wigner function to describe the system.

A study of non-equilibrium quantum tunneling, such as the systems studied above, is necessary to understand arrays of Josephson junctions. A Wigner function-based approach to macroscopic quantum tunneling using a corresponding master equation that incorporates the tunneling coupled to an environment has been developed [78]. Again, the total system is projected onto a reduced Wigner function that is based upon the eigenstates of the closed system (no coupling to the reservoir). Hence, the coupling to the reservoir can be treated via a WKB approach.

12.3.1 Coupling to a resonator

In a variant of the above discussions, the incorporation of a small Josephson junction, coupled to a superconducting ring, in a microwave cavity has been discussed [79]. For a single small Josephson junction in the cavity, one may write the Hamiltonian as

$$\begin{aligned} H &= H_0 + H_{\text{int}} \\ H_0 &= \hbar\omega\left[\left(a^\dagger a + \frac{1}{2}\right) + (1 + 2\lambda)\left(b^\dagger b + \frac{1}{2}\right)\right] \\ H_{\text{int}} &= \hbar\omega(a^\dagger b + ab^\dagger) + \lambda(a^\dagger b^\dagger + ab), \end{aligned} \tag{12.18}$$

where $\lambda = e^2 E_J L$, L is the inductance of the ring, and the bosonic operators a and b refer to the cavity and the junction operators, respectively. The Josephson junction emits its own microwaves when a voltage bias is applied to the junction [80]. Since the ring makes a form of cavity, this can be treated by harmonic oscillator wave functions just as for a regular electromagnetic cavity. Here, $\omega = 1/\sqrt{LC}$ describes the resonator (as a resonant circuit), and another frequency is introduced for the junction as $\omega' = \sqrt{(1 + 4\lambda)/LC}$. Now, as is well known, we may describe conjugate variables for the resonant circuit and junction in terms of the flux and charge as

$$\begin{aligned} \Phi &= (a^\dagger + a)\sqrt{2\hbar/L\omega} \\ Q &= i(a^\dagger - a)\sqrt{\hbar L\omega/2} \end{aligned} \tag{12.19}$$

for the microwave circuit and

$$\begin{aligned} \varphi &= (b^\dagger + b)\sqrt{2\hbar/\omega' L} \\ q &= i(a^\dagger - a)\sqrt{\hbar\omega' L/2} \end{aligned} \tag{12.20}$$

for the junction circuit. In each case, the uncertainty relation gives a value of $\hbar$ simply because the effective charge of the tunneling Cooper pair is twice the electron charge. The inductance for this can be inferred by the discrete change in flux is given by the flux quantum. Thus, adding a flux quantum to the Josephson ring circuit requires a Cooper pair tunneling. Hence, one can now develop a four-variable phase space representation with the Wigner function. There are, in fact, certain similarities here with the Jaynes–Cummings model, discussed in section 11.1, but with the two-state atom being replaced by a multi-level Josephson junction and ring circuit. Of course, this also changes slightly the model used for the interaction with the microwave cavity. This set of equations is useful for many different studies and is generic in that sense. For example, one group has gone somewhat further by coupling two electromagnetic cavities to a Josephson junction [81]. Here, the circuit is a series connection of the two cavities with the junction. In this situation, H_0 just becomes the bosonic representations of two cavities and the interaction is through the Josephson junction itself, with the flux variables of the two cavities coupling to the Josephson phase.

A variation on the above concept is using two cavities which are coupled by a single Josephson junction [82]. In this situation, it is observed that the tunneling of a Cooper pair through the Josephson junction, from one cavity to the other, excites two photons, one in each of the cavities, or resonators. These microwave photons

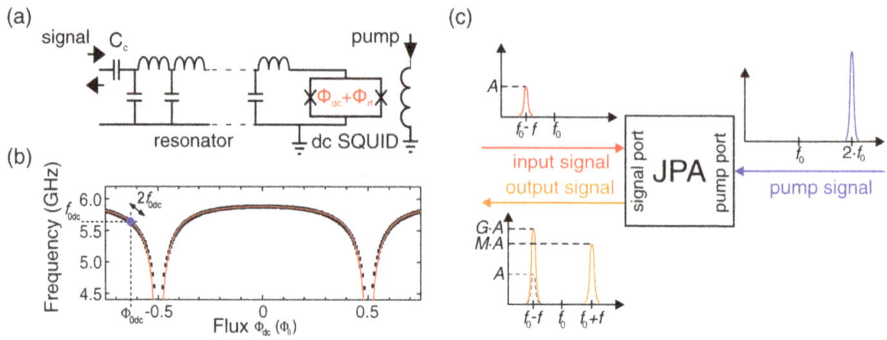

Figure 12.8. Flux-driven Josephson parametric amplifier (JPA). (a) Circuit diagram. The transmission line resonator is terminated by a dc SQUID (loop with crosses symbolizing Josephson junctions) at one end. A magnetic flux Φdc + Φrf penetrating the dc SQUID modulates the resonant frequency. (b) Dependence of the resonant frequency on the dc flux. The red line is a fit of a distributed circuit model [30] to the data (■). Blue dot: the operation point used in our experiments. (c) Schematic of the operating principle of the JPA. Reproduced from Zhong *et al* [86] under the Creative Commons attribute 3.0 license from the Institute of Physics.

leak out and can be observed experimentally. The role of nonlinearities, arising from the nonlinear current through the junction, can be important in the system. The coupling parameter is determined by the effective parameters of the individual oscillators, and is defined by the square root factors in (12.19). This parameter can be adjusted somewhat by the use of an external scanning tunneling microscope [83, 84]. If the coupling to both loops is reduced to zero, one gets a nondegenerate parametric amplifier, while if it is coupled to only a single resonator, the authors suggest they achieve an anti-Jaynes–Cummings model, in that the driving field de-excites both the resonator and the junction. In the latter case, the Wigner function demonstrates negative regions of correlation between the resonator and the junction.

12.3.2 SQUIDS

The idea of the parametric amplifier is useful, also, in microwave applications. When two Josephson junctions are incorporated into the ring, with connections made between the two junctions, the circuit is known as a superconducting quantum interference device, or SQUID [85]. This is a more complicated circuit than the single junction-ring combination, but can be used in the same manner as discussed above. The SQUID can be coupled to a resonator, consisting of a quarter-wave transmission line through which the signal (to be amplified) is passed. The resonant frequency of the SQUID can be adjusted by coupling it to a source of ac magnetic field that, in turn, provides the pump signal for the parametric amplifier [86]. In figure 12.8, the circuit and principles of operation are illustrated. Periodically varying the resonant frequency of the SQUID at twice the frequency of the operating point, f_0, results in parametric amplification: a signal applied at $f_0 - f$ impinging at the signal port is amplified. If the incoming signal consists of vacuum fluctuations, this process is the analogue of parametric down-conversion in optics, and correlation between the signal and the idler, at $f_0 + f$, is established, which creates a squeezed state, as discussed in chapter 11. The correlation and the properties of the squeezed state are best described by a phase space Wigner function, just as in the case of optics. In the case of squeezing, the Wigner function exhibits a clear narrowing in the appropriate variables.

12.3.3 Qubits

As we discussed above, a Josephson junction coupled to a superconducting ring creates two states that can be monitored and entangled. Hence, it became clear that this structure could be used to create a qubit, the quantum bit for quantum computing, which is an analog bit rather than a digital one in that it can take complex values whose magnitude remains unity [87]. One of the earliest measurements was with quantum state tomography, as revealed in the Wigner function constructed after measurements [88]. In this work, several different possible qubit configurations, all of which involved superconducting states, were discussed. The authors, however, chose to use SQUIDS in which a small superconducting island coupled via two ultrasmall, but identical, Josephson junctions. Here, they worked in the charge state basis in which the single-electron charging energy was much larger than the Josephson coupling constant; then the state of the qubit could be read by a single-electron transistor [89, 90] coupled capacitively to the charge qubit. If we take the spin-up case as $|0\rangle$ and the spin-down state as $|1\rangle$, then the qubit density matrix can be described by the density matrix in this basis as

$$\rho = \begin{bmatrix} \rho_{00} & \rho_{01} \\ \rho_{01} & \rho_{11} \end{bmatrix} = \rho_{00}|0\rangle\langle 0| + \rho_{01}|0\rangle\langle 1| + \rho_{10}|1\rangle\langle 0| + \rho_{11}|1\rangle\langle 1|. \tag{12.21}$$

Then, four real parameters can be described as

$$\begin{aligned} r_0 &= \rho_{00} + \rho_{11}, & r_x &= \rho_{01} + \rho_{10}, \\ r_y &= i(\rho_{01} - \rho_{10}), & r_z &= \rho_{00} - \rho_{11}. \end{aligned} \tag{12.22}$$

The dissipative current flowing through the single-electron transistor is proportional to the probability of a projective operator measurement $|1\rangle\langle 1|$ on the qubit state, and this is the experimental approach [88]. This measurement is the equivalent of a σ_z measurement on the qubit, as

$$p_1 = \mathrm{Tr}\{\rho|1\rangle\langle 1|\} = \frac{1}{2}[1 - \mathrm{Tr}\{\rho\sigma_z\}] = \rho_{11}. \tag{12.23}$$

The four parameters (12.22) can be determined from such projective measurements on the qubit, and these can then be used to tomographically describe the Wigner function for the qubit. In a reversal of the process, the same authors have proposed a method of creating a superposition of two macroscopic quantum states in a single-mode microwave cavity couple to a superconducting charge qubit [91]. Such an interaction would demonstrate a simple cavity quantum-electrodynamics effect via the qubit. The density matrix is similar in form to (12.21) except that the states are defined as a product of the qubit state and the cavity mode, rather than that of a single spin. The superposition of the two cavity states, as measured by the charge

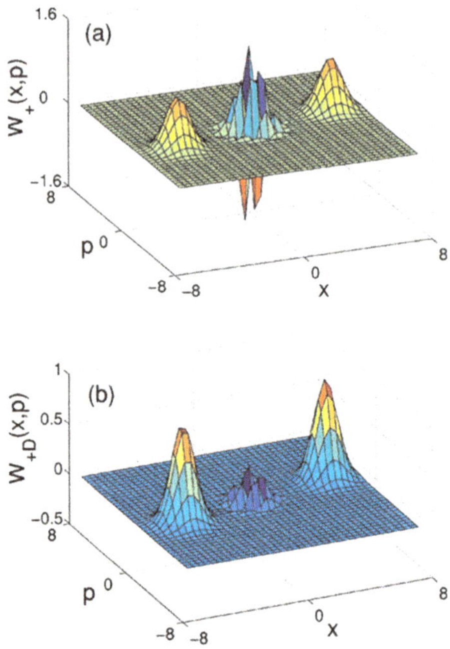

Figure 12.9. Wigner functions for the cavity modes without energy dissipation (a) and with energy dissipation (b). Figure reprinted with permission from Y X Liu *et al* [91]. Copyright 2005 by the American Physical Society.

qubit, are shown in figure 12.9. Unsurprisingly, the images bear a strong resemblance to figure 12.4 and those in chapter 8.

Normally, one can sense the state of a SQUID used as a qubit by determining the magnetic flux enclosed within the qubit. Here, the SQUID is current biased by the connections between the two junctions, so that the latter are in parallel branches of the ring. A more complicated circuit works in the charge qubit state, with a third junction spanning the first two and the addition of small capacitors. A pair of capacitors and another current source allow one to read out the state of the three-junction circuit [92]. This measurement and readout process has been modeled through the addition of a dissipated mechanism which ensures the irreversibility of the additional current, and this works to decohere the qubit [93]. Here, the qubit results from the three-junction circuit acting as a two-level atom. The third junction is considered to be the readout junction and this state may be clearly illustrated by constructing the Wigner function. As expected, it clearly demonstrates the two possible states and the entanglement between them.

A variant on the junction-ring qubit can be made by placing the Josephson junction as a shunt across a parallel inductor–capacitor circuit. This allows the junction to be biased by coupling through the inductor. Then, the additional resonator can be constructed from a coplanar waveguide (analogous to the transmission line discussed previously). This structure has been shown to be capable of generating arbitrary quantum states, each of which demonstrates an

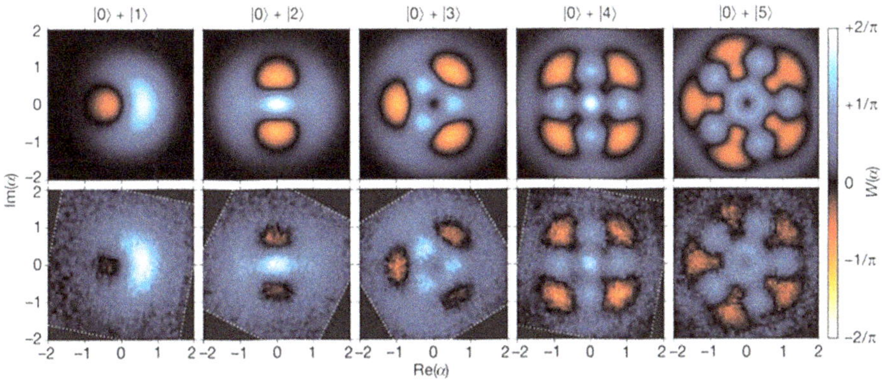

Figure 12.10. The Wigner tomography of superpositions of resonator Fock states $|0\rangle + |N\rangle$ for the first five excited states. The Wigner function is plotted as a function of the complex resonator amplitude in photon number units. Figure is reprinted by permission from *Nature* **459** 546 (2009), M Hofheinz *et al*, copyright 2009.

interesting Wigner function, as expected [94]. The Wigner function itself is computed from the parity function, as described above, and in chapter 11. In a general form, it becomes

$$W(\alpha) = \frac{2}{\pi}\mathrm{Tr}\{D(-\alpha)\rho D(\alpha)\Pi\}, \tag{12.24}$$

where α is the complex amplitude of the resonator. To measure the Wigner function, the resonator is first prepared in a desired state, which then defines the density matrix. During the analysis process, the resonator is then driven by the microwaves to displace the state by

$$D(-\alpha) = D^{\dagger}(\alpha) = e^{\alpha^{*}a-\alpha a^{\dagger}}, \tag{12.25}$$

where the a are the usual oscillator operators. The diagonal elements of the shifted density matrix are measured by a swap operation. In figure 12.10, we show the Wigner functions as a function of the complex amplitude of the resonator in photon number units for several different combinations of Fock states. That is, the Wigner function demonstrates its correlation between the $|N\rangle$ state and the $|0\rangle$ state. But, one can go further, and to higher eigenstates of the resonator. In figure 12.11, the so-called 'Voodoo cat' state is shown for a Schrödinger cat state. Here, the state $|\alpha = 2\rangle$ is used for the 'alive' state, $|\alpha = 2e^{i2\pi/3}\rangle$ represents the 'dead' state, and $|\alpha = 2e^{i4\pi/3}\rangle$ is the 'zombie' state, in the terminology of [94]. The figure compares the theoretical Wigner function with that measured experimentally.

The study of dissipation in the qubit coupled to a resonator is a variant of the Jaynes–Cummings model in that the qubit and the interaction are slightly modified from a simple two-level model due to the extreme nonlinearity of the Josephson current producing the interaction between the two. One of the important aspects is that it has been found that dissipation will lead to the so-called death of both the entanglement and the corresponding negativity of the Wigner function [95]. This effect has also been illustrated in the case of strong coupling [96]. Again, the qubit is

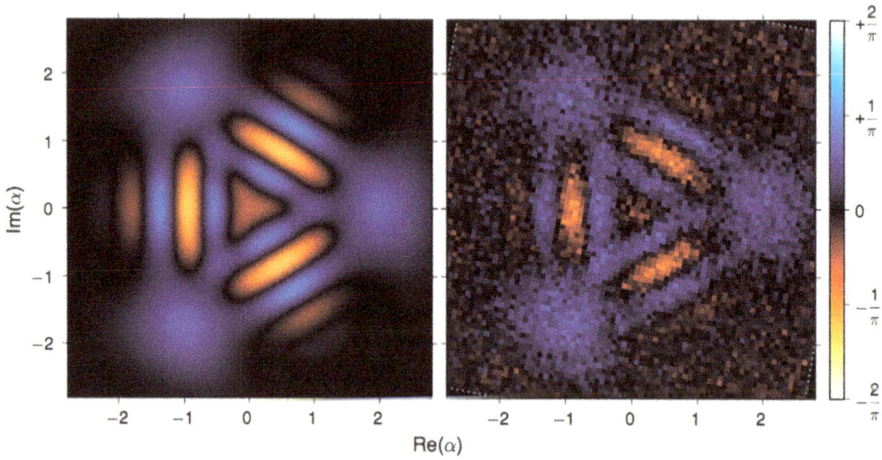

Figure 12.11. Wigner topography of the 'Voodoo cat' state. The left panel shows the theoretical Wigner function, while the right panel shows the experimental results for comparison. The 'Voodoo cat' state is an equal superposition of the alive, dead, and 'zombie' states, discussed in the text. The experimental result has been truncated at $n = 9$. The fidelity of the results is 0.83, showing that states up to nine photons can be created accurately. Figure is reprinted by permission from *Nature* **459** 546 (2009), M Hofheinz *et al*, copyright 2009.

treated as a two-level atom and the resonator as an harmonic oscillator system, with the interaction being treated somewhat differently. Then, a Q function is generated, with this function being somewhat similar to the smoothed Wigner function, as

$$Q(X, P) = \frac{1}{\pi}\langle X + iP|\rho_{osc}|X + iP\rangle, \tag{12.26}$$

where ρ_{osc} is the reduced density matrix obtained by projecting out the qubit variables form the ground state of the system, and the variables are the reduced harmonic oscillator variables that are related to the corresponding operators for the harmonic oscillator. This density matrix is also used in the Wigner function. The reduced coupling strength is given by

$$\lambda = g\sqrt{\frac{\hbar}{2m\omega_0}}, \tag{12.27}$$

where g is the qubit-oscillator coupling strength. The Q and Wigner functions are shown in figure 12.12, where it is clear that both the entanglement and the negativity of the Wigner function disappear as the coupling (and dissipation) is increased. Nevertheless, the authors demonstrate that a variety of states can be created in this system, including the Schrödinger cat state, squeezed states and entangled states [96, 97]. The coupled qubit-resonator has been extended to anharmonic resonators, as well [98].

As remarked above, the resonator-qubit system can lose its entanglement and negativity of the Wigner function in the presence of dissipation. However, the system can be stabilized by applying an active feedback from an ancillary system, or by reservoir engineering [99, 100]. The study of these systems has also been extended to

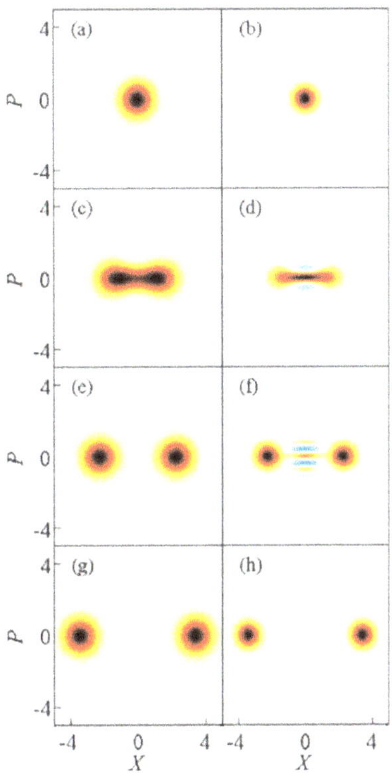

Figure 12.12. The Q function (12.26), on the left, and the Wigner function, on the right, of the oscillator's state in the ground state of the combined system. Here, the ratio of the oscillator quantized energy and the superconducting gap is 0.1 and the qubit is unbiased. The reduced coupling strength (12.27), in units of $\hbar\omega_0$ are given as 0.5 (a,b), 2 (c,d), 2.5 (e,f), and 3.5 (g,h). The color scheme is adjusted so the maximum value of the functions is always black, and yellow and red are positive values, white is zero value and blue and green are negative values. As one moves down the panels, the oscillator goes from a coherent state with no photons to a squeezed state and then to an entangled state. Reprinted figure with permission from Ashhab and Nori [96], copyright 2010 by the American Physical Society.

an array of qubits in an electron ensemble [101]. Finally, the reader is directed to some excellent reviews of the coupled microwave resonator and the superconducting qubit [102–104] that contain far more references than we have room for in this effort.

12.4 Plasmas

One of the earliest applications of the Wigner function to plasma physics paralleled developments in semiconductor transport, in that effort was devoted to deriving an equation of motion for the Wigner function [105, 106]. In this work, the two-particle Coulomb interaction and a magnetic field were both introduced, although the former is simplified by using the BBGKY hierarchy to limit the equation to only the second order two-body distribution function. The magnetic field affects the collisions, but true quantum effects were neglected in the equation of motion, beyond those arising from the nonlocality of the Wigner function. Their major goal was to extend the normal classical approach to effects of high magnetic fields. A more general look at the collisional effects was followed at about the same time [107]. In this latter work, the inhomogeneity that can arise in the plasma was also considered. In chapter 4, the expansion of the potential from the general form of the Wigner equation was discussed. This expansion for the effective potential has often been called the Wigner–Kirkwood expansion, and this has been studied for its role in the two-particle distribution function [108–110]. Here, it is found that higher-order terms lead to quantum corrections to this two-particle distribution. A Feynman path integral approach to finding the Wigner function for a canonical plasma has also been developed [111].

12.4.1 Kinetics

At high plasma densities, the densities become degenerate, particularly in condensed matter. The nonlinear kinetic equations for the degenerate Fermi–Dirac distribution in an electron plasma were also derived [112, 113]. The general requirements for the spin-1/2 quantum plasma of electrons were subsequently reviewed [114, 115]. The transverse currents in the plasma that arise when the magnetic field is present were studied for a quantum plasma in the presence of collisions [116]. In this latter work, the magnetic field is considered via the vector potential, and this provides some additional complications to the corresponding nonlocal potential term in the Wigner equation of motion.

The quantum Brownian motion of particles in a plasma, in the presence of damping and the diffusion of the particles, has also been studied [117]. One application of this work is to dilute impurities which are embedded in an ultracold degenerate quantum gas. Obviously, the inhomogeneity of the plasma is important in this context, but the authors still predict Gaussian stationary states in the plasma. A two-dimensional, one-component plasma has been studied in the cell model with cylindrical symmetry [118]. In this latter work, the counter ions are placed at the center of the cylindrical space with the electrons in the larger area around them. The role of strong correlations in a two-component degenerate plasma have also been studied [119].

While we mentioned fluctuations and diffusion above, the effects of the scattering of electromagnetic waves from these fluctuations becomes important in plasma systems, particularly in the study of possible fusion reactors [120]. The single-scattering process was derived for a compact technique based upon the Wigner function, but it was found that multiple scattering was necessary to accurately provide for the angular and spatial spreading of the scattered waves from the turbulence in the plasma. In multi-stream plasmas, it is found that Landau damping of the waves by the particles can be treated by a Wigner function approach, and the damping leads naturally to broadening of this Wigner function [121]. In these inhomogeneous plasmas, it has become natural to use the coupled Wigner and Poisson equations to describe both the plasma kinetics and its electrostatics. Such an approach has been used to treat the stability of waves in the plasma [122]. In the presence of waves in the plasma and a drift process, it is found that Landau damping can lead to photon acceleration when the wave velocity is comparable to the particle velocity, with the photons being generated by a corresponding optical wave [123].

For sure, the waves can be coupled to the plasma motion, and indeed the plasma self-oscillations, often referred to a Langmuir waves, can become significant, particularly with a nonlinear coupling [124–126]. As the electromagnetic waves often have a transverse component, this can affect the longitudinal current, as well [127].

In solving the coupled Wigner and Poisson equations, one approach uses an approximation to an 'exact' solution, written as [128]

$$f_W = [A(q)H + B(q)]e^{C(q)H}, \tag{12.28}$$

where H is the Hamiltonian in a quartic/quadratic combined potential. The factors A, B, and C are functions only of the position q. If the Hamiltonian is written as

$$H = \frac{p^2}{2} + \mu\left(\frac{q^2}{2} - \frac{q^4}{24}\right) \tag{12.29}$$

in reduced units, the derivative of the Wigner function introduces an additional parameter Γ as [124]

$$\frac{\partial f_W}{\partial q} = \Gamma\mu q\left\{\left[2H - \mu\left(q^2 - \frac{q^4}{12}\right)\frac{\partial^3 f_W}{\partial H^3} + 3\frac{\partial^2 f_W}{\partial H^2}\right]\right\}. \tag{12.30}$$

Then solving the coupled Wigner–Poisson system allows one to find the parameters in the equilibrium case, if one assumes the initial values for these parameters. In

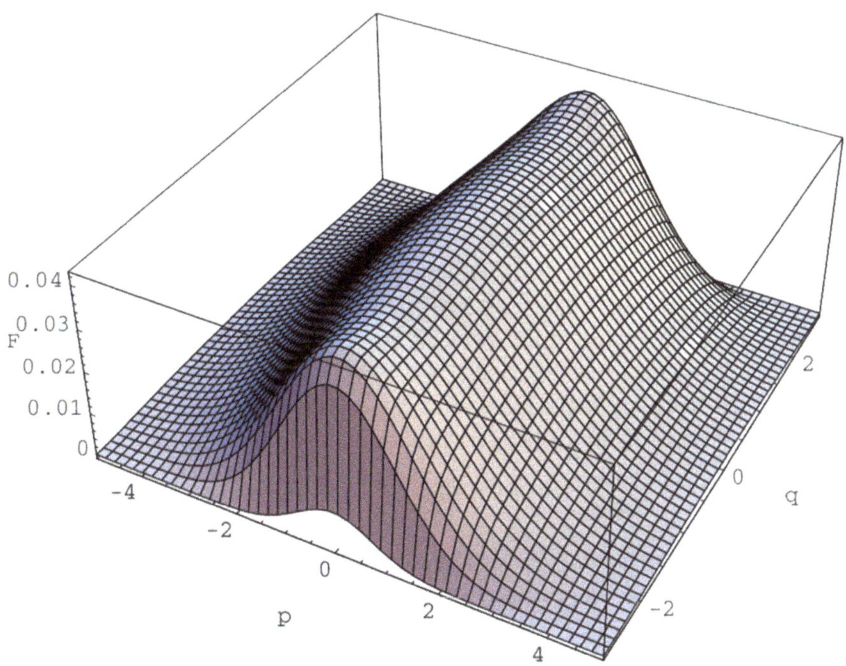

Figure 12.13. The exact Wigner function according to (12.28) for $\Gamma = 1$, $\mu = 1$, and $A_0 = 0$, $B_0 = 0.42$, and $C_0 = 16$. Reprinted from F Hass and P K Shukla [128], with the permission of AIP Publishing.

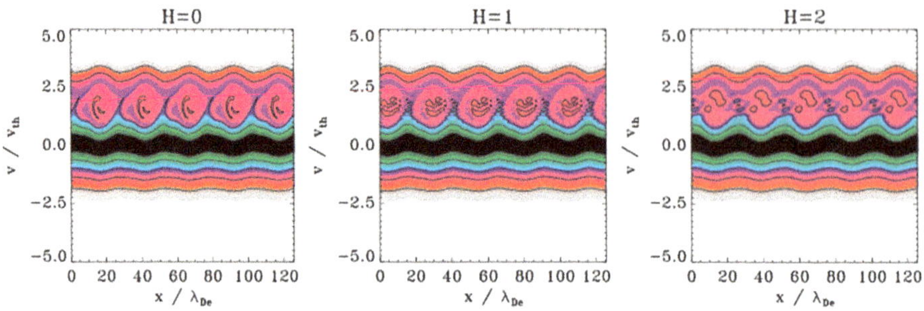

Figure 12.14. Simulation of the phase distribution function found from the Wigner–Poisson equations at a normalized time of $\omega_P t = 200$, in terms of the plasma frequency. The parameter H is the ratio of the plasma energy to the thermal energy of the particle, the velocity is normalized to the thermal velocity, and the position is normalized to the plasma wavelength. Reprinted with permission from F Haas *et al* [129], copyright 2008 by the American Physical Society.

figure 12.13, we show the resulting exact Wigner function for the case of $\Gamma = 1$, $\mu = 1$, and $A_0 = 0$, $B_0 = 0.42$, and $C_0 = 16$. In the presence of turbulence, the density can become quite inhomogeneous, but also fairly regular due to the wavelength of the wave involved [129]. In figure 12.14, we show the Wigner distribution function in phase space, as determined from the coupled Wigner–Poisson equations for a turbulent plasma. Here a perturbative solution building upon the approach given just above is used to find the wave interactions. Here, the parameter is the normalized Planck constant $H = \hbar\omega_P/mv_{th}^2$, that is, the ratio of the plasma energy to the thermal energy of the particles.

The study of the propagation of electromagnetic waves through the ionosphere is one of the oldest studies in radio physics. Of course, some interest lies in the fact that some layers of the ionosphere are conducting and this tends to trap low-frequency waves in the Earth-ionosphere duct. Then, atmospheric ducting arises from gradients in the index of refraction of various layers of the atmosphere and waves trapped in this region tend to propagate around the earth at constant altitude. These different effects can lead to over the horizon transmission of radio signals for thousands of miles. This can also occur in the high-frequency regime by wave guided by the geomagnetic field [130]. The two points where the field line touches the earth are known as *conjugate points*. One approach to the study of such propagation treats the radio wave as a distribution of quasi-particles described by a Wigner distribution function [131, 132]. The collisions of these quasi-particles with irregularities in the propagation medium leads to modifications of the distribution function and give rise to wave scattering.

12.4.2 High-density plasmas

Many forms of high-density plasmas can be generated, for example by laser excitation of condensed matter. Collisional absorption or inverse Bremstrahlung absorption is a common absorption method of laser light in plasmas [133]. This has been studied by using a Wigner representation of the density matrix to describe the quantum-mechanical effects through the Lindhard dielectric model. In this effort, it is found that almost all the quantum effects are contained within the dielectric function. Another interest in this excitation is the role of the transverse light field on the motion of the particles. Using the equations of motion for the Wigner function to describe the distribution of charge-carrier populations, it has been shown that the effect of these transverse light fields on the kinetic behavior of the electron–hole plasma is different than that found in atomic or molecular systems [134]. Here, it is shown that while the momentum of the light is absorbed by the carriers, there is no direct current linear in the intensity of the light, i.e. the light does not induce general current to lowest order.

Wigner function analysis has been shown to be useful for the analysis of time–frequency signals in fusion reactors, as well [135]. This follows due to the simplicity of the Wigner formulation and the computational ease of the implementation. Often there is a high energy density in these systems, and this can lead to Landau damping of a wave in such plasmas [136]. Using the Wigner formulation, these latter authors have also incorporated the Bohm potential and Fermi statistics to account further for the high carrier density. Finally, it was found that a Wigner kinetic theory was useful to treat the coupled 'kinetic theory and molecular dynamics' in treating the dense plasma [137]. The effects of radiation transfer from the fusion reactor have also been studied with the Wigner function [138, 139].

Still another type of plasma, which also can be characterized as dense, is the arc or jet plasma. The dynamic behavior of a double-arc nitrogen plasma has been characterized with the aid of the Wigner distribution [140]. Such a plasma can give rise to a variety of oscillatory behavior. This same double-arc nitrogen plasma was studied by similar techniques, in which it was found that some of the fluctuations are caused by power supply undulations as well as motion of the plasma around the anode [141].

12.4.3 Hydrogen

The thermodynamic properties of a deuterium plasma were investigated with a quantum dynamics method based upon the Wigner function phase space approach for a dense plasma of deuterium [142]. Here, it was found that the particle trajectories in a Monte Carlo simulation dominated the momentum–momentum correlation function. The approach improved upon earlier theoretical approaches which did not use the Wigner function, in that quantum exchange-correlation effects were included for the dense plasma. Similarly, the momentum distribution function for a weakly degenerate was investigated with the Wigner approach [143].

Another approach with the Wigner function was the determination of the hydrogen emission line in a low-density hydrogen plasma typical of interstellar situations [144]. In a similar vein, a weakly ionized, hydrogen-based plasma was studied for the quantum situation, and the wave dispersion relation and statistical properties determined with a Wigner distribution approach [145].

12.5 Relativistic systems

The need to develop a relativistic form of quantum mechanics was realized almost at the beginning, at least by Schrödinger, who proposed an extension of his famous equation that meets the requirements of special relativity. While one often thinks of relativity only in the realm of astrophysics, in fact it is important across all scales of science. At the largest scale we do have astrophysics, but then there is the realm of our everyday experience in life, and at the smaller scale come atomic and molecular studies, and then subatomic particles. Relativity and quantum physics are important across this entire spectrum, as we will demonstrate shortly.

Traditionally, in quantum mechanics and in classical mechanics, we usually write the energy in terms of the momentum as

$$E = \frac{p^2}{2m}. \tag{12.31}$$

Of course, this relationship establishes the energy shell in momentum. Life off the shell arises from the quantization of the system, as discussed in many previous chapters. The relativistic form of this equation was given by Einstein to be [146]

$$E^2 = p^2c^2 + m_0^2c^4, \tag{12.32}$$

where m_0 is the rest mass, i.e. the mass that the particle takes when it has zero momentum. Dirac, however, approached the problem in a different manner [147]. To see this, we will follow the approach of Schiff [148]. Dirac wanted to have a Hamiltonian that would make his equations linear in the spatial dimensions. If we rewrite (12.32) as

$$\hbar\omega = \pm\sqrt{\hbar^2c^2k^2 + m_0^2c^4}, \tag{12.33}$$

then one can approach his equations, which we will give below. First, however, we want to explore relativity in an unexpected place.

It is not usual for people in the condensed matter community, besides many other fields of physics, to worry significantly about relativistic transport. It is true that there are some processes such as non-parabolic bands and spin–orbit interactions that most will admit have their origins in relativistic effects. But relativistic effects occur already at the first, naive studies of band structure. That is, when we create a periodic structure such as a lattice in a condensed material, the Bloch wave function for the electrons is not unique, as one finds that the wave momentum k ($=p/m_0$) is only uniquely defined in the first Brillouin zone. When we use a value of k which runs through $-\pi/a < k < \pi/a$ to provide this first Brillouin zone, we are using what is called a Wigner–Seitz cell, which are those values of k closer to the Γ point ($k = 0$) than to any point shifted from this one by a reciprocal lattice vector $G = n \cdot 2\pi/a$, where n is any integer. This means that the momentum vector k is only defined up to a reciprocal lattice vector G, so that (12.31) must also be satisfied for any value of the shifted momentum vector, as

$$E = \frac{\hbar^2(k - G)^2}{2m}. \tag{12.34}$$

This means that the free carrier energy bands are also replicated throughout the reciprocal lattice. A simple example is shown in figure 12.15. The central curve

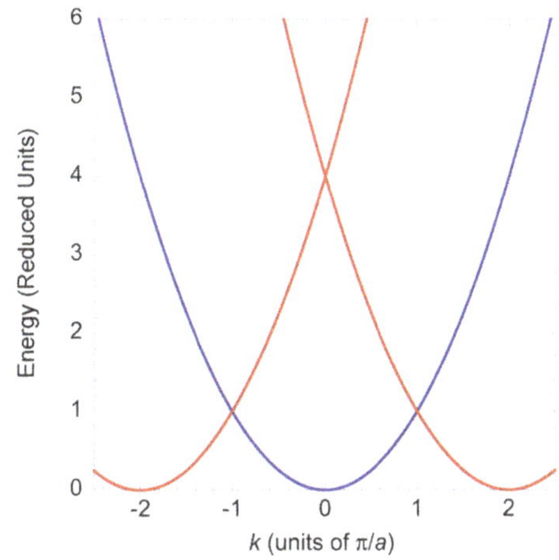

Figure 12.15. Free electron energy bands in a periodic, one-dimensional lattice. The atomic potentials open gaps at the crossing, as discussed in the text.

represents (12.31), while the left and right curves represent (12.34) for $G = 2\pi/a$ and $G = -2\pi/a$, respectively. The energy is degenerate at ± 1 as well as at 0 in this limited plot. It must be noted that parabolas arise from all values of G, not just those shown, but this figure is for the empty lattice; that is, this figure arises when we neglect the atomic potentials that come from the atoms in the lattice. It arises only because of the periodicity that has been imposed upon the lattice. If we now add the atomic potentials, these will cause a perturbation that opens gaps at each of the crossings in figure 12.15. Using simple perturbation theory based upon the Schrödinger equation leads to a new equation to describe the bands around, say, the point at 1, which is given by solving the matrix

$$\begin{vmatrix} (E_k - E) & U_G \\ U_G & (E_{k-G} - E) \end{vmatrix} = 0. \tag{12.35}$$

Here, U_G is the perturbing potential from the atomic potentials. To proceed, we define the reduced wave number as $k' = k - \pi/a$, and then expand the two bands around the point 1 in the figure, as

$$\begin{aligned} E_k &= \frac{\hbar^2}{2m}(k)^2 = \frac{\hbar^2}{2m}\left(k' + \frac{\pi}{a}\right)^2 \simeq \frac{\hbar^2}{2m}\left(\frac{\pi}{a}\right)^2 + \frac{\hbar^2}{2m}2k'\frac{\pi}{a} + \ldots \\ E_{k-2\pi/a} &= \frac{\hbar^2}{2m}\left(k - \frac{2\pi}{a}\right)^2 = \frac{\hbar^2}{2m}\left(k' - \frac{\pi}{a}\right)^2 \\ &\simeq \frac{\hbar^2}{2m}\left(\frac{\pi}{a}\right)^2 - \frac{\hbar^2}{2m}2k'\frac{\pi}{a} + \ldots. \end{aligned} \tag{12.36}$$

If we now use these expansions in the determinant of the matrices (12.35), we find the energies as given by

$$E - E_{\pi/a} = \pm\sqrt{E_{\pi/a}\left(\frac{2\hbar^2 k'^2}{m}\right) + U_G^2}, \tag{12.37}$$

where $E_{\pi/a}$ is the first term in the last expressions of (12.36). It is clear that $E_{\pi/a}$ plays the role of the $m_0c^2/2$ in (12.33). The left-hand side of (12.37) tells us that the renormalized zero of energy lies in the center of the gap. The lower band is completely full at low temperature in semiconductors, but excitations lead to the presence of holes in this lower band, and these holes are the analogues of Dirac's positrons. Thus, we see that relativistic-like effects arise from the very first principles of band theory in condensed matter physics. Hence, it is not surprising that deviations from the simple parabolic bands of (12.31) are termed relativistic effects.

Normally, the consideration of relativistic effects in transport in semiconductors is mainly after the fact and expressed in terms of simple analytic functions. This, however, came to a crashing halt when graphene was isolated from graphite layers in 2004 [149]. Graphene is a single atomic layer of carbon atoms that are hexagonally coordinated. More importantly, it does not have a band gap, and the bands are linear as a result of this fact. This means that the band structure in the neighborhood of the band crossing, which is termed the Dirac point, behave like relativistic Dirac bands (discussed below), but with a zero rest mass. The carriers, be they electrons or holes, still are fermions, so they tend to be characterized with the phrase *massless Dirac fermions*. Of course, these carriers are not really massless, as they have a dynamic mass given the momentum and the linear bands.

Now, if one directly introduces quantization to (12.32) by replacing the momentum and energy with the corresponding operators, one obtains

$$-\hbar^2\frac{\partial^2\psi}{\partial t^2} = -\hbar^2c^2\nabla^2\psi + m_0^2c^4\psi. \tag{12.38}$$

This equation has come to be known with a different name, and that is the Klein–Gordon (or Klein–Gordon–Foch) equation, as these authors came up with at about the same time. One problem with (12.38) is that it contains no spin variables, and so it has become associated with bosons. Nevertheless, it has some useful properties, in that the probability charge and probability current densities can be written as

$$\begin{aligned} Q &= \frac{i\hbar}{2m_0c^2}\left(\psi^*\frac{\partial\psi}{\partial t} - \psi\frac{\partial\psi^*}{\partial t}\right) \\ J &= \frac{\hbar}{2im_0}(\psi^*\nabla\psi - \psi\nabla\psi^*). \end{aligned} \tag{12.39}$$

In order to gain the spin variables, Dirac began with the form (12.33), and sought to find an equation linear in the momentum. One can take the positive root and insert it into Schrödinger's equation, for which we can achieve Dirac's goal. Dirac actually approached this with the simplest Hamiltonian that is linear in the momentum and mass terms (nevertheless, it was a masterstroke to realize this equation would work), which is

$$H = c\widetilde{\alpha}\cdot\mathbf{p} + \beta m_0 c^2. \tag{12.40}$$

To follow the full development of the resulting Dirac equations is beyond the aim here, but this development can be found in several textbooks [148, 150]. In order to accommodate all the constraints on the Dirac equation, the quantities $\widetilde{\alpha}$ and β must be 4×4 matrices. In addition, $\widetilde{\alpha}$ is a vector, so that each component of this vector is a 4×4 matrix. Dirac correspondingly found that there were four solutions to the resulting equations, two of which had positive energy and two of which had negative energy. One of each of these solutions had spin-up and the other spin-down. The negative-energy solutions came to be known as positrons, and were found experimentally.

12.5.1 Waves and particles

One of the earliest efforts to utilize relativistic, covariant Wigner functions dealt with the presence of large electric fields and large magnetic fields found in the relativistic quantum electrodynamic plasmas near quasars [151]. Hence, we are beginning here with our astronomical scale. Here, the basic derivation incorporated a density matrix, which naturally leads to the Wigner function. These authors then discussed the nature of the dispersion relations for waves in these very energetic plasmas. While the astrophysics considerations have been clear, some of these plasma properties have become of interest for intense laser-generated plasmas. Here, the kinetic properties of dense plasmas without magnetic fields have been formulated for spin-less particles, basing the Wigner functions on the Klein–Gordon equation [152], although the case for including all the electromagnetic fields has also been considered [153–155].

It has been proposed that the phase space description provided by Wigner functions provides a more intuitive view of the quantum-electrodynamics vacuum [156]. Here, it was felt that to obtain an effective description of the local vacuum, coupled to full relativistic electric and magnetic fields, one needed to use only a single time parameter, thus suggesting that the Wigner function was more useful than the two-time Green's functions. These authors developed the formalism for the phase space description of the Dirac vacuum in an approximation suitable for strong fields, and considered several simple examples.

The relaxation of relativistic dense matter through the use of a relaxation time approximation was considered [157]. Here, self-diffusion within a two-fluid plasma (the spin-up and spin-down particles) besides some modifications in the structure of the transport parameters were found to be significant. Somewhat later, the use of strong electromagnetic fields could still be handled by the Wigner function approach, which allowed for the formation of Lorentz-invariant distribution functions for the particles and anti-particles [158]. In this latter case, the Wigner function satisfied closed kinetic equations, at least in the semi-classical limit.

In previous sections and chapters, we have discussed the nature of non-Gaussian behavior for the Wigner function. Quite early on, Hudson demonstrated that the

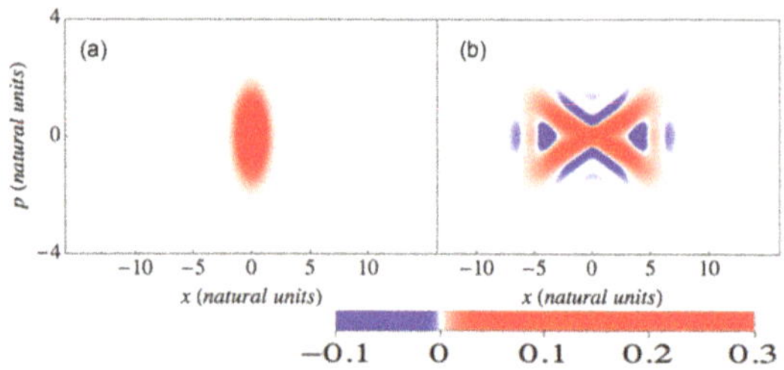

Figure 12.16. (a) The relativistic Wigner function of the spinor (12.39) at $t = 0$ in reduced units. (b) The relativistic Wigner function at $t = 7.7$ in reduced units. As indicated, the red and blue colors represent positive and negative values, respectively. Reprinted with permission from A G Campos *et al* [160]. Copyright 2014 by the American Physical Society.

only pure state for which a positive definite Wigner function existed was a Gaussian state [159], for which the presence of non-Gaussian behavior was meant to indicate more complex behavior. However, it has been demonstrated that this result can be violated in relativistic quantum mechanics [160]. Here, these latter authors begin with the four-variable wave function

$$\psi = \frac{Ce^{-x^2/2\sigma^2}}{\sqrt{2m_0c(m_0c + p^{(0)})}}\begin{pmatrix} p^{(0)} + m_0c \\ 0 \\ i(p^{(0)} + m_0c) \\ 0 \end{pmatrix}, \tag{12.41}$$

where $p^{(0)}$ is the reduced energy in the momentum four-vector. This wave function has a strictly positive relativistic Wigner function at $t = 0$, as it is the ground state for a system. Yet, at positive times, it develops a significantly non-Gaussian Wigner function, as shown in figure 12.16. The x-shaped packet at positive times is a result of the interference between the particles and the anti-particles that appear in the wave function. As these particles and anti-particles have different dynamics, the interference arises as a result of this fact.

In other work, the Wigner formalism has been applied to the study of the quantum cosmology of the universe, where the concept of deformation quantization is applied [161]. Here, the authors study a Friedman–Lemátre–Robertson–Walker model of the universe that is filled with radiation and dust or cosmic strings. For this system, the classical super-Hamiltonian can be represented as

$$\Pi_x^2 - \Omega_{cs}x^2 = 0, \tag{12.42}$$

which looks like a classical oscillator, where Π_x is the proper momentum and Ω_{cs} is the renormalized coefficients of the quadratic potential. The relativistic equation of motion for the equivalent quantum system are given by

$$\left[\left(\Pi_x - i\frac{\partial}{\partial x}\right)^2 - \Omega_{cs}\left(x + i\frac{\partial}{\partial \Pi_x}\right)^2\right]W = 0, \tag{12.43}$$

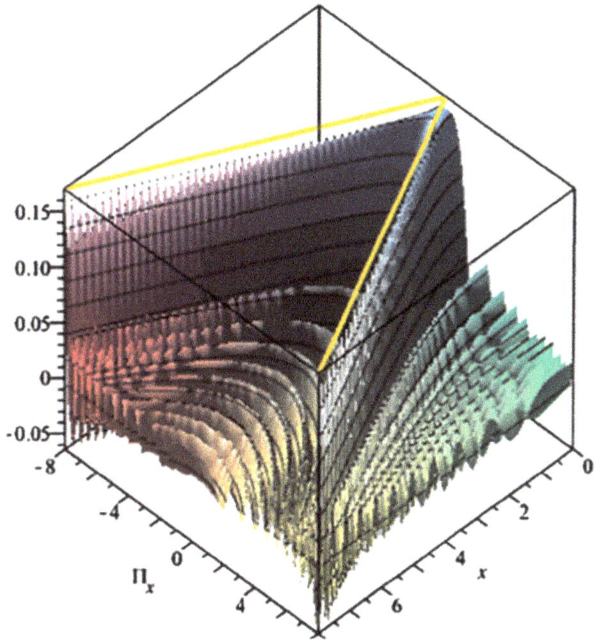

Figure 12.17. The Wigner function of a cosmic-string-filled universe, according to the equation (12.43). The corresponding classical trajectory is denoted by the gold line. This figure is plotted for $\Omega_{cs} = 1$. Reprinted by permission from Springer, 'The quantum state of the universe from deformation quantization and classical-quantum correlation', M Rashki and S Jalalzadeh [161], copyright 2017.

where W is the Wigner function in these reduced coordinates. In figure 12.17, the Wigner function is shown for a cosmic-string-filled universe. There is apparently a turning point at $x = 0$ in these coordinates, as one can see from the gold line for classical behavior. The Wigner function found from (12.43) has significant oscillatory behavior that arises both inside and outside the classical trajectory. Surprisingly, exactly this behavior was described for semi-classical states by Berry and Balazs [162]. The relativistic situation for the universe model does not seem to add much to the non-relativistic approach to semi-classical quantization. In other work, the role of the chiral anomaly and vorticity in the condensation of massive fermions has also been discussed using the Wigner functions [163, 164].

More closely to the everyday world in which we live is the idea of scalar fields. These have value in that it is easier to understand their quantization in terms of harmonic oscillators, yet they can still include the effects of relativity. It has been shown that, in this situation, one may develop a gauge-invariant quantum kinetic theory based upon a corresponding gauge-invariant Wigner function. This provides a basis for quantum plasmas, whose properties can be determined from moments of the kinetic equations for the Wigner functions as previously [165]. The electromagnetic fields from both the scalar and vector potentials were included in this work. On the other hand, a kinetic equation for only the scalar fields has also been developed, and the non-equilibrium transport considered [166].

12.5.2 The strong force

So far, in the chapters and sections preceding this one, we have dealt with particles and electromagnetic fields, if only with the Coulombic electron–electron interaction in some cases. In general, the quantum treatment of this set of particles and waves, including relativistic effects, is referred to as quantum electrodynamics (QED). In QED, the particles are the electron, proton, neutron, and photon, where the photon is the carrier of the electromagnetic field and interaction. Of course, there are the corresponding anti-particles, some of which are charged and some are not. Some are fermions (electron and proton) and some are bosons (neutron and photon). These are handled well within Dirac's theory of relativistic quantum mechanics. However, there was another solution to Dirac's equation found which gave a particle at zero energy that was its own anti-particle as well as being a fermion. This was the so-called Majorana fermion [167]. Even so, there was a problem with understanding how the nucleus was held together. This persisted until well after World War II, when ideas of the strong force began to appear. Several ideas came together in the following decades into what is now known as quantum chromodynamics (QCD), including a new set of fundamental particles from which the earlier 'fundamental particles' are composed, and the strong force. At one scale, about 10^{-15} m, the strong force holds protons and neutrons together to form nuclei. At a slightly smaller scale, this force holds the new particles, the quarks, together to form the protons and neutrons. These latter two particles are known as hadrons, which are composite particles of two kinds: baryons, composed of three quarks, and mesons, composed of one quark and one antiquark. The proton and neutron are baryons. The electron, on the other hand, is a form of lepton, which is a fermion that does not respond to the strong force. If this all seems complicated, then one can understand why it is sometimes referred to as a zoo.

Needless to say, there are still more layers of what is called today the standard model, and one layer relates to the so-called 'flavor' of the quark. This is often treated by a pseudospin, or isospin, variable in analogy to the normal spin which some quarks possess. By decomposing a Wigner function into multiple functions in spin and isospin space, one common model for the transport of quarks has been generated, described as a two-flavor model [168]. Then, if one can generate a Wigner phase space distribution for the quark transverse position and its momentum in a proton, it has been shown that this is sufficient to evaluate their cross product, the orbital angular momentum [169].

In QED, the photon carries the electromagnetic force. In a similar manner, in QCD there is a particle, the gluon, that carries the strong force. Just as one can have two photon correlations in quantum optics, so one can have two gluon correlations. These correlations have been studied in the collisions between heavy ions and light ions, in which the two gluons are characterized by one- and two-gluon Wigner function distributions for both the transverse momenta and impact parameters of the collisions [170].

Since the quarks are held together, so to speak, by the strong force, and the gluons are the carrier of this force, it is natural to think about quark–gluon interactions, just as one talks about electron–photon interactions. Then, the kinetic theory of the quark–gluon plasma can be treated by means of gauge-covariant one-particle Wigner distribution functions for the quarks, anti-quarks, and gluons. Such a description has been used to treat the collisions that can occur in such a plasma [171]. This quark–gluon plasma is the result of a phase transition in which individual hadrons are destroyed and replaced by a highly energetic system of quarks and gluons (and anti-quarks). The method by which this process is achieved is the use of ultra-relativistic heavy-ion collisions in high-energy accelerators, such as CERN. Here, any heavy ion is an ionized ion whose nuclear mass is larger than that of helium. In the plasma, the chiral symmetry is restored to the system, and chirally invariant transport equations can be written in terms of the Wigner function [172–177], as mentioned. Decoherence by a possible entanglement of length scales of the system has been studied, as well [178]. A coarse-graining of the Wigner function into a Husimi function has been used to determine the entropy production from these collisions in the case where the dynamics has gone chaotic [179].

A Wigner function description of the electromagnetic fields has been used to describe the behavior of low-energy photons radiated during deceleration of the colliding heavy ions [180]. Another aspect of the heavy-ion collisions is the generation of so-called parton jets. Here, the partons are simply the free hadrons that are emitted during the collision event of the two heavy ions. These impact the so-called vacuum state surrounding the colliding ions, and Wigner functions are useful here to describe the nature of the vacuum [181] and the parton shower [182]. In figure 12.18, the Wigner distribution function for quark–antiquark pairs is plotted as a function of the normalized position and momentum for a 100 GeV parton jet shower. The Wigner function has also been used to study the production of quark pairs during the plasma formation [183], as well as the specific role of the bottom quarks (one type of quark) [184].

Relativistic effects also occur within the nucleus of atoms. Here, Wigner functions have been usefully applied to study the dynamics of mesons with this approach [185]. The non-equilibrium situation has also been studied to determine the kinetics of a possible meson condensate and the quasi-particle excitations from this condensate [186].

12.5.3 Other forces

Of course, the electromagnetic and the strong forces are not the only forces. There are the weak force and gravity, but these have not been studied as much with Wigner functions. Now, the neutrino is a subatomic fermion that interacts only with the weak force and gravity. The mass of the neutrino is much smaller than that of other elementary particles, and this has been the source of much argument and speculation. The basic idea of thc neutrino dates to Pauli, but it is now known that it comes in three varieties (referred to as flavors). The prevailing (although still argumentative) view is that the neutrino is massless, which allows it to travel at the speed of light, a view consistent with recent experiments with accuracy of about 1 part per million [187]. Somewhat earlier the transport of neutrinos was studied,

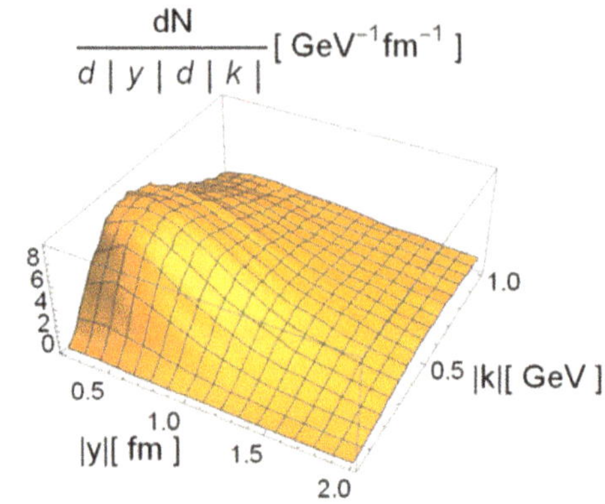

Figure 12.18. The statistical distribution of quark–antiquark pairs in a 100 GeV parton jet shower in terms of the relative spatial and momentum coordinates. The coordinates are defined in the common rest frame of the pair at the time the latter parton is created in the shower. Reprinted from R J Fries *et al* [182], with permission from Elsevier.

because the manner in which they interact with their surrounding media can be important to the early universe. In this earlier work, a Wigner function was taken to study the neutrino propagation in matter [188]. This latter group then studied the self-interaction of the neutrinos with the Wigner function approach [189]. Somewhat later, the coherence of neutron oscillations was shown to be explainable in a Wigner representation [190].

Quantum gravity has been studied with Wigner functions only slightly more often than neutrinos. The earliest work appears to be a study of the decoherence that occurs in QED when coupled to quantum gravity, where the decoherence is thought to arise from back-reaction on the scalar field [191]. Then, a quantum theory in curved space–time, including quantum gravity, was developed with the Wigner function [192]. Then, the semi-classical movement of a particle in the massive but homogeneous scalar field of quantum gravity was studied, where it was shown that a back-reaction on the particle occurred [193]. More recently, quantum gravity and collapse was studied as a mechanism for the collapse of the normal quantum wave function [194]. A theoretical formulation of Wigner function kinetics was developed and applied to formulating the generalized uncertainty principle for quantum gravity [195]. Then, the wave kinetics of acoustic-gravity waves were studied with the use of the Wigner function approach [196]. Finally, the effect that quantum gravity has on the normal quantum harmonic oscillator was studied from the approach of the generalized uncertainty principles mentioned just above. The differences may or may not be measureable, depending upon the size of the Planck mass [197].

12.6 Quantum cascade laser

The quantum cascade laser (QCL) has a long and rich history, as the basic ideas were described by Kazarinov and Suris almost half a century ago [198]. The basic

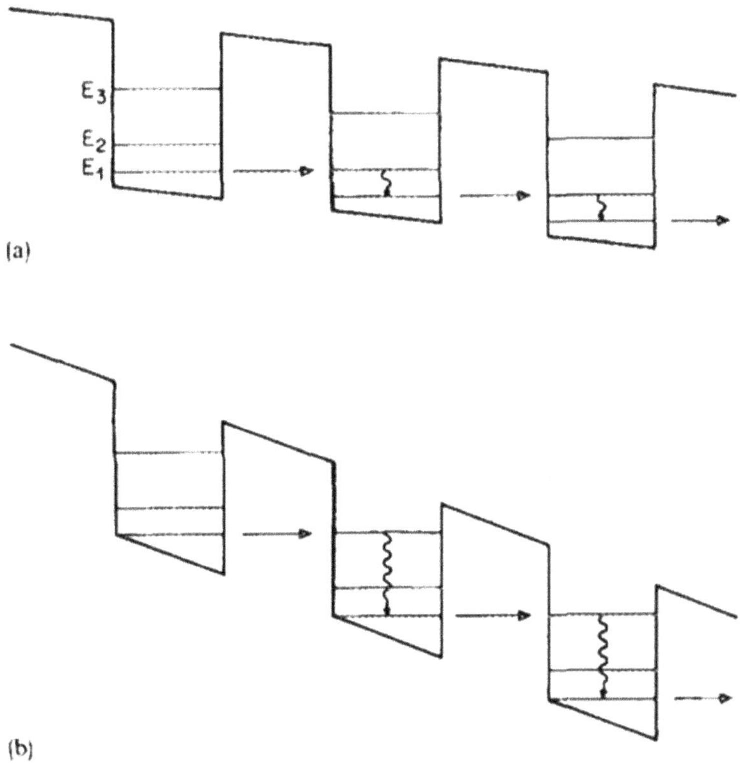

Figure 12.19. Schematic illustration of sequential resonant tunneling of electrons for a potential energy drop across the superlattice period equal respectively (a) to the energy difference between the first excited state and the ground state of the wells, and (b) to the energy difference between the second excited state and the ground state of the wells. Reprinted from F Cappaso *et al* [199] with the permission of AIP Publishing.

idea is that, through the use of molecular beam epitaxy, one can carry out band gap engineering and create a series of quantum wells, particularly in the conduction band. These quantum wells will have a series of bound states within each well. Then, under bias, the lowest bound state in one well will line up with an upper bound state in the neighboring well, so that electrons can tunnel through from the first well to the second well [199]. The basic idea is shown in figure 12.19. Once in the second well, the electron can relax to the lower well either radiatively or nonradiatively; if the former, a photon can be emitted, and one can conceive of an intra-band infrared laser by this process [200]. Between proposing a device and its realization, some time can be required, and it was in 1994 that the first QCL was demonstrated [201]. Not only has this device become an important infrared source, it has become one of the few options for generating THz radiation.

Of course, the interest here is the use of the Wigner function to study the QCL is described in some detail by Iotti and Rossi [202], and they mention the use of the Wigner function for transport in open optical devices, but they do not apply it to the QCL. However, the advantages of the Wigner function were used in the study of such open-boundary transport in superlattices [203]. Then, it was used to study electrothermal effects in QCLs, as well [204].

References

[1] Zachos C K, Fairlie D B and Cartwright T L 2005 *Quantum Mechanics in Phase Sparce* (Singapore: World Scientific)

[2] Heisenberg W 1925 *Z. Phys.* **33** 879

[3] Schrödinger E 1926 *Ann. Phys.* **60** 437

[4] Wigner E 1932 *Phys. Rev.* **40** 749

[5] Feynman R P 1987 *Quantum Implications* ed B J Hiley and F D Peat (London: Routledge) p 235

[6] Schrödinger E 1935 *Naturwiss.* **23** 807 823, 844
Tr. by Trimmer J D 1980 *Proc. Am. Phil. Soc.* **124** 323

[7] Groenewold H J 1946 *Physica* **12** 405

[8] Blaszak M and Domanski Z 2012 *Ann. Phys.* **327** 167

[9] Van der Jeugt J 2013 *J. Phys.* A **46** 475302

[10] Campos D 2017 *Pramana—J. Phys.* **88** 54

[11] Sellier J M and Dimov I 2015 *J. Comp. Electron.* **14** 907

[12] Anastopoulos C, Kechribaris S and Mylonas D 2010 *Phys. Rev.* A **82** 047119

[13] Álvarez-Estrada R F 2014 *Entropy* **16** 1426

[14] Rasinariu C 2013 *Fortshcr. Phys.* **61** 4

[15] Sels D, Brosens F and Magnus W 2013 *Physica* A **392** 326

[16] de Castro A S M and Dodonov V V 2013 *J. Phys.* A **46** 395304

[17] Stöckmann H J 1999 *Quantum Chaos* (Cambridge: Cambridge Univ. Press)

[18] Mehmani B and Aiello A 2012 *Eur. J. Phys.* **33** 1367

[19] Laguna H G and Sagar R P 2014 *Ann. Phys.* **526** 555

[20] Santos J F G and Bernardini A E 2016 *Physica* A **445** 75

[21] Caldeira A O and Leggett A J 1983 *Physica* A **121** 587

[22] Kohler S, Dittrich T and Hänggi P 1997 *Phys. Rev.* E **55** 300

[23] Ankerhold J 2003 *Europhys. Lett.* **61** 301

[24] Feynman R P and Hibbs A R 1965 *Quantum Mechanics and Path Integrals* (New York: McGraw-Hill) sec 12.8

[25] Grabert H, Schramm P and Ingold G-L 1988 *Phys. Rep.* **168** 115

[26] Liu H, Zhu L, Bai S and Shi Q 2014 *J. Chem. Phys.* **140** 134106

[27] Prado Reynoso M Á, López Vázquez P C and Gorin T 2017 *Phys. Rev.* A **95** 022118

[28] Rose P A, McClung A C, Keating T E, Steege A T C, Egge E S and Pattanayak A K 2016 *Pramana—J. Phys.* **87** 32

[29] Pachón L A, Ingold G-L and Dittrich T 2010 *Chem. Phys.* **375** 209

[30] Marchukov O V, Volosniev A G, Federov D V, Jensen A S and Zinner N T 2014 *J. Phys.* B **47** 195303

[31] Julsgaard B and Mølmer K 2014 *Phys. Rev.* A **89** 012333

[32] Ding S, Maslennikov G, Hablützel R, Loh H and Matsukevich D 2017 *Phys. Rev. Lett.* **119** 150404

[33] Gnonni M G, Palma M L, Tufarelli T, Olivares S, Kim M S and Paris M G A 2013 *Phys. Rev.* A **87** 062104

[34] Corney J F and Olsen M K 2015 *Phys. Rev.* A **91** 023824

[35] Balbutsev E B, Molodtsova I V and Schuck P 2011 *Nucl. Phys.* A **872** 42
[36] Balbutsev E B, Molodtsova I V and Schuck P 2013 *Phys. Rev.* C **88** 014306
[37] Dobrowolski A, Mazurek K and Gózdz A 2016 *Phys. Rev.* C **94** 054322
[38] Balbutsev E B, Molodtsova I V and Schuck P 2017 *Phys. Atomic Nucl.* **80** 86
[39] Dowker H F and Halliwell J J 1992 *Phys. Rev.* D **46** 1580
[40] Osborn T A, Kondart'eva M F, Tabisz G C and McQuarrie B R 1999 *J. Phys.* A **32** 4149
[41] Cartwright T, Fairlie D and Zachos C 1998 *Phys. Rev.* D **58** 025002
[42] Martin-Fierro E and Llorente J M 2006 *Gomez. Chem. Phys.* **322** 13
[43] Weiss C, Cornish S L, Gardiner S A and Breuer H-P 2016 *Phys. Rev.* A **93** 013605
[44] Nedjalkov M, Selberherr S, Ferry D K, Vasileska D, Dollfus P, Querlioz D, Dimov I and Schwaha P 2013 *Ann. Phys.* **328** 220
[45] Ferry D K, Akis R and Brunner R 2015 *Phys. Scr.* **T165** 014010
[46] Jacoboni C, Brunetti R and Monastra S 2003 *Phys. Rev.* B **68** 125205
[47] Kuz'menkov L S and Maksimov S G 2005 *Theor. Math. Phys.* **143** 621
[48] Berg B, Pimak L I, Polkovnikov A, Olsen M K, Fleischhauer M and Schleich W P 2009 *Phys. Rev.* A **80** 033624
[49] A. Poulsen J 2011 *J. Chem. Phys.* **134** 034118
[50] Bonella S and Ciccotti G 2014 *Entropy* **16** 86
[51] Kauffmann S K 2011 *Found. Phys.* **41** 805
[52] de Gosson M A 2016 *Phys. Rep.* **623** 1
[53] Tilma T, Everitt M J, Samson J H, Munro W J and Nemoto K 2016 *Phys. Rev. Lett.* **117** 180401
[54] Yang H *et al* 2016 *Nature Commun.* **17** 12532
[55] Bocquillon E, Parmentier F D, Grenier C, Berroir J-M, Degiovanni P, Glattli D C, Plaçais B, Cavanni A, Jin Y and Fève G 2012 *Phys. Rev. Lett.* **108** 196803
[56] Bocquillon E, Freulon V, Berroir J-M, Degiovanni P, Plaçais B, Cavanni A and Fève G 2013 *Science* **339** 1054
[57] Ferraro D, Roussel B, Cabart C, Thibierge E, Fève G, Grenier C and Degiovanni P 2014 *Phys. Rev. Lett.* **113** 166403
[58] Levitov L S, Lee H and Lesovik G B 1996 *J. Math. Phys.* **37** 4845
[59] Ronetti F, Vannucci L, Ferraro D, Jonckheere T, Rech J, Martin T and Sassatti M 2017
[60] Ferraro D, Jonckheere T, Rech J and Martin T 2017 *Phys. Status Solidi* B **254** 1600531
[61] Marguerite A, Bocquillon E, Berroir J-M, Plaçais B, Cavanna A, Jin Y, Degiovanni P and Fève G 2017 *Phys. Status Solidi* B **254** 1600618
[62] Rundle R P, Mills P W, Tilma T, Samson J H and Everitt M J 2017 *Phys. Rev.* A **96** 022117
[63] Deléglise S, Sayrin C, Bernu J, Brane M, Raimond J M and Haroche S 2008 *Nature* **455** 510
[64] Devitt S J 2016 *Phys. Rev.* A **94** 032329
[65] Alsina D and Latorre J I 2016 *Phys. Rev.* A **94** 012314
[66] Tian Y, Wang Z, Zhang P, Li G, Li J and Zhang T 2018 *Phys. Rev.* A **97** 013840
[67] Oh S and Choi S-I 1996 *Phys. Rev.* B **54** 4440
[68] Riedel E 1964 *Z. Naturforsch.* **19a** 1634
[69] Hamilton C A and Shapiro S 1971 *Phys. Rev. Lett.* **26** 426
[70] Joshi A 2000 *Phys. Lett.* A **270** 249
[71] Zou J, Shao B and Su W-Y 2001 *Phys. Lett.* A **285** 401
[72] Fan H-Y, Wang J-S and Liu S-G 2006 *Phys. Lett.* A **359** 580
[73] Opatrny T, Miranowicz A and Bajer J 1996 *J. Mod. Opt.* **43** 417

[74] Miranowicz A, Leonski W and Imoto N 2001 *Modern Nonlinear Optics, Part 1* 2nd edn ed M W Evans (New York: Wiley) pp 155–193
[75] Grigorio L S, Guimares M S, Rougemont R, Wotzasek C and Zarro C A D 2013 *Phys. Rev.* D **88** 065009
[76] Widom A, Megaloudis G, Clark T D, Prance R J and Prance H 1982 *J. Phys.* A **15** 3877
[77] Konstadopoulou A, Hollingsworth J M, Everitt M, Vourdas A, Clark T D and Ralph J F 2003 *Euro. Phys. J.* B **32** 270
[78] Calzetta E and Verdaguer E 2006 *J. Phys.* A **39** 9503
[79] Chang P, Shao B, Wang Z-M and Zou J 2008 *Commun. Theor. Phys.* **49** 1622
[80] Josephson B D 1962 *Phys. Lett.* **1** 251
[81] Armour A D, Kubala B and Ankerhold J 2015 *Phys. Rev.* B **91** 184508
[82] Dambach S, Kubala B and Ankerhold J 2017 *Fortschr. Phys.* **65** 1600061
[83] Ruby M, Pientka F, Peng Y, von Oppen F, Heinrich B W and Franke K J 2015 *Phys. Rev. Lett.* **115** 087001
[84] Jäck B, Eltschka M, Assig M, Etzkorn M, Ast C R and Kern K 2016 *Phys. Rev.* B **93** 020504
[85] Jaklevic R C, Lambe J, Silver A H and Mercereau J E 1964 *Phys. Rev. Lett.* **12** 159
[86] Zhong L *et al* 2013 *New J. Phys.* **15** 125013
[87] Marinescu D C and Marinescu G M 2005 *Approaching Quantum Computing* (Upper Saddle, NJ: Prentice Hall)
[88] Liu Y-X, Wei L F and Nori F 2005 *Phys. Rev.* **B 72** 014547
[89] Ferry D K, Goodnick S M and Bird J P 2009 *Transport in Nanostructures* 2nd edn (Cambridge: Cambridge Univ. Press) sec 6.2
[90] Ferry D K 2015 *Transport in Semiconductor Mesoscopic Devices* (Bristol: Institute of Physics Publishing) sec 8.2
[91] Liu Y-X, Wei L F and Nori F 2005 *Phys. Rev.* A **72** 033818
[92] Vion D, Aassime A, Cottet A, Joyez P, Pothier H, Urbina C, Esteve D and Devoret M H 2002 *Science* **296** 886
[93] Hutchinson G D, Holmes C A, Stace T M, Spiller T P, Milburn G J, Barrett S D, Hasko D G and Williams D A 2006 *Phys. Rev.* A **74** 062302
[94] Hofheinz M *et al* 2009 *Nature* **459** 546
[95] Mohamed A-B A 2010 *Phys. Lett.* A **374** 4115
[96] Ashhab S and Nori F 2010 *Phys. Rev.* A **81** 042311
[97] Balamurugan M, Chakrabarti R and Virgin Jenisha B 2017 *Physica* A **473** 428
[98] DiVincenzo D P and Smolin J A 2012 *New J. Phys.* **14** 013051
[99] Holland E T *et al* 2015 *Phys. Rev. Lett.* **115** 180501
[100] Ateto M S 2017 *Quantum Inf. Process.* **16** 267
[101] Reboiro M, Civitarese O, Ramirez R and Tielas T 2017 *Phys. Scr.* **92** 094004
[102] Bulata I, Ashhab S and Nori F 2011 *Rep. Prog. Phys.* **74** 104401
[103] Xiang Z L, Ashhab S, You J Q and Nori F 2013 *Rev. Mod. Phys.* **85** 623
[104] Gu X, Kockum A F, Miranowicz A, Liu Y X and Nori F 2017 *Phys. Rep.* **718-719** 1
[105] Yang B-J and Yao S-G 1991 *Phys. Rev.* A **43** 1983
[106] Kuz'menkov L S and Maksimov S G 2002 *Theor. Math. Phys.* **131** 641
[107] Trigger S A and Schram P P J M 1991 *Phys. Lett.* A **158** 458
[108] Gombert M-M and Leger D 1994 *Phys. Lett.* A **185** 417
[109] Leger D and Gombert M-M 1996 *Phys. Lett.* A **222** 182

[110] Gombert M-M and Leger D 1998 *Phys. Rev.* E **57** 3962
[111] Larkin A S, Filinov V S and Fortov V E 2016 *Contrib. Plasma Phys.* **56** 187
[112] Eliasson B and Shukla P K 2008 *Phys. Scr.* **78** 025503
[113] Brodin G, Ekman R and Zamanian J 2017 *Plasma Phys. Contr. Fusion* **59** 014043
[114] Zamanian J, Marklund M and Brodin G 2010 *New J. Phys.* **12** 043019
[115] Larkin A S and Filinov V S 2018 *Contrib. Plasma Phys.* **58** 107
[116] Latyshev A V and Yushkanov A A 2012 *Plasma Phys. Rep.* **38** 899
[117] Massignan P, Lampo A, Wehr J and Lewenstein M J 2015 *Phys. Rev.* A **91** 033627
[118] Mallarino J P and Téllez G 2015 *Phys. Rev.* E **91** 062140
[119] Larkin A S, Filinov V S and Fortov V E 2018 *J. Phys.* A **51** 035002
[120] Smith G R, Cook D R, Kaufman A N, Kritz A H and McDonald S W 1993 *Phys. Fluids* B **5** 4299
[121] Anderson D, Hall B, Lisak M and Marklund M 2002 *Phys. Rev.* E **65** 046417
[122] Haas F, Manfredi G and Goedert J 2003 *Brazilian J. Phys.* **33** 128
[123] Mendonça J T 2006 *Phys. Scr.* **74** C61
[124] Jovanovic D and Fedele R 2007 *Phys. Lett.* A **364** 304
[125] Tyshetskiy Y, Vladimirov S V and Kompaneets R 2011 *Phys. Plasmas* **18** 112104
[126] Bose A and Janaki M S 2013 *Phys. Plasmas* **20** 032104
[127] Latyshev A V and Yushkanov A A 2015 *Fluid Dyn.* **50** 820
[128] Haas F and Shukla P K 2008 *Phys. Plasmas* **15** 112302
[129] Haas F, Eliasson B, Shukla P K and Manfredi G 2008 *Phys. Rev.* E **78** 056407
[130] Hargreaves J K and Ecklund W L 1968 *Radio Sci.* **3** 698
[131] Bremmer H 1973 *Radio Sci.* **8** 511
[132] Ho A Y, Kuo S P and Lee M C 1994 *Radio Sci.* **29** 1179
[133] Kul H-J and Plagne L 2001 *Phys. Plasmas* **8** 5244
[134] Lindberg M and Binder R 2003 *J. Phys. Condens. Matter* **15** 1119
[135] Bizarro J P S and Figueiredo A C A *et al* 2006 *Nucl. Fusion* **46** 645
[136] Zhu J and Ji P 2012 *Plasma Phys. Control Fusion* **54** 065004
[137] Graziani F R, Bauer J D and Murillo M S 2014 *Phys. Rev.* E **90** 033104
[138] Rosato J 2014 *Phys. Lett.* A **378** 2586
[139] Rosato J 2017 *Ann. Phys.* **383** 130
[140] Chéron B G, Bultel A and Delair L 2007 *IEEE Trans. Plasma Sci.* **35** 498
[141] Tu X, Yan J, Yu L, Can K and Chéron B G 2007 *Appl. Phys. Lett.* **91** 131501
[142] Filinov V S, Levashov P R, Botan A V, Bonitz M and Forlov V E 2009 *J. Phys.* A **42** 214002
[143] Larkin A S and Filinov V S 2017 *Math. Montisnigri* **40** 55
[144] Rosato J 2017 *Eur. Phys. J.* D **71** 28
[145] Slyusarenko Y V and Silusarenko O Y 2017 *J. Math. Phys.* **58** 113302
[146] Schrödinger E 1926 *Ann. Phys.* **81** 109
[147] Dirac P A M 1928 *Proc. R. Soc.* A **117** 610
[148] Schiff L I 1955 *Quantum Mechanices* 2nd edn (New York: McGraw-Hill) sec 43
[149] Novoselov K S, Geim A K, Morozov S V, Jiang D, Zhang Y, Dubonos S V, Grigorieva I V and Firsov A A 2004 *Science* **306** 666
[150] Ferry D K 2018 *An Introduction to Quantum Transport in Semiconductors* (Singapore: Pan Stanford Publishing) ch 11
[151] Hakim R and Heyvaerts J 1978 *Phys. Rev.* A **18** 1250

[152] Mendonça J T 2011 *Phys. Plasmas* **18** 062101
[153] Zhu J and Ji P 2010 *Phys. Rev.* E **81** 036406
[154] Mahajan S M and Asenjo F A 2016 *Phys. Plasmas* **23** 056301
[155] Sheng X-L, Rischke D H, Vasak D and Wang Q 2018 *Eur. Phys. J.* A **54** 21
[156] Bialynicki-Bitula I, Górnicki P and Rafelski J 1991 *Phys. Rev.* D **44** 1825
[157] Hakim R, Mornas L, Peter P and Sivak H D 1992 *Phys. Rev.* D **46** 4803
[158] Morozov V G, Röpke G and Höll A 2002 *Theor. Math. Phys.* **132** 1029
[159] Hudson R L 1974 *Rep. Math. Phys.* **6** 249
[160] Campos A G, Cabrera R, Bondar D I and Rabitz H A 2014 *Phys. Rev.* A **90** 034102
[161] Rashki M and Jalalzadeh S 2017 *Gen. Relativ.-Gravit.* **49** 14
[162] Berry M V and Balazs N I 1979 *J. Phys.* A **12** 625
[163] Fang R-H, Pang J-Y, Wang Q and Wang X-N 2017 *Phys. Rev.* D **95** 014032
[164] Gao J-H, Pu S and Wang Q 2017 *Phys. Rev.* D **96** 016002
[165] Haas F, Zamanian J, Marklund M and Brodin G 2010 *New J. Phys.* **12** 073027
[166] Garbrecht B and Garny M 2012 *Ann. Phys.* **327** 914
[167] Majorana E 1937 *Il Nuovo Cimento* **14** 171
[168] Florkowski W 1998 *Eur. Phys. J.* A **2** 77
[169] Engelhardt M 2017 *Phys. Rev.* D **95** 094505
[170] Kovchegov Y V and Wertepny D E 2014 *Nucl. Phys.* A **925** 254
[171] Selikhov A V 1991 *Phys. Lett.* B **268** 263
[172] Florkowski W, Hüffner J, Klevansky S P and Neise L 1996 *Ann. Phys.* **245** 445
[173] Ochs S and Heinz U 1998 *Ann. Phys.* **266** 351
[174] Rehberg P 1999 *J. Phys.* G **25** 373
[175] Filinov V S, Bonitz M, Ivanov Y B, Skokov V V, Levashov P R and Fortnov V E 2011 *Contrib. Plasma Phys.* **51** 322
[176] Filinov V S, Ivanov Y B, Fortov V E, Bonitz M and Levashov P R 2013 *Phys. Rev.* C **87** 035207
[177] Guo C-Q, Zhang C-J and Xu J 2017 *Eur. Phys. J.* A **53** 233
[178] Akkelin S V and Sinyukov Y M 2014 *Phys. Rev.* C **89** 034910
[179] Tsukji H, Iida H, Kunihiro T, Ohnishi A and Takahasi T T 2016 *Phys. Rev.* D **94** 091502
[180] Koide T and Kodama T 2016 *J. Phys.* G **43** 095103
[181] Jeon S and Epelbaum T 2016 *Ann. Phys.* **364** 1
[182] Fries R J, Han K and Ko C M 2016 *Nucl. Phys.* A **956** 601
[183] Song T, Aichelin J and Bratkovskaya E 2017 *Phys. Rev.* C **96** 014907
[184] Chen B and Zhao J 2017 *Phys. Lett.* B **772** 819
[185] Alonso J D and Canvellas A P 1991 *Nucl. Phys.* A **526** 623
[186] Matsui T and Matsuo M 2008 *Nucl. Phys.* A **809** 211
[187] Adamson P *et al* 2015 *Phys. Rev.* D **52** 052005
[188] Sirera M and Pérez A 1999 *Phys. Rev.* D **59** 125011
[189] Sirera M and Pérez A 2004 *J. Phys.* G **30** 1173
[190] Herranen M, Kainulainen K and Rahkila P M 2009 *Nucl. Phys.* B **810** 389
[191] Kiefer C 1992 *Phys. Rev.* D **46** 1658
[192] Antonsen F 1997 *Phys. Rev.* D **56** 920
[193] Wissowski H and Kastrup H A 1999 *Nucl. Phys. B (suppl.)* **57** 299
[194] De Unánue A and Sudarsky D 2008 *Phys. Rev.* D **78** 043510
[195] Marchiolli M A and Ruzzi M 2012 *Ann.Phys.* **327** 1538

[196] Mendonça J T and Stenflo L 2015 *Phys. Scr.* **90** 055001
[197] Das S, Robbins M P G and Walton M A 2016 *Can. J. Phys.* **94** 139
[198] Kazarinov R F and Suris R A 1974 *Sov. Phys. Semicond.* **5** 207
[199] Capasso F, Mohammed K and Cho A Y 1986 *Appl. Phys. Lett.* **48** 478
[200] Capasso F, Mohammed K and Cho A Y 1986 *IEEE J. Quantum Electron.* **22** 1853
[201] Faist J, Capasso F, Sivco D L, Sirtori C, Hutchinson A L and Cho A Y 1994 *Science* **264** 553
[202] Iotti R C and Rossi F 2005 *Rep. Prog. Phys.* **68** 2533
[203] Jonasson O and Knezevic I 2015 *J. Comput. Electron.* **14** 879
[204] Mei S, Shi Y B, Jonasson O and Knezevic I 2017 *Handbook of Optoelectronic Device Modeling and Simulation* ed J Piprek vol 2 (Boca Raton, FL: CRC Press)

www.ingramcontent.com/pod-product-compliance
Ingram Content Group UK Ltd.
Pitfield, Milton Keynes, MK11 3LW, UK
UKHW051901100926
8556IPUK00005B/718

9 780750 316699